Green Energy and Technology

Climate change, environmental impact and the limited natural resources urge scientific research and novel technical solutions. The monograph series Green Energy and Technology serves as a publishing platform for scientific and technological approaches to “green”—i.e. environmentally friendly and sustainable—technologies. While a focus lies on energy and power supply, it also covers “green” solutions in industrial engineering and engineering design. Green Energy and Technology addresses researchers, advanced students, technical consultants as well as decision makers in industries and politics. Hence, the level of presentation spans from instructional to highly technical.

Indexed in Scopus.

Indexed in Ei Compendex.

More information about this series at https://link.springer.com/bookseries/8059

Vahid Vahidinasab • Behnam Mohammadi-Ivatloo
Editors

Active Building Energy Systems

Operation and Control

Springer

Editors
Vahid Vahidinasab
Department of Engineering
School of Science and Technology
Nottingham Trent University
Nottingham, UK

Behnam Mohammadi-Ivatloo
Department of Electrical and Electronics Engineering
Muğla Sıtkı Koçman University
Muğla, Turkey

ISSN 1865-3529 ISSN 1865-3537 (electronic)
Green Energy and Technology
ISBN 978-3-030-79741-6 ISBN 978-3-030-79742-3 (eBook)
https://doi.org/10.1007/978-3-030-79742-3

This Springer imprint is published by the registered company Springer Nature Switzerland AG
The registered company address is: Gewerbestrasse 11, 6330 Cham, Switzerland

Preface

Thanks to public consciousness of global environmental problems, ambitious targets were set over the last years by different countries all around the world to reach net-zero carbon emissions. Globally, buildings are accountable for about 40% of carbon emissions, and in most countries, they consume around 40% of the produced energy. As a result, to cope with the energy crisis, any solutions provided in this regard would have to consider the energy issues of buildings.

Active buildings (ABs) are able to support the broader energy system by intelligent integration of renewable-based energy technologies for heating, cooling, electricity, and transport. ABs are designed to be energy efficient, with novel ways of creating, controlling, and releasing energy [R1]. ABs can work standalone as a self-sufficient energy entity or can connect with the other ABs in a district area and trade energy via the network. They have the potential of interacting with local as well as national-level energy grids, and by behaving as an isolated or a positive energy agent, they can provide energy services to reduce the stress on the upstream grids and defer investment costs.

In this regard, the operation and control of ABs are being developed as a fundamental research and outreach part of the future smart energy systems. The ABs concept is an intelligent integration of renewable energy sources like rooftop photovoltaic (PV) and wind units, different types of loads and control (centralized and decentralized), and energy management systems. They can be operated as a controllable subsystem which can operate in grid-connected as well as in an islanded mode.

These new systems called for a (re)thinking on the definition of the control, operation, and optimization of energy systems of ABs. The key issues for the operation, control, and optimization of these systems include integrated energy technologies, energy management systems, and control techniques (including distributed, coordinated, centralized, and decentralized), and optimization methods should be carefully updated focusing on different building-level control layers and in both of the islanded and grid-connected modes. Analysis of the impacts of the building-level renewable energy sources, electric vehicles, heat and electricity storage technologies, demand-side response (DSR), and market policies are the most important features that need to be considered in the control and operation of ABs' energy systems.

This book introduces an overview of the concept, technologies, control strategies, operation and market participation, and optimization of ABs while covering a comprehensive and in-depth review of building-level energy systems technologies, different control, operation, and market mechanisms.

Chapter 1 provides an overview of promising active buildings by discussing the concept and definition of such buildings as well as the energy services that can be provided by them. Different enabling technologies and challenges of active buildings are also discussed and a literature review on the subject is provided.

Chapter 2 studies the active buildings from a social science perspective. In this chapter, the interaction between people, homes, and energy is discussed. Different case sites with different characteristics, themes, energy specifics, occupant behavior, and architectural aesthetic are analyzed in this chapter.

Chapter 3 discusses the application of static energy storage systems and mobile energy storage in active buildings. The energy storage systems can alleviate the impact of uncertainties in forecasted demand and renewable energy generations of active buildings and can also act as a backup increasing the reliability of active buildings. Different technologies, requirements, and the application of different energy storage systems are presented in this chapter.

Chapter 4 discusses hybrid AC/DC electrical power grid applications in active buildings from power electronics perspectives. The chapter studies the main structures of hybrid AC/DC electrical power grids and the associated power electronics topologies, and power quality aspects from the main AC grid point of view. In this chapter, experimental results of laboratory prototypes are presented, highlighting the advantages of the proposed hybrid AC/DC electrical grids.

Chapter 5 provides a systematic workflow for the application of model predictive control in active buildings. The introduced workflow consists of physics-based modeling methods for analysis and evaluation, and model-based and data-driven techniques for developing low-complexity and control-oriented models. Their case study has shown that using the proposed model the cost of space heating could be reduced by 11%.

Chapter 6 presents a co-design framework for simultaneous optimization of the design and operation of residential buildings using model predictive control (MPC). This chapter considers the impact of seasonal storage on the proposed co-design framework. The system flexibility and self-sufficiency are increased by incorporating the proposed co-design approach and total cost and emission is reduced considerably.

Chapter 7 provides a comprehensive overview of the concepts, frameworks, and methods involved in active building management and control. Different energy systems structures like as centralized, decentralized, and distributed structures are discussed in this chapter. This chapter also reviews the application of different control techniques like PID, model predictive control (MPC), multi-agent system approaches, and artificial intelligence-based and data-driven methods.

Chapter 8 studies the control of a population of active buildings at the network level. The aggregated active buildings can provide different services to the network such as frequency control and coordinated volt/VAR control. This chapter considers active buildings as multi-vector nano energy hubs and proposes proper control techniques. The chapter also provides a detailed case study that considers demand-side frequency regulation, power tracking, and formal control synthesis methods using energy storage systems (ESSs) and thermostatically controlled loads (TCLs).

Chapter 9 discusses the cybersecurity issues of active buildings. As active buildings are using the Internet of Things (IoT) for sensing and communication and are relying on ICT, they are required to provide security, privacy, safety, reliability, and, resilience against attacks and abnormal conditions. This chapter provides a roadmap for tackling cyber-physical security in active buildings by focusing on risk assessment, security metrics, intrusion detection, threat modeling, and simulation.

Chapter 10 studies the energy management systems of active buildings by answering three questions to how energy management could be applied in active buildings? What are the requirements and constraints? And what techniques could be deployed to make energy management more efficient and user-friendly?

Chapter 11 studies the possibility of flexibility provided by active buildings. This chapter identifies the flexible resources (appliances/devices) in active buildings and their corresponding degree of flexibility. The requirement of the different flexibility services in the network is determined and the connection of the flexibility potential in active buildings and the flexibility required of the network is addressed.

Chapter 12 studies the participation of active buildings in peer-to-peer (P2P) and local transactive energy markets. Active buildings can act as prosumers in the local energy markets, and by using the local energy markets, different benefits could be obtained from the active building and network perspectives. This chapter analyzes different technologies and potentials in a community of active buildings and describes how these potentials can be activated through local transactive energy markets.

Chapter 13 studies non-intrusive load monitoring (NILM) as a technique to provide useful information for energy management of active buildings. By using NILM, the power consumption of each appliance can be extracted from the given total consumption of the active buildings. This chapter provides the basic concepts about NILM, an overview of various NILM algorithms, and their challenges.

Chapter 14 studies the demand response service provision by active buildings and the role of aggregators in providing such services. This chapter outlines the structure of aggregators and retailers. Then, the benefits, barriers, motivators, and challenges of demand response programs are discussed. The existing demand response programs and enabling technologies for implementing such programs are also described in this chapter. To conclude, we would like to sincerely thank all of the authors who contributed to this book. Also, the editors would like to extend their deep gratitude to the reviewers for their thoughtful comments on all the submitted book chapters including those that were accepted and published in this book.

In creating this book, the editors, authors, and reviewers dedicated their time and enthusiasm with the hope that it will be beneficial to academics, industry professionals, and students interested in this field.

Nottingham, UK Vahid Vahidinasab
Muğla, Turkey Behnam Mohammadi-Ivatloo

Contents

1 **Active Buildings: Concept, Definition, Enabling Technologies, Challenges, and Literature Review** 1
Omid Sadeghian, Vahid Vahidinasab, and Behnam Mohammadi-Ivatloo

2 **Why Active Buildings? Realising the Potentials of Energy Networked Homes: A Social Scientific Perspective** 25
Kate O'Sullivan, Fiona Shirani, Nick Pidgeon, and Karen Henwood

3 **Electrical Energy Storage Devices for Active Buildings** 51
Mohamad Amin Rajabinezhad, Hooman Firoozi, Hosna Khajeh, and Hannu Laaksonen

4 **Hybrid AC/DC Electrical Power Grids in Active Buildings: A Power Electronics Perspective** 71
Vitor Monteiro, Catia Oliveira, Sergio Coelho, and Joao L. Afonso

5 **A Modelling Workflow for Predictive Control in Residential Buildings** 99
E. O'Dwyer, E. Atam, P. Falugi, E. C. Kerrigan, M. A. Zagorowska, and N. Shah

6 **MPC and Optimal Design of Residential Buildings with Seasonal Storage: A Case Study** 129
P. Falugi, E. O'Dwyer, M. A. Zagorowska, E. Atam, E. C. Kerrigan, G. Strbac, and N. Shah

7 **Control and Management of Active Buildings** 161
Ben Wooding, Vahid Vahidinasab, Milad Kazemi, and Sadegh Soudjani

8 **Operation and Control of a Population of Active Buildings at Network Level** 193
Ben Wooding, Vahid Vahidinasab, and Sadegh Soudjani

9 **Cybersecurity Roadmap for Active Buildings** 219
Ricardo M. Czekster, Charles Morisset, Aad van Moorsel, John C. Mace, Walter A. Bassage, and John A. Clark

10 **Energy Management Systems of Grid-Connected Active Buildings** 251
Mohammad Shaterabadi, Hosna Khajeh, Hooman Firoozi, and Hannu Laaksonen

11 **Active Building as an Electricity Network Service Provider** 273
Hooman Firoozi, Hosna Khajeh, Hannu Laaksonen, and Mohammad Shaterabadi

12 **Participation of Active Buildings in Peer-to-Peer and Local Transactive Energy Markets** 295
Vahid Vahidinasab, Mohammad Nasimifar, and Behnam Mohammadi-Ivatloo

13 **Non-intrusive Load Monitoring and Its Application in Energy Flexibility Potential Extraction of Active Buildings** 331
Elnaz Azizi, Mohammad T. H. Beheshti, and Sadegh Bolouki

14 **Active Buildings Demand Response: Provision and Aggregation** 355
Omid Sadeghian, Arash Moradzadeh, Behnam Mohammadi-Ivatloo, and Vahid Vahidinasab

Index 381

Chapter 1
Active Buildings: Concept, Definition, Enabling Technologies, Challenges, and Literature Review

Omid Sadeghian, Vahid Vahidinasab, and Behnam Mohammadi-Ivatloo

Status Quo, Challenges, and Outlook
Electricity systems should confront to the emerging energy crisis and environmental issues by instigating the demand flexibility. To overcome these issues, the buildings as a large energy consumer in the today's world are moving toward active buildings. Demand flexibility and self-generation of electricity are two main features of such buildings. Building activation could be challenging due to enabling technologies and technical restrictions of the grid. Therefore, the first stage is to answer the question what kind of technologies in viewpoint of communication infrastructures, energy storage systems, and distributed generators is required for building activation? and what energy services could be provided by active buildings for the power grid?

1.1 Introduction

Buildings are a large energy consumer in the world by the responsibility of consuming 40% of total energy consumption (Zhan et al., 2018). In addition, buildings are responsible for 36% of total CO_2 emissions (Langevin et al., 2019).

O. Sadeghian (✉)
Faculty of Electrical and Computer Engineering, University of Tabriz, Tabriz, Iran

V. Vahidinasab
Department of Engineering, School of Science and Technology, Nottingham Trent University, Nottingham, UK
e-mail: vahid.vahidinasab@ntu.ac.uk

B. Mohammadi-Ivatloo
Department of Electrical and Electronics Engineering, Muğla Sıtkı Koçman University, Muğla, Turkey

V. Vahidinasab, B. Mohammadi-Ivatloo (eds.), *Active Building Energy Systems*, Green Energy and Technology, https://doi.org/10.1007/978-3-030-79742-3_1

Recently, energy-saving in the building has been focused on due to the energy crisis and global warming (Sher et al., 2019). In the future, by increasing the load demand, when no control strategies are implemented for the demand side, a significant strain occurred in the electricity grid (Verbruggen & Driesen, 2015). In general, there are two approaches for energy reduction of buildings and meeting the energy challenges, including energy efficiency and integration of renewable energy sources with buildings (Fosas et al., 2020; Wang et al., 2017). The newly defined active building concept as the solution to such challenges is an improved version of the traditional envelope, which is called grid-responsive or grid-aware buildings, which can provide energy services for the network by the production of energy or timing their use (Luo et al., 2016). Energy crisis and emission issues have accelerated the movement toward active buildings and activation of energy systems in recent years, and therefore, they have received a lot of attention from researchers (Luo et al., 2016).

1.1.1 Concept and Definition of Active Buildings

The active building concept is known to be related to load flexibility (Bulut et al., 2016) and defined as a building which has two major characteristics, including demand flexibility or self-generation of electricity (Luo et al., 2016). The demand flexibility of buildings is achieved by participation in demand-side response (DSR) programs (Chen et al., 2018). Through participation in such programs, three actions, including load shifting (Sadeghian et al., 2020a), load reduction/curtailment (Sadeghian et al., 2019a), and load interruption (Wang et al., 2020), may be accomplished by participation in different price-based and incentive-based programs (Aalami et al., 2010). The amount of load flexibility depends on the price elasticity of the understudy demands (Sadeghian et al., 2019b).

Self-generation of electricity is another major characteristic of active buildings (Luo et al., 2019). The most efficient way for self-generation is to use renewable energy sources, and buildings are expected to be more integrated with such promising energy sources (Schirrer et al., 2013). However, the renewable generations in buildings are subjected to technical limitations such as congestion and voltage limit (Luo et al., 2019). The operation of both the individual buildings and the grid should be optimized for the efficient use of renewable energies. The overall control aim of such a smart building is to maximize resident comfort with minimum energy consumption (Wang et al., 2012a).

As mentioned, demand flexibility and self-generation are two major features of active buildings, but as claimed in (Bulut et al., 2016), this concept is more related to demand flexibility rather than self-generation. However, simultaneous investigation on both the renewable and responsive solutions is ongoing to achieve the highest activation level of buildings (Dewidar et al., 2013).

In Fig. 1.1, the two major types of energy services that can be provided by active buildings, including self-generation and demand flexibility, are shown.

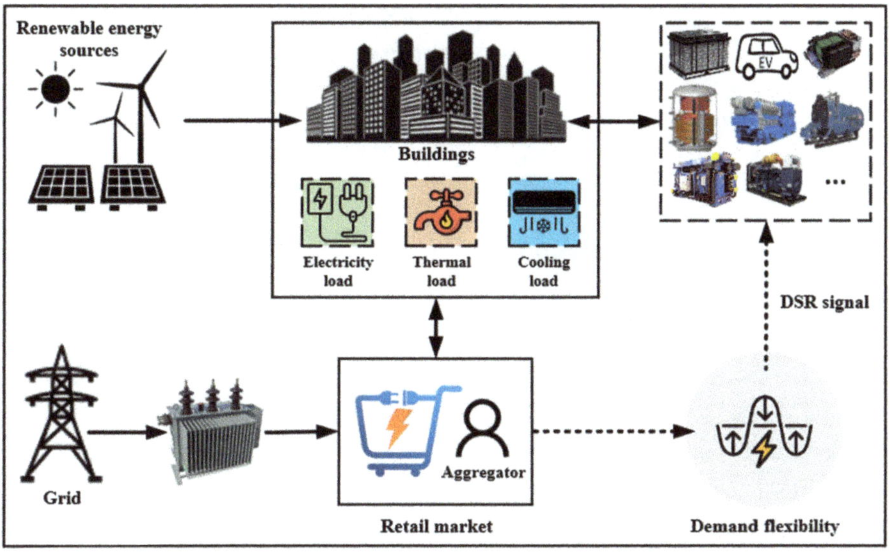

Fig. 1.1 Schematic diagram of the active building and its contribution for providing energy services

Despite the similarities between active buildings and smart buildings, there is a difference between them. Smart building is an overall term compared to an active building. In a smart building, the energy is appropriately managed to prevent energy waste by energy efficiency and energy-saving approaches. In smart buildings, the capability of energy exchange with the grid to respond to the grid conditions is optional, and therefore, it can be available or not available. This is why the active building should respond to the grid and impact on it by load management or providing electricity generated by distributed generation sources, especially renewable energy sources, energy storage systems, and electric vehicles. The energy storage systems and DSR programs are critical to shut down the energy sources in peak load period and providing grid services, when a DSR signal or a request from the grid is received.

In Fig. 1.2, the main keywords of the previous works related to active buildings are illustrated.

1.1.2 Purpose and Organization

In this chapter, the concept, definition, and challenges of the active building are outlined. In addition, the capability of active building for providing energy services such as DSR programs, ancillary services, and energy sharing with other buildings is described. Furthermore, distributed generation in active buildings such as energy

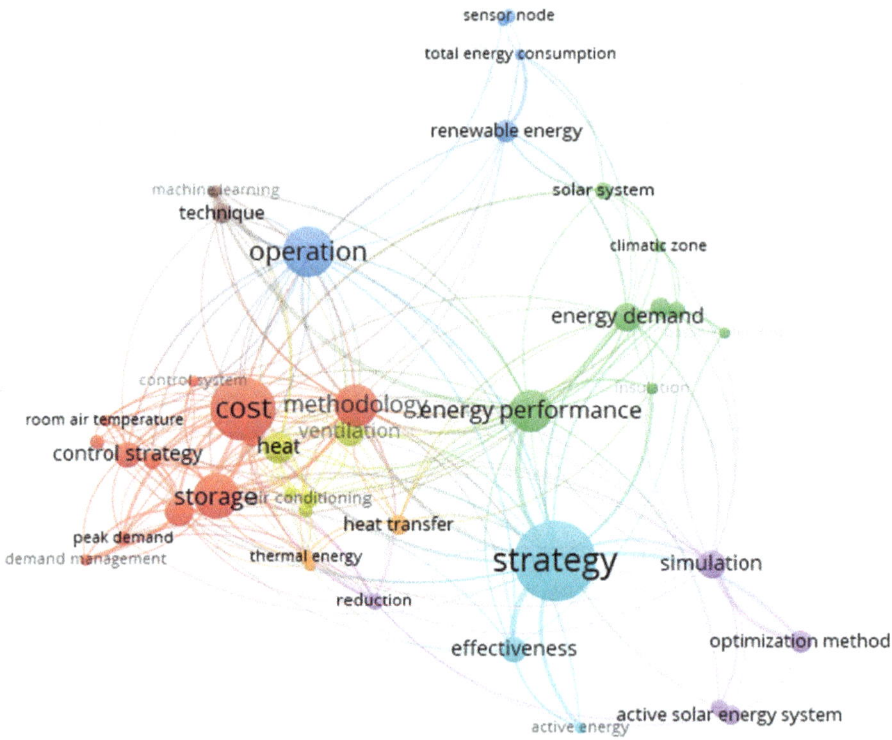

Fig. 1.2 Main keywords of the previous works related to active buildings

storage systems and renewable and non-renewable generators are discussed. This chapter also deals with role of electrified active buildings in decarbonization. Afterward, the literature on the understudied topic is reviewed. Different works have been conducted in this regard, containing demand and behavior forecasting and enabling technologies. Furthermore, the literature about integration of buildings with energy storage systems, DSR strategies, and renewable energy sources is reviewed. The previous works related to providing ancillary services are also reviewed in this chapter.

With this regard, this chapter is continued by the energy services that can be provided by active buildings. The role of electrified active buildings in decarbonization is explained in Sect. 1.3. Distributed generators in active buildings for self-generation of electricity are enumerated in Sect. 1.4. Section 1.5 deals with energy storage systems, which are appropriate to be used in active buildings. Challenges associated with development of active buildings are discussed in Sect. 1.6. Literature review about the active buildings is presented in Sect. 1.7. Finally, the chapter is concluded in the last section.

1.2 Providing Energy Services by Active Buildings

As mentioned, demand flexibility and power generation are two features of active buildings, but in a more comprehensive definition, in addition to these two features, having energy storage systems and sharing electricity among buildings are also the features of active buildings. These characteristics make a building to be flexible or active. Accordingly, the main features that make a building to be active are illustrated in Fig. 1.3, which are extracted based on the characteristics of active building presented in (Vahidinasab et al., 2021).

1.2.1 Demand-Side Response Programs

Instigating demand flexibility needs to move from passive consumers to active players to impact the operation of the network and participate in the real-time power market (Bulut et al., 2016). Therefore, one of the prerequisites for the creation of active buildings is the activation of energy systems in preconstructed or newly constructed buildings (Fosas et al., 2020). During the load flexibility, the weather conditions, user's behavior, energy production available, and market energy price should be considered (Lauro et al., 2015). For instance, the power used by the air

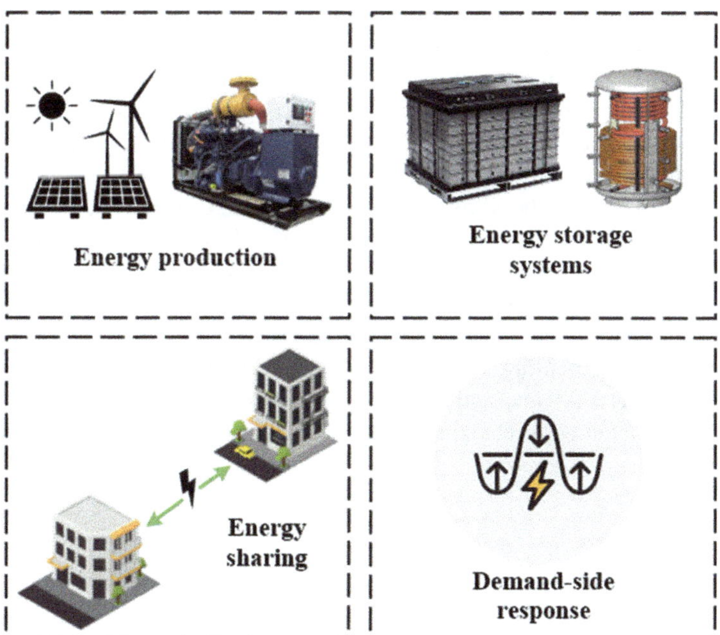

Fig. 1.3 Major features that enable buildings to be active

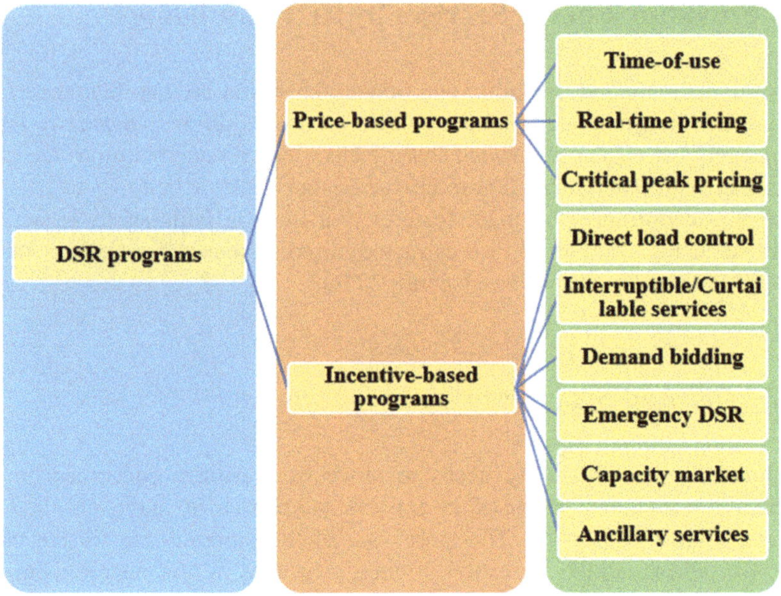

Fig. 1.4 Demand-side response programs and related subcategories

conditioning system actually is to offset the cooling or heating load generated from mainly three sources: building envelope, indoor equipment, and human body (Luo et al., 2016), which should be considered in demand flexibility programs.

In addition to the mentioned features, insulation and thermal mass inertia can also be the feature for active buildings (Verbruggen & Driesen, 2015; Long & Ye, 2015). In (Zeiler & Boxem, 2009), the active building is mentioned as the opposite of conventional passive buildings. Different demand-side programs can be integrated by active buildings that are categorized in two main categories of price-based and incentive-based programs. In price-based programs, the electricity price is changed to impact the electricity consumption pattern, whereas in incentive-based programs, an incentive is paid to the customers in exchange for modifying their consumption pattern based on incentive-based programs (Chen et al., 2018). However, for implementing such programs, some prerequisites, including modeling, enabling technologies, and drivers, are required. Figure 1.4 illustrates these programs associated with related subcategories (Albadi & El-Saadany, 2008; Yan et al., 2018).

1.2.2 Providing Ancillary Services by Active Buildings

In addition to participation in DSR programs, active buildings are also effective to provide ancillary services. Such services are some operational reserves, which are

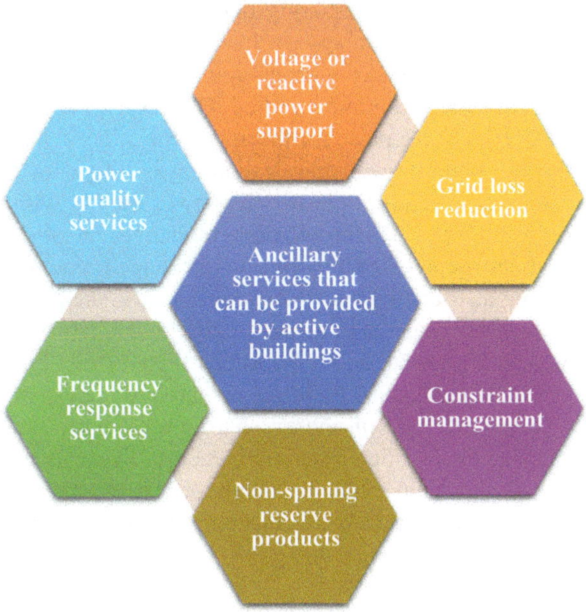

Fig. 1.5 A number of ancillary services that can be provided by active buildings

employed by system operator to ensure the power grid reliability and operational characteristics of the grid like the power quality. Ancillary services play the safeguard role against uncertainties associated with the power grid (Lymperopoulos et al., 2015). By using such services, the system operator improves the operational parameters of the grid by keeping a balance between real-time supply and load. Active buildings are able to provide different kinds of ancillary services related to voltage and reactive power, grid losses, power quality, frequency response, and so forth (Pirbazari, 2011). Providing these energy services can be feasible by distributed generation to control the active and reactive power (Dominguez-Garcia & Hadjicostis, 2010). Although capacitor banks cannot be completely replaced by distributed generation, at least the capacitor's size can be reduced (Dominguez-Garcia & Hadjicostis, 2010). The ancillary services that can be provided by active buildings are illustrated in Fig. 1.5. These services are depended to the different features of the building such as the technology used (such as renewable energy sources, energy storage systems, electric vehicles, etc.) and the related sector (such as residential, commercial, industrial, etc.).

1.2.3 Energy Sharing

Active buildings are able to share electrical energy with neighboring buildings using the energy from local distributed generators and energy storage systems (Xu et al., 2020). Energy sharing depends on the energy surplus in buildings (Kayo et al., 2014). This energy exchange action should benefit both the buildings that are exchanging energy. In energy sharing, the objective of the buildings (cost, reliability, etc.) cannot be unified. Otherwise, the energy exchange is not beneficial and does not take place for those buildings. For instance, the origin building shares its surplus power with the destination building for profit. On the other hand, the destination building improves its power quality with the received power from the origin building. Therefore, their objectives in power sharing cannot be unified. Different kinds of energy can be exchanged among active buildings based on the existing distributed generators. For instance, the presence of the combined heat and power (CHP) system in an active abuilding enables that building to share its power and heat with other active buildings (Kayo et al., 2014), but the presence of diesel generator enables that building to share only power with other building.

A specific kind of energy sharing among buildings is the peer-to-peer energy exchange or so-called transactive energy. This energy network is for energy sharing among individual customers without a centralized server (Siano et al., 2019). A challenge about this kind of energy sharing is related to its security perspective (Tushar et al., 2020).

1.3 Decarbonization by Electrified Active Buildings

Decarbonization is one of the important goals in moving toward the active buildings since such buildings are able to efficiently utilize the energy for existing demands such as heating and lighting demands. Using renewable energy sources is a sample for decarbonization of buildings. For the most decarbonization, electrification is the best solution (Tarroja et al., 2018). Electrification is the utilization of electric-based energy units (like heat pumps) instead of the energy unit that operate based on the non-electric energy carriers (such as CHP and diesel systems). Electrification of heating systems is seen as one of the prerequisites for development of renewable energy sources. In addition to installing renewable energy sources, electrification of conventional vehicles and using heat pumps are existing options for electrification in buildings (Vahidinasab et al., 2021). Utilizing electric boilers is another approach for electrification (Bühler et al., 2019). Figure 1.6 shows the existing options for electrification in buildings. Among them, using heat pumps is the well-known option that has been widely investigated in the literature for electrification of heating systems (Fawcett et al., 2014; Neirotti et al., 2020).

1.4 Distributed Generation in Active Buildings

Distributed generation is the solution for self-generation of electricity in buildings. There are two major kinds of distributed generators, including renewable and non-

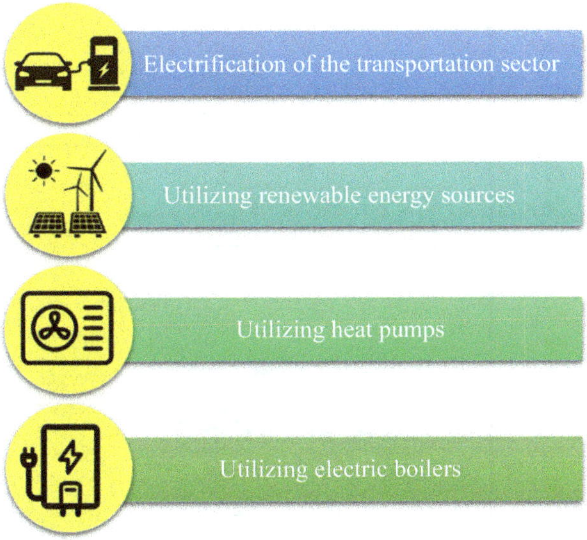

Fig. 1.6 Existing options toward electrification

renewable generators (Theo et al., 2017). Renewable-based distributed generators especially solar photovoltaic and wind turbine are more developing rather than non-renewable-based generators. However, some non-renewable generators such as fuel cell are developing due to its advantages such as environmental friendly operation. The two major kinds of distributed generators associated with related subcategories are shown in Fig. 1.7.

1.5 Energy Storage Systems in Active Buildings

Energy storage systems are an important component of active buildings, which can effectively enhance the flexibility of building loads. Generally, energy storage systems are charged in low-demand hours due to lower prices of energy. In contrast, they are discharged in high-demand hours because of higher prices of energy. Using such systems are also effective for decreasing the operation costs of buildings (Niu et al., 2019). There are different kinds of energy storage systems such as electrical, heating, cooling, and hydrogen storages, which have different operation characteristics. In Fig. 1.8, existing types of energy storage systems that can be utilized in active buildings are illustrated.

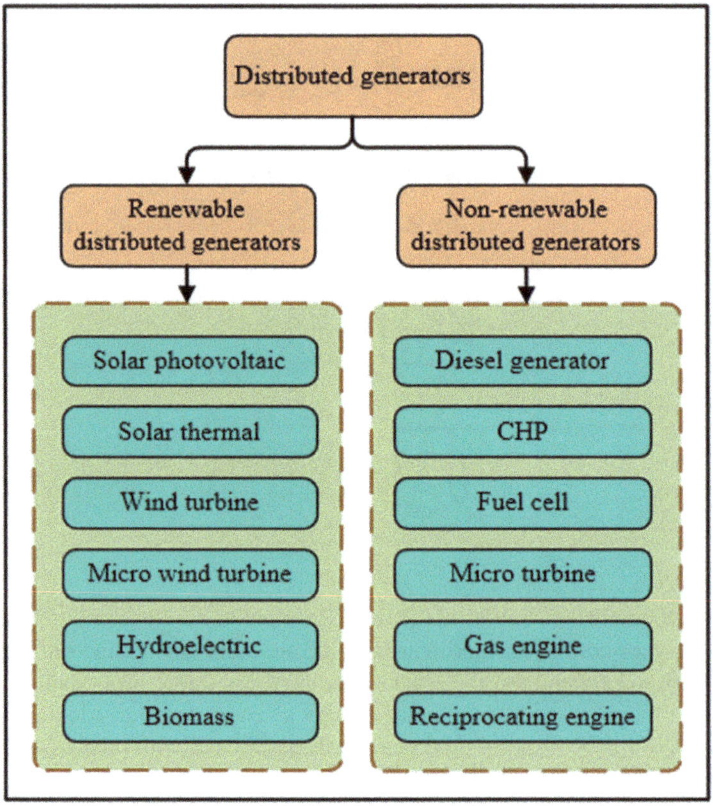

Fig. 1.7 Distributed generators and related subcategories that can be used in active buildings

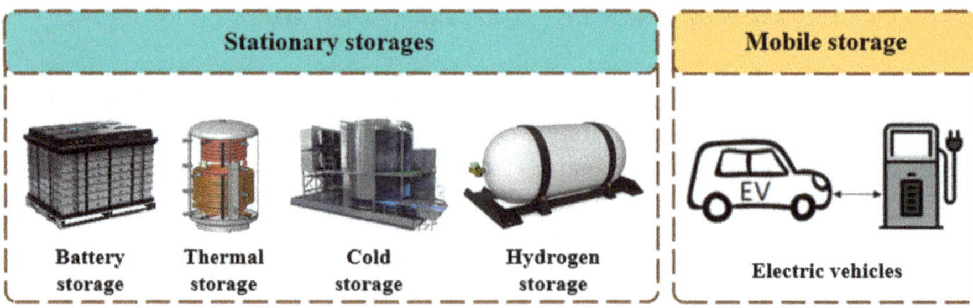

Fig. 1.8 Energy storage systems suitable for using in active buildings

1.6 Challenges Toward Active Buildings

Although the active buildings are considerably focused by researchers due to the numerous benefits of such buildings, there are some challenges to such a concept.

Active buildings need well-specified information and communication technology (ICT) for favorite operation about interaction with grid considering comfort requirements, which include indoor air quality, visual comfort, and thermal comfort (Aduda et al., 2014). Therefore, one of the major barriers to activate the buildings is enabling technologies. However, by developments of sensors and wireless technologies, this challenge is solving. Active buildings need information technology (IT) for power management (Weng & Agarwal, 2012). Some other communication technologies may be used in active buildings, including Big Data, Cloud Computing, and sensor technologies (Plageras et al., 2018). Furthermore, ICT is required for energy and comfort in active buildings (Aduda et al., 2014). The Internet of Things (IoT) is another technology for energy management in active buildings (Minoli et al., 2017). The IoT technology is rapidly developing because of the high share of buildings in energy consumption.

In active buildings, all energy systems, including lighting, heating, cooling, and ventilation systems, should be optimally controlled (Sun et al., 2013). For this aim, load forecasting is an effective factor in energy flexibility as well as energy saving of buildings (Gayeski et al., 2012; Roy et al., 2020). In spite of the advantages of load forecasting, there are some issues about it like load data preparation (Fallah et al., 2018).

The lack of a clear definition or building code for designing is another problem about active buildings concept. However, a code is defined in (Fosas et al., 2020) for active buildings as (1) minimum energy consumption, (2) self-generation (3) grid services provision, and (4) minimum carbon emission. Although buildings could become an active part of the power system by using the existing strategies and technologies, no comprehensive and integrated scheme has been presented for the development of active buildings yet.

Another barrier to this development is the relatively high cost of technologies to build active buildings (Bulut et al., 2016). However, currently, wireless sensor networks are the most flexible and promising technologies for creating energy-intelligent buildings (Nguyen & Aiello, 2013).

1.7 Literature Review

A wide number of research studies have been studied on the active building topic. The researches in this field have been accomplished in different viewpoints such as load forecasting, enabling technologies and energy storage systems, and comparing passive and active energy systems, renewable energy sources, ancillary services, and DSR programs. In the following, the previous works are reviewed.

1.7.1 Load Forecasting

As mentioned, load forecasting is an effective factor for the activation of buildings, which has been investigated in some research studies. The suitable forecasting of load demand is effective in the adaptation of load with generation planning. Accordingly, the existing electric demand forecasting approaches have been discussed in (Hernandez et al., 2014) due to the importance of load prediction in recent years by increasing the load demand and also the development of emerging smart grids. In another research (Ma et al., 2018), an approach is introduced to detect the activity of lightweight appliances of buildings, in which a single metering device is used to establish the detection and learning process for all appliances. Furthermore, in (Zhang et al., 2019a), a deep learning approach is adopted to model and control the thermal comfort of buildings, in which the IoT is used to generate the data. The obtained results showed a linear relationship between the air conditioning setpoint and comfort level, which linear relationship contributes to the accuracy and speed of searching the setpoint. In the following, the literature about utilizing two widely used approaches appropriate for load forecasting in active buildings is reviewed.

1.7.1.1 Model Predictive Control Approaches

Some research works have employed strategies based on the model predictive control approach, which is an effective approach to appropriately use available resources based on the forecasted information (e.g., RES production, weather, and load demand) and subjects to constraints. In (Schirrer et al., 2013), an approach based on model predictive control is presented to do a trade-off between minimum grid congestion and maximum self-coverage of energy demand by self-generation of buildings. In addition, in (Dhulipala et al., 2019), the authors have introduced a distributed model predictive control approach for Volt-Var control of smart-building-integrated distribution system with the aim of voltage profile improvement, loss reduction, and conservation voltage reduction. This reference has considered the thermal inertia of buildings and also the ability to provide reactive power (Sadeghian et al., 2020b) by smart buildings. Both the power and heat resources of buildings are coordinately controlled using an economic model predictive control approach in (Liberati et al., 2019) with the aim of participation in DSR by considering the users' requirements. Different loads, including flexible loads, electric vehicles, battery storage, and heat pump, are managed through the proposed approach, and the control objective is energy cost minimization.

1.7.1.2 Load Disaggregation Techniques

A number of other researches have adopted load disaggregation techniques. In (Azizi et al., 2020), a novel event-based optimization algorithm based on a load disaggregation technique. i.e., non-intrusive load monitoring has been proposed. This approach is inexpensive for actual cases with small training datasets since it has a low computation burden. In another work (Moradzadeh et al., 2020), an unsupervised approach based on dimensionality reduction has been employed for pattern recognition of power consumption in electrical appliances. Dimension reduction of data has been accomplished by principal component analysis in this reference. Moreover, in (Zeinal-Kheiri et al., 2020), a disaggregated approach using low-frequency data has been adopted, in which the solution is an appropriate coordination of power consumption of the appliances, by minimizing the least square errors.

1.7.2 Enabling Technologies

Enabling technologies of active buildings has been investigated in some researches. In (Nugur et al., 2019), the design and implementation of IoT gateway on the basis of the most up-to-date technologies has been described for cloud-based building energy management systems, which are the solutions to issues caused by the conventional energy management systems. In addition, optimal communication technology scheduling for smart buildings with the aim of minimizing the energy consumption of such communication devices, which have been ignored by researchers and engineers, has been investigated in (Zhang et al., 2019b). In another attempt (Zhang et al., 2020), a learning strategy for the thermal model of the smart building based on the IoT gateway has been investigated. Through this framework, without manual configuration, the thermal models of thermal zones are identified. Furthermore, design and implementation of an emergency smart lighting system based on the IoT gateway is studied in (Xu et al., 2019) for 400 emergency lights situated in nine different building types.

1.7.3 Ancillary Services

Some research works have investigated the role of active buildings in proving ancillary services. In (Keskar et al., 2020), an experimental study to evaluate the energy efficiency of three university buildings, when providing ancillary services, has been accomplished. Further, in (Kiliccote et al., 2011), providing non-spinning reserve product by commercial buildings through a price-based automated DSR has been studied, in which the understudy building participates in the ancillary service market. In some other researches (Pavlak et al., 2014; Lin et al., 2017), the frequency regulation capability of medium and large office buildings through participation in the ancillary service market has been investigated. Moreover, in (Lymperopoulos et al., 2015), ancillary service provision by commercial buildings in Switzerland based on real-time prices has been investigated. In this reference, the effect of spot price on the comfort level of customers has been studied.

1.7.4 Energy Storage Systems

Some research studies have studied the integration of energy storage systems with buildings due to the positive influence of them to mitigate the surplus energy of renewable generations and enabling the building demand to be flexible (Xu et al., 2012). In addition to the mentioned advantages, energy storage systems are useful from the viewpoint of economic and reliability improvement (Sadeghian et al., 2020c). In literature, the integration of electrical, thermal, and hybrid storage systems with active buildings has been studied. In the following, these researches are reviewed.

1.7.4.1 Stationary Energy Storage Systems

Integration of different energy storage systems with buildings has been accomplished in some works. In (Xu et al., 2012), the application of different energy storage systems, including ice storage, hot water tank, and battery storage, has been evaluated by considering the uncertainty in solar radiation and load demand. The obtained results show that ice storage and hot water tank are more effective in energy saving compared to battery storage systems. The reason for this is the high investment cost of battery storage and its short lifetime. Operation management based on forecasted energy consumption cost has been presented for optimal scheduling of a cluster of homes in (Zhang et al., 2013) for two cases, including a 30-home case and a 90-home case. In this reference, wind turbines, electric vehicles, electrical and thermal storages, boiler, and CHP system have been considered. The obtained results show the effectiveness of the approach in cost-saving and energy demand reduction.

1.7.4.2 Electric Vehicles

Integration of electric vehicles as a large-scale mobile storage with buildings has been considered in some researches. In (Wang et al., 2012a), plug-in hybrid electric vehicles are integrated with buildings to provide both power and energy and make the building more economical and flexible. The obtained results showed that by integrating a proper amount of such vehicles, the efficiency and reliability of the power supply and also the resident comfort level are preserved. In addition, in (Aliasghari et al., 2018), electric vehicles are integrated with renewable-based microgrids for participation in power and reserve markets, in which incentives are paid to the electric vehicle's owners to participate in DSR programs by paying profit.

1.7.4.3 Thermal Energy Storage Systems

Some other researchers have studied the integration of thermal energy storage systems with buildings. In (Xu & Van Dessel, 2008; Xu et al., 2007), a real active building integrated with solar energy and thermal storage is studied, in which thermoelectric modules are used to control the indoor air temperature. In addition, in (Chen et al., 2014a), an active building is integrated with thermal energy storage, in which a control strategy is adopted to optimally control the charging mechanism. In another work (Wei et al., 2019), the authors have investigated the application of thermal energy storage systems, in which the considered thermal storage is based on phase change material, since such storage systems can provide a considerable amount of thermal energy at a fixed temperature. The influence of solar radiation and the external environment has been taken into account. In this reference, the peak load and power fluctuation are improved by the proposed approach based on the obtained results from numerical simulation. Moreover, the authors in (Chen et al., 2014b) have investigated the optimal design and operation of an active building integrated with the thermal energy storage system.

1.7.5 Passive and Active Energy Management Systems

Some research studies have compared the passive and active energy-saving strategies. In (Sun et al., 2018), both the active and passive energy-efficient technologies have been evaluated for efficiency improvement of a building situated in Southern Asia at Singapore with hot and humid climates. The obtained results showed that the active technologies such as high-performance air-conditioning and energy-efficient lighting systems are more efficient since they have a short payback period and achieve 23–28% and 12–17% energy saving, respectively. On the other hand, passive systems such as window film have a long payback period and only achieve 5% energy saving. The satisfying payback period in this reference is considered to be 10 years. In another attempt (Ma et al., 2016), several passive and active approaches are considered for an active office building in Tianjin. The passive approaches included the changeable external shading technology, natural light utilization, etc., whereas the active approaches included the free cooling technology, heat pump system, air conditioning, etc. By using the adopted approaches, 15% of energy consumption has been saved for the understudied case. A fast DSR scheme for the grid-connected active buildings using both active and passive cold storages has been investigated in (Cui et al., 2015). In this approach, the chillers are turned off when needed. The indoor temperature and the power demand reduction are forecasted in this reference.

1.7.6 Energy Sharing among Buildings

Energy sharing has been studied in the literature, where energy sharing of different components are considered. In a research, the energy generated by a CHP system is shared among buildings from different sectors (Kayo et al., 2014). In addition, renewable energy sources and battery have been shared among five apartment buildings in (Roberts et al., 2019) and between two smart buildings in an off-grid mode (Hakimi & Hasankhani, 2020). Furthermore, trigeneration and energy storage systems have been shared among building clusters with the aim of cost minimization in (Dai et al., 2015).

The peer-to-peer transactive energy as a special form of energy sharing among buildings has also been researched in the literature. In (Cui et al., 2020), optimal energy sharing among six pairs of buildings is studied in a distributed approach with the aim of minimizing total social energy cost. In another research, the peer-to-peer energy transaction has been considered to share the production of renewable energy sources among different types of buildings each one including a flexibility source (Cui et al., 2019). Further, a two-stage strategy for peer-to-pear energy sharing among smart homes located in a microgrid is studied, in which a photovoltaic/battery system been shared for this aim (Alam et al., 2019). Moreover, in (Mehrjerdi, 2020), a hybrid photovoltaic/hydrogen storage system is integrated with buildings that enable the pear-to-peer energy trading among them.

1.7.7 Demand-Side Response Programs

DSR programs have received a lot of attention, which are useful programs for the activation of buildings. In (Samad et al., 2016), the authors reviewed the motivation for DSR and highlight the enabling technologies, control and communication protocols, and architectural models, which are currently used. In this references, four projects for commercial buildings and microgrids in China, the United Kingdom, and the United States have been discussed. In addition, in (Zare Oskouei et al., 2020), the combination of electric vehicles and photovoltaic has been considered as responsive components of an active building with the aim of peak reduction and profit maximization for customers.

Some DSR programs ignoring the renewable energy sources are introduced for buildings in some researches. In (Bilgin et al., 2016), the regulation service is provided by flexible loads of buildings by responding to price signals broadcasted by the independent system operator. In this reference, the smart building operator modulates the aggregate loads of the building to respond to the price signal, in which the historical data are used to forecast the behavior of the loads. In addition, a decentralized energy management approach based on price-driven DSR has been presented in (Moradzadeh & Tomsovic, 2013) for cost improvement of both the utility and customers, whereas the customers' satisfaction is retained at a good level. In this reference, the uncertainty in demand estimation of non-schedulable loads is modeled using the normal distribution function. Further, a time-varying DSR program for residential appliances with the aim of minimizing the cost of individual customers and load demand flattening is presented in (Safdarian et al., 2014), in which a distributed algorithm is used to coordinate the DSR provided by buildings residents. A Finnish distribution network is used in this reference as a case study with three cases, including base case, non-coordinated DSR, and coordinated DSR. In another work (Razmara et al., 2018), an approach is presented for the building-to-grid strategy with two objectives, including load factor maximization of the grid and energy cost reduction of the office buildings' electricity consumers. In this reference, an office building at Michigan Technological University has been considered as a case study, while the grid included the capacitor banks and transformer load tap changers. The obtained results showed that 25% cost saving is achieved through the proposed approach. Furthermore, in (Diekerhof et al., 2018), an approach based on hierarchical distributed robust optimization has been adopted for DSR of electro-thermal heating units with the aim of load balance, in which the proposed approach is able to accommodate different customer types. The obtained results indicated that the aggregator could improve the load peak by 37%, while the electricity bill of customers could be reduced by 11%.

1.7.7.1 Energy Storage-Based Demand-Side Response Programs

Utilizing energy storage systems for demand flexibility programs has been investigated in some researches. An approach is introduced for mitigating load peak of smart buildings using a DSR program based on vehicle-to-building by satisfying the vehicles' load (Dagdougui et al., 2019), in which the building is integrated with a hybrid photovoltaic/battery system. In another work (Gabbar & Othman, 2018), smart building energy automation is integrated with a battery storage system, in which the battery lifetime is retained by limiting the charging/discharging of batteries through incorporating its operation with DSR.

1.7.7.2 Autonomous Demand-Side Response Programs

In some other works, autonomous DSR programs are considered for active buildings to enable them for demand flexibility. The authors in (Costanzo et al., 2012) have investigated an autonomous demand-side management approach for heating, ventilation, and air conditioning (HVAC) system, refrigeration, and water heating system, in which the proposed scheme containing three modules included DSR management, load balancing, and admission control so that different loads can participate in the demand flexibility program. In this reference, the goal is to minimize the total operational cost considering the capacity constraints and operational requirements for individual appliances. The autonomous demand has also been investigated with the aim of cost minimization in a hybrid photovoltaic/wind/battery system in (Forouzandehmehr et al., 2015) for some responsive loads, including HVAC system, refrigeration, and water heating system.

1.7.8 Renewable Energy Sources

In a number of other researches, DSR programs have been considered in renewable-integrated buildings. A DSR program for the responsive loads of a photovoltaic/batter-based building is considered by local distribution companies in (Alrumayh & Bhattacharya, 2019). In this approach, while the consumers optimize their economic benefit, the local distribution company tries to improve the network operation by implementing DSR with direct communication with consumers. The network objectives were reducing the peak demand, minimizing the power losses and retaining the voltage in a good level. In (Thomas et al., 2018), an approach is presented for energy management of a grid-connected office building integrated with a hybrid photovoltaic/electric vehicles/battery system considering the uncertainty in the driving schedule of electric vehicles and photovoltaic generation. The obtained results showed the lower cost of the stochastic approach compared to the deterministic approach. An energy management approach based on information fusion for energy and comfort management of buildings has been introduced in (Wang et al., 2012b), in which a hybrid wind/photovoltaic/battery system is considered as the primary source, whereas the grid is considered as the backup power supply. In this reference, a high comfort level with minimal energy use is investigated through two individual comfort models. Some interruptible loads are also considered to contribute to the demand flexibility based on the predetermined priority of loads. Further, in (Wang & Wang, 2013), to enable the photovoltaic/wind/battery-building for bi-directional trading with the utility, an adaptive negotiation agent is considered to improve the building intelligence. By using the proposed framework, the building dynamically adjusts its behavior in response to varying behaviors of the opponent. The advantage of the proposed approach is maximizing the payoffs of the trader with minimum negotiation time. Moreover, a price-responsive model based on minimizing the net energy cost is presented in (Ostadijafari et al., 2020) for HVAC systems integrated with inflexible loads and hybrid photovoltaic/battery system, in which the comfort level of residents is maintained at a good level. This approach uses the occupancy information to effectively optimize the problem.

1.8 Conclusions

In this chapter, the active building concept and definition as a promising solution for energy sustainability was discussed, and the literature, in this regard, was reviewed. The characteristics and requirements needed to build activate buildings were described. The ability to respond to the grid via demand flexibility and electricity generation was mentioned as the major features of active buildings. Capability of active buildings for different energy services was also discussed in this chapter. The previous works in different aspects of the active building topic were reviewed. The research studies were about load forecasting, enabling technologies and DSR strategies, and comparing passive and active energy systems, renewable energy sources, ancillary services, energy sharing, and energy storage systems. As the literature showed, the activation of buildings is critical for energy and cost-saving and also for improving the grid reliability. The comfort level of residents is another advantage for the activation of buildings. The previous works also revealed that although the activation is investigated mostly for residential buildings, other building types such as office and commercial buildings are important for activation, whereas they have been somewhat ignored.

References

Aalami, H. A., Moghaddam, M. P., & Yousefi, G. R. (2010). Modeling and prioritizing demand response programs in power markets. *Electric Power Systems Research, 80*(4), 426–435.

Aduda, K. O., Zeiler, W., Boxem, G., & Labeodan, T. (2014). On defining information and communication technology requirements and associated challenges for 'energy and comfort active' buildings. *Procedia Computer Science, 32*, 979–984.

Alam, M. R., St-Hilaire, M., & Kunz, T. (2019). Peer-to-peer energy trading among smart homes. *Applied Energy, 238*(October 2018), 1434–1443.

Albadi, M. H., & El-Saadany, E. F. (2008). A summary of demand response in electricity markets. *Electric Power Systems Research, 78*(11), 1989–1996.

Aliasghari, P., Mohammadi-Ivatloo, B., Alipour, M., Abapour, M., & Zare, K. (2018). Optimal scheduling of plug-in electric vehicles and renewable micro-grid in energy and reserve markets considering demand response program. *Journal of Cleaner Production, 186*, 293–303.

Alrumayh, O., & Bhattacharya, K. (2019). Flexibility of residential loads for demand response provisions in smart grid. *IEEE Transactions on Smart Grid, 10*(6), 6284–6297.

Azizi, E., Shotorbani, A. M., Hamidi-Beheshti, M. T., Mohammadi-Ivatloo, B., & Bolouki, S. (2020). Residential household non-intrusive load monitoring via smart event-based optimization. *IEEE Transactions on Consumer Electronics, 3063*(c), 1–9.

Bilgin, E., Caramanis, M. C., Paschalidis, I. C., & Cassandras, C. G. (2016). Provision of regulation service by smart buildings. *IEEE Transactions on Smart Grid, 7*(3), 1683–1693.

Bühler, F., Zühlsdorf, B., Van Nguyen, T., & Elmegaard, B. (2019). A comparative assessment of electrification strategies for industrial sites: Case of milk powder production. *Applied Energy, 250*(April), 1383–1401.

Bulut, M. B., Odlare, M., Stigson, P., Wallin, F., & Vassileva, I. (2016). Active buildings in smart grids – Exploring the views of the Swedish energy and buildings sectors. *Energy and Buildings, 117*, 185–198.

Chen, Y., Athienitis, A. K., & Galal, K. E. (2014a). A charging control strategy for active building-integrated thermal energy storage systems using frequency domain modeling. *Energy and Buildings, 84*, 651–661.

Chen, Y., Galal, K. E., & Athienitis, A. K. (2014b). Design and operation methodology for active building-integrated thermal energy storage systems. *Energy and Buildings, 84*, 575–585.

Chen, Y., Xu, P., Gu, J., Schmidt, F., & Li, W. (2018). Measures to improve energy demand flexibility in buildings for demand response (DR): A review. *Energy and Buildings, 177*. Elsevier B.V., 125–139.

Costanzo, G. T., Zhu, G., Anjos, M. F., & Savard, G. (2012). A system architecture for autonomous demand side load management in smart buildings. *IEEE Transactions on Smart Grid, 3*(4), 2157–2165.

Cui, B., Wang, S., Yan, C., & Xue, X. (2015). Evaluation of a fast power demand response strategy using active and passive building cold storages for smart grid applications. *Energy Conversion and Management, 102*, 227–238.

Cui, S., Wang, Y. W., & Xiao, J. W. (2019). Peer-to-peer energy sharing among smart energy buildings by distributed transaction. *IEEE Transactions on Smart Grid, 10*(6), 6491–6501.

Cui, S., Wang, Y. W., Shi, Y., & Xiao, J. W. (2020). A new and fair peer-to-peer energy sharing framework for energy buildings. *IEEE Transactions on Smart Grid, 11*(5), 3817–3826.

Dagdougui, H., Ouammi, A., & Dessaint, L. A. (2019). Peak load reduction in a smart building integrating microgrid and V2B-based demand response scheme. *IEEE Systems Journal, 13*(3), 3274–3282.

Dai, R., Hu, M., Yang, D., & Chen, Y. (2015). A collaborative operation decision model for distributed building clusters. *Energy, 84*, 759–773.

Dewidar, Y., Mohamed, K. M., Ashour, N. M., Dewidar, K. M., Mohamed, N. M., & Ashour, Y. S. (2013). Living skins: A new concept of self active building envelope regulating systems. In *Advancing the green agenda; technology, practices and policies conference–BUID*, pp. 1–8.

Dhulipala, S. C., Monteiro, R. V. A., Teixeira, R. F. D. S., Ruben, C., Bretas, A. S., & Guimaraes, G. C. (2019). Distributed model-predictive control strategy for distribution network volt/VAR control: A smart-building-based approach. *IEEE Transactions on Industry Applications, 55*(6), 7041–7051.

Diekerhof, M., Peterssen, F., & Monti, A. (2018). Hierarchical distributed robust optimization for demand response services. *IEEE Transactions on Smart Grid, 9*(6), 6018–6029.

Dominguez-Garcia, A. D., & Hadjicostis, C. N. (2010). Coordination and control of distributed energy resources for provision of ancillary services. In *2010 1st IEEE international conference on smart grid communications, SmartGridComm 2010*, pp. 537–542.

Fallah, S. N., Deo, R. C., Shojafar, M., Conti, M., & Shamshirband, S. (2018). Computational intelligence approaches for energy load forecasting in smart energy management grids: State of the art, future challenges, and research directions. *Energies, 11*(3).

Fawcett, T., Layberry, R., & Eyre, N. (2014). Electrification of heating: The role of heat pumps. In *BIEE Conference*, September, pp. 1–13.

Forouzandehmehr, N., Esmalifalak, M., Mohsenian-Rad, H., & Han, Z. (2015). Autonomous demand response using stochastic differential games. *IEEE Transactions on Smart Grid, 6*(1), 291–300.

Fosas, D., Nikolaidou, E., Roberts, M., Allen, S., Walker, I., & Coley, D. (2020). Towards active buildings: Rating grid-servicing buildings. *Building Services Engineering Research and Technology*, 1–27.

Gabbar, H. A., & Othman, A. M. (2018). Resilient battery management for buildings. *Energy Conservation Residential and Commercial Industrial Facility*, 249–266.

Gayeski, N. T., Armstrong, P. R., & Norford, L. K. (2012). Predictive pre-cooling of thermo-active building systems with low-lift chillers. *HVAC R Research, 18*(5), 858–873.

Hakimi, S. M., & Hasankhani, A. (2020). Intelligent energy management in off-grid smart buildings with energy interaction. *Journal of Cleaner Production, 244*.

Hernandez, L., et al. (2014). A survey on electric power demand forecasting: Future trends in smart grids, microgrids and smart buildings. *IEEE Communication Surveys and Tutorials, 16*(3), 1460–1495.

Kayo, G., Hasan, A., & Siren, K. (2014). Energy sharing and matching in different combinations of buildings, CHP capacities and operation strategy. *Energy and Buildings, 82*, 685–695.

Keskar, A., Anderson, D., Johnson, J. X., Hiskens, I. A., & Mathieu, J. L. (2020). Do commercial buildings become less efficient when they provide grid ancillary services? *Energy Efficiency, 13*(3), 487–501.

Kiliccote, S., Piette, M. A., Koch, E., & Hennage, D. (2011). Utilizing Automated Demand Response in commercial buildings as non-spinning reserve product for ancillary services markets. *Proceedings of the IEEE Conference on Decision and Control*, 4354–4360.

Langevin, J., Harris, C. B., & Reyna, J. L. (2019). Assessing the potential to reduce U.S. building CO2 emissions 80% by 2050. *Joule, 3*(10), 2403–2424.

Lauro, F., Moretti, F., Capozzoli, A., & Panzieri, S. (2015). Model predictive control for building active demand response systems. *Energy Procedia, 83*, 494–503.

Liberati, F., Di Giorgio, A., Giuseppi, A., Pietrabissa, A., Habib, E., & Martirano, L. (2019). Joint model predictive control of electric and heating resources in a smart building. *IEEE Transactions on Industry Applications, 55*(6), 7015–7027.

Lin, Y., Mathieu, J. L., Johnson, J. X., Hiskens, I. A., & Backhaus, S. (2017). Explaining inefficiencies in commercial buildings providing power system ancillary services. *Energy and Buildings, 152*, 216–226.

Long, L., & Ye, H. (2015). Effects of thermophysical properties of wall materials on energy performance in an active building. *Energy Procedia, 75*, 1850–1855.

Luo, Y., Zhang, L., Liu, Z., Wang, Y., Meng, F., & Wu, J. (2016). Thermal performance evaluation of an active building integrated photovoltaic thermoelectric wall system. *Applied Energy, 177*, 25–39.

Luo, Y., Zhang, L., Bozlar, M., Liu, Z., Guo, H., & Meggers, F. (2019). Active building envelope systems toward renewable and sustainable energy. *Renewable and Sustainable Energy Reviews, 104*(January), 470–491.

Lymperopoulos, I., Qureshi, F. A., Nghiem, T., Khatir, A. A., & Jones, C. N. (2015). Providing ancillary service with commercial buildings: The Swiss perspective. *IFAC-PapersOnLine, 28*(8), 6–13.

Ma, H., Zhou, W., Lu, X., Ding, Z., & Cao, Y. (2016). Application of low cost active and passive energy saving technologies in an ultra-low energy consumption building. *Energy Procedia, 88*, 807–813.

Ma, M., Lin, W., Zhang, J., Wang, P., Zhou, Y., & Liang, X. (2018). Toward energy-awareness smart building: Discover the fingerprint of your electrical appliances. *IEEE Transactions on Industrial Informatics, 14*(4), 1458–1468.

Mehrjerdi, H. (2020). Peer to peer home energy management incorporating hydrogen storage system and solar generating units. *Renewable Energy, 156*, 183–192.

Minoli, D., Sohraby, K., & Occhiogrosso, B. (2017). IoT Considerations, requirements, and architectures for smart buildings-energy optimization and next-generation building management systems. *IEEE Internet of Things Journal, 4*(1), 269–283.

Moradzadeh, B., & Tomsovic, K. (2013). Two-stage residential energy management considering network operational constraints. *IEEE Transactions on Smart Grid, 4*(4), 2339–2346.

Moradzadeh, A., Sadeghian, O., Pourhossein, K., Mohammadi-Ivatloo, B., & Anvari-Moghaddam, A. (2020). Improving residential load disaggregation for sustainable development of energy via principal component analysis. *Sustainability, 12*(8), 3158.

Neirotti, F., Noussan, M., & Simonetti, M. (2020). Towards the electrification of buildings heating – Real heat pumps electricity mixes based on high resolution operational profiles. *Energy, 195*, 116974.

Nguyen, T. A., & Aiello, M. (2013). Energy intelligent buildings based on user activity: A survey. *Energy and Buildings, 56*, 244–257.

Niu, J., Tian, Z., Lu, Y., & Zhao, H. (2019). Flexible dispatch of a building energy system using building thermal storage and battery energy storage. *Applied Energy, 243*(March), 274–287.

Nugur, A., Pipattanasomporn, M., Kuzlu, M., & Rahman, S. (2019). Design and development of an IoT gateway for smart building applications. *IEEE Internet of Things Journal, 6*(5), 9020–9029.

Ostadijafari, M., Dubey, A., & Yu, N. (2020). Linearized price-responsive HVAC controller for optimal scheduling of smart building loads. *IEEE Transactions on Smart Grid, 11*(4), 3131–3145.

Pavlak, G. S., Henze, G. P., & Cushing, V. J. (2014). Optimizing commercial building participation in energy and ancillary service markets. *Energy and Buildings, 81*, 115–126.

Pirbazari, A. M. (2011). Ancillary services definitions, markets and practices in the world. In *2010 IEEE/PES transmission and distribution conference and exposition: Latin America, T and D-LA 2010*, pp. 32–36.

Plageras, A. P., Psannis, K. E., Stergiou, C., Wang, H., & Gupta, B. B. (2018). Efficient IoT-based sensor BIG Data collection–processing and analysis in smart buildings. *Future Generation Computer Systems, 82*, 349–357.

Razmara, M., Bharati, G. R., Shahbakhti, M., Paudyal, S., & Robinett, R. D. (2018). Bilevel optimization framework for smart building-to-grid systems. *IEEE Transactions on Smart Grid, 9*(2), 582–593.

Roberts, M. B., Bruce, A., & MacGill, I. (2019). Impact of shared battery energy storage systems on photovoltaic self-consumption and electricity bills in apartment buildings. *Applied Energy, 245*(March), 78–95.

Roy, S. S., Samui, P., Nagtode, I., Jain, H., Shivaramakrishnan, V., & Mohammadi-ivatloo, B. (2020). Forecasting heating and cooling loads of buildings: A comparative performance analysis. *Journal of Ambient Intelligence and Humanized Computing, 11*(3), 1253–1264.

Sadeghian, O., Shotorbani, A. M., & Mohammadi-Ivatloo, B. (2019a). Generation maintenance scheduling in virtual power plants. *IET Generation Transmission and Distribution, 13*(12), 2584–2596.

Sadeghian, O., Nazari-Heris, M., Abapour, M., Taheri, S. S., & Zare, K. (2019b). Improving reliability of distribution networks using plug-in electric vehicles and demand response. *Journal of Modern Power System and Clean Energy, 7*(5), 1189–1199.

Sadeghian, O., Moradzadeh, A., Mohammadi-Ivatloo, B., Abapour, M., & Marquez, F. P. G. (2020a). Generation units maintenance in combined heat and power integrated systems using the mixed integer quadratic programming approach. *Energies, 13*(11), 2840.

Sadeghian, O., Oshnoei, A., Kheradmandi, M., & Mohammadi-Ivatloo, B. (2020b). Optimal placement of multi-period-based switched capacitor in radial distribution systems. *Computers and Electrical Engineering, 82*, 106549.

Sadeghian, O., Oshnoei, A., Khezri, R., & Muyeen, S. M. (2020c). Risk-constrained stochastic optimal allocation of energy storage system in virtual power plants. *Journal of Energy Storage, 31*(January), 101732.

Safdarian, A., Fotuhi-Firuzabad, M., & Lehtonen, M. (2014). A distributed algorithm for managing residential demand response in smart grids. *IEEE Transactions on Industrial Informatics, 10*(4), 2385–2393.

Samad, T., Koch, E., & Stluka, P. (2016). Automated demand response for smart buildings and microgrids: The state of the practice and research challenges. *Proceedings of the IEEE, 104*(4), 726–744.

Schirrer, A., Konig, O., Ghaemi, S., Kupzog, F., & Kozek, M. (2013). Hierarchical application of model-predictive control for efficient integration of active buildings into low voltage grids. In *2013 working modelling simulation on cyber-physical energy systems, MSCPES 2013*, pp. 1–6.

Sher, F., Kawai, A., Güleç, F., & Sadiq, H. (2019). Sustainable energy saving alternatives in small buildings. *Sustainable Energy Technologies and Assessments, 32*(February), 92–99.

Siano, P., De Marco, G., Rolan, A., & Loia, V. (2019). A survey and evaluation of the potentials of distributed ledger technology for peer-to-peer transactive energy exchanges in local energy markets. *IEEE Systems Journal, 13*(3), 3454–3466.

Sun, B., Luh, P. B., Jia, Q. S., Jiang, Z., Wang, F., & Song, C. (2013). Building energy management: Integrated control of active and passive heating, cooling, lighting, shading, and ventilation systems. *IEEE Transactions on Automation Science and Engineering, 10*(3), 588–602.

Sun, X., Gou, Z., & Lau, S. S. Y. (2018). Cost-effectiveness of active and passive design strategies for existing building retrofits in tropical climate: Case study of a zero energy building. *Journal of Cleaner Production, 183*, 35–45.
Tarroja, B., et al. (2018). Translating climate change and heating system electrification impacts on building energy use to future greenhouse gas emissions and electric grid capacity requirements in California. *Applied Energy, 225*(April), 522–534.
Theo, W. L., Lim, J. S., Ho, W. S., Hashim, H., & Lee, C. T. (2017). Review of distributed generation (DG) system planning and optimisation techniques: Comparison of numerical and mathematical modelling methods. *Renewable and Sustainable Energy Reviews, 67*, 531–573.
Thomas, D., Deblecker, O., & Ioakimidis, C. S. (2018). Optimal operation of an energy management system for a grid-connected smart building considering photovoltaics' uncertainty and stochastic electric vehicles' driving schedule. *Applied Energy, 210*, 1188–1206.
Tushar, W., Saha, T. K., Yuen, C., Smith, D., & Poor, H. V. (2020). Peer-to-peer trading in electricity networks: An overview. *IEEE Transactions on Smart Grid, 11*(4), 3185–3200.
Vahidinasab, V., Ardalan, C., Mohammadi-Ivatloo, B., Giaouris, D., & Walker, S. L. (2021). Active building as an energy system: Concept, challenges, and outlook. *IEEE Access*, 1–1.
Verbruggen, B., & Driesen, J. (2015). Grid impact indicators for active building simulations. *IEEE Transactions on Sustainable Energy, 6*(1), 43–50.
Wang, Z., & Wang, L. (2013). Adaptive negotiation agent for facilitating bi-directional energy trading between smart building and utility grid. *IEEE Transactions on Smart Grid, 4*(2), 702–710.
Wang, Z., Wang, L., Dounis, A. I., & Yang, R. (2012a). Integration of plug-in hybrid electric vehicles into energy and comfort management for smart building. *Energy and Buildings, 47*, 260–266.
Wang, Z., Wang, L., Dounis, A. I., & Yang, R. (2012b). Multi-agent control system with information fusion based comfort model for smart buildings. *Applied Energy, 99*, 247–254.
Wang, Y., Shukla, A., & Liu, S. (2017). A state of art review on methodologies for heat transfer and energy flow characteristics of the active building envelopes. *Renewable and Sustainable Energy Reviews, 78*(September 2016), 1102–1116.
Wang, B., Li, Y., Ming, W., & Wang, S. (2020). Deep reinforcement learning method for demand response management of interruptible load. *IEEE Transactions on Smart Grid, 11*(4), 3146–3155.
Wei, F., et al. (2019). A novel thermal energy storage system in smart building based on phase change material. *IEEE Transactions on Smart Grid, 10*(3), 2846–2857.
Weng, T., & Agarwal, Y. (2012). From buildings to smart buildings-sensing and actuation to improve energy efficiency. *IEEE Design and Test of Computers, 29*(4), 36–44.
Xu, X., & Van Dessel, S. (2008). Evaluation of a prototype active building envelope window-system. *Energy and Buildings, 40*(2), 168–174.
Xu, X., Van Dessel, S., & Messac, A. (2007). Study of the performance of thermoelectric modules for use in active building envelopes. *Building and Environment, 42*(3), 1489–1502.
Xu, Z., Guan, X., Jia, Q. S., Wu, J., Wang, D., & Chen, S. (2012). Performance analysis and comparison on energy storage devices for smart building energy management. *IEEE Transactions on Smart Grid, 3*(4), 2136–2147.
Xu, W., et al. (2019). The design, implementation, and deployment of a smart lighting system for smart buildings. *IEEE Internet of Things Journal, 6*(4), 7266–7281.
Xu, X., et al. (2020). Data-driven game-based pricing for sharing rooftop photovoltaic generation and energy storage in the residential building cluster under uncertainties. *IEEE Transactions on Industrial Informatics, 17*(7), 4480–4491.
Yan, X., Ozturk, Y., Hu, Z., & Song, Y. (2018). A review on price-driven residential demand response. *Renewable and Sustainable Energy Reviews, 96*(August), 411–419.
Zare Oskouei, M., Mohammadi-Ivatloo, B., Abapour, M., Anvari-Moghaddam, A., & Mehrjerdi, H. (2020). Practical implementation of residential load management system by considering vehicle-for-power transfer: Profit analysis. *Sustainable Cities and Society, 60*(May), 102144.

Zeiler, W., & Boxem, G. (2009). Geothermal active building concept. In *Sustainability in energy and buildings – Proceedings of the international conference in sustainability in energy and buildings, SEB'09*, pp. 305–314.

Zeinal-Kheiri, S., Shotorbani, A. M., & Mohammadi-Ivatloo, B. (2020). Residential load disaggregation considering state transitions. *IEEE Transactions on Industrial Informatics, 16*(2), 743–753.

Zhan, J., Liu, W., Wu, F., Li, Z., & Wang, C. (2018). Life cycle energy consumption and greenhouse gas emissions of urban residential buildings in Guangzhou city. *Journal of Cleaner Production, 194*, 318–326.

Zhang, D., Shah, N., & Papageorgiou, L. G. (2013). Efficient energy consumption and operation management in a smart building with microgrid. *Energy Conversion and Management, 74*, 209–222.

Zhang, W., Hu, W., & Wen, Y. (2019a). Thermal comfort modeling for smart buildings: A fine-grained deep learning approach. *IEEE Internet of Things Journal, 6*(2), 2540–2549.

Zhang, L., Kerrigan, E. C., & Pal, B. C. (2019b). Optimal communication scheduling in the smart grid. *IEEE Transactions on Industrial Informatics, 15*(9), 5257–5265.

Zhang, X., Pipattanasomporn, M., Chen, T., & Rahman, S. (2020). An IoT-based thermal model learning framework for smart buildings. *IEEE Internet of Things Journal, 7*(1), 518–527.

Chapter 2
Why Active Buildings? Realising the Potentials of Energy Networked Homes: A Social Scientific Perspective

Kate O'Sullivan, Fiona Shirani, Nick Pidgeon, and Karen Henwood

2.1 Introduction

Active Buildings present an ambitious, contemporary conceptualisation that is situated at the intersection between a nexus of interrelated environmental, technical, societal and policy problems currently focussed on reducing carbon emissions. Incorporating low carbon building fabric design, renewable energy production, energy storage capacity and intelligent digitalisation, it has at its heart a core technocratic aim: to facilitate the scale-up of single buildings to neighbourhoods and beyond. At the same time, the very idea of Active Buildings has an imaginative side in that it is seeking to realise the potential of energy networked homes in ways that are both people/system and contemporary/future orientated.

In this chapter, we begin by outlining our understanding of what constitutes an Active Home as a particular type of Active Building (see Fig. 2.1). This highlights the changes to the built, energy and social landscapes that may be expected through the realisation of Active Buildings and also informs the justification of our research design. Following this, we outline the methodological approach we are taking in this research project to empirically study a set of Active Buildings that are under construction in the UK(2020–2023).

We highlight our efforts to bring to light factors that are underrepresented in some existing concepts and that are necessary to understanding the many diverse and valued roles that Active Buildings will play in society, as material places of commerce, education, healthcare or home. Buildings, in all forms, have subjective

K. O'Sullivan (✉) · F. Shirani · K. Henwood
School of Social Sciences, Cardiff University, Cardiff, UK
e-mail: OSullivanKM1@cardiff.ac.uk

N. Pidgeon
School of Psychology, Cardiff University, Cardiff, UK

V. Vahidinasab, B. Mohammadi-Ivatloo (eds.), *Active Building Energy Systems*, Green Energy and Technology, https://doi.org/10.1007/978-3-030-79742-3_2

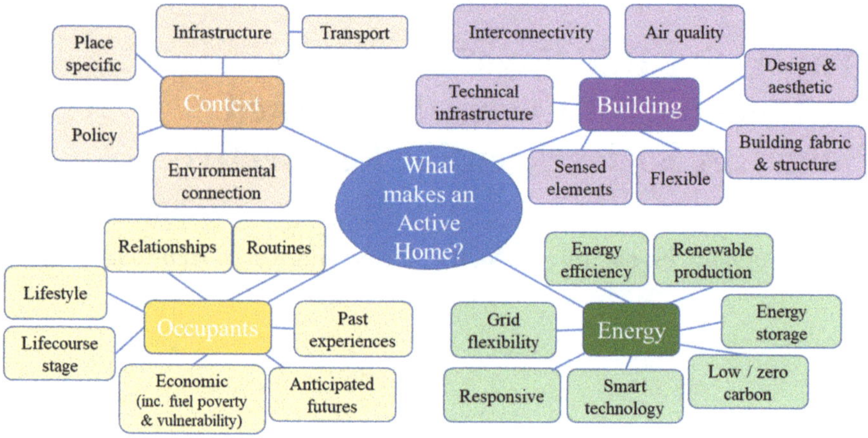

Fig. 2.1 Active homes as dynamic socio-technical configurations

and powerful values and meanings attached to them. For homes, these are formed through a multitude of factors: people's past and anticipated futures, life course transitions, social relationships as well as wider social, economic and political contexts and structures, all of which vary in how they assemble in space and time. Such factors together also influence how people carry out their everyday and energy practices. Our concluding argument is for investigating the interplay between people and energy both *in situ* and *in and through time,* as this reveals that how buildings are imagined (i.e. subjectively perceived/wanted/desired, as well as more abstractly conceptualised) can play a dynamic role in the making of future energy infrastructure.

2.1.1 Active Buildings as Active Homes

In the UK, the Active Building Centre presents a contemporary conceptualisation of Active Buildings, which directly seek to be efficient energy producers, have zero carbon emissions (including embodied) and provide grid flexibility. Elliott et al. (2020) provide the following definition of an Active Building:

> Active Buildings support the wider energy system by intelligently integrating renewable energy technologies for heat, power and transport. Active Buildings are designed to be energy efficient, with novel ways of creating, controlling and releasing energy. Active Buildings have the potential to be energy self-sufficient. When connected with other Active Buildings in a network, they could have the ability to trade energy.

Active Buildings may present a logical development of predecessor sustainable and low carbon building concepts such as Green Buildings, energy-efficient and Passivhaus buildings, net zero buildings and smart buildings. Table 2.1. outlines

Table 2.1 Connection between Active Buildings and predecessor conceptualisations, standards and certificates

Predecessor concept/certificate/standard	Influence on Active Building	Key references
Green Buildings (GBs)	Sustainable resource use throughout building life cycle	Zuo and Zhao (2014) and Mattoni et al. (2018)
Passivhaus	Fabric first building design, ventilation systems and layouts that minimise energy demand	Mitchell and Natarajan (2019) and Passivhaus Trust (Undated)
Energy Performance Certificate (EPC)	Energy efficiency of building structure, in-building energy systems and appliances	Buildings Performance Institute Europe (2014) and Semple and Jenkins (2020)
Net/Nearly Zero Carbon/Energy Buildings (NZEB)	Integration of energy-efficient buildings with renewable energy production	D'Agostino and Mazzarella (2019) and Liu et al. (2019)
Smart buildings	Intelligent digitalisation that facilitates communication between other digitally intelligent buildings, infrastructures and organisations	European Technology Innovation Plan [ETIP] (2020) and Ofgem (2020)

some characteristics of these pre-existing conceptualisations that link into the Active Building conceptualisation; for an in-depth review of this, see O'Sullivan et al. (2020).

This Active Building conceptualisation is different to other adoptions of the word 'active' in relation to low carbon and sustainable buildings, such as 'active houses' in Denmark and France that use solar energy and storage as a means of increasing energy efficiency (Liu et al., 2019). The Active Building conceptualisation is also distinct from when 'active' is used to differentiate buildings with technology that produces, uses and exports renewable energy, from 'passive' buildings such as Passivhaus, which greatly reduce energy consumption and emissions due to in situ energy efficiency properties such as thermal insulation and manual ventilation. Other concepts of 'Active House' such as that of a city-wide sustainable pilot in Stockholm, Sweden (Nilsson et al., 2018), are where the homes are 'active' due to the smart technology used by occupants of the building in line with environmental and time of use (ToU) prompts from the localised smart grid. Finally, there is the use of 'Active House' by the International Active House Alliance networking and partner platform for energy and building industry, academia and planners. The Active House label they have developed is based on three core principles – 'comfort', 'energy' and 'environment' – and adopts a holistic approach to develop a healthy and comfortable indoor climate, 'without negative impact on the climate –

measured in terms of energy, fresh water consumption and the use of sustainable materials' (Active House, Undated, para. 3).

Active Homes are essentially Active Buildings that are lived in. This difference is important as it impacts their materiality, technology mix and operation, their expected day to day use, their expected intrinsic values and meanings and the emotional attachments made to them – changing from an Active Building or Active House to an Active Home. In addition, their Activeness and how this manifests into the built design and then interplays with their occupants' subjectivities also means that they hold differences to traditional (non-active) homes. These differences may be more obvious, for example some Active Homes require new occupant behaviours are adopted to achieve their low/zero carbon and grid flexibility outputs. Other Active Homes may instead influence daily routines and energy behaviours more subtly through utilizing different energy sources, technologies and services that impact how, when and to what effect energy is consumed.

As daily routines and energy behaviours are more than their tangible end product (Henwood et al., 2016a; Shirani et al., 2017; Ozaki, 2018), i.e. heating is turned on to get heat, but also to fulfil practices of caring, health or hospitality, elucidating how and why people carry out such practices is essential to understanding the functions an Active Home must strive to maintain. In combination with the changes to daily life that Active Homes may bring is the alteration to the home's physical materiality and also possibly to the intrinsic values and meanings that are attached to the home. These three areas – daily life, materiality, value and meaning – are in constant interplay and themselves subjectively affected by the personal narratives, social relationships, anticipated futures and current life course of the occupants (Shirani & Henwood, 2011; Groves et al., 2016) in addition to multi-scalar sociocultural, economic and political contexts and energy geographies (Ozaki & Shaw, 2014; Middlemiss et al., 2019; Roberts & Henwood, 2019; Golubchikov & O'Sullivan, 2020).

Considering again predecessor concepts to Active Building, each has anticipated or assumed that the building and its occupants will interplay in different ways to achieve a concept's designed outcomes, with varying success. All have succumbed to varying degrees to the 'performance gap', whereby during their operation a gap emerges between the predicted and actual performance of the building. While several reasons have been cited for this and there are differences in cause and effect between building concepts (c.f. Ade & Rehm, 2019; Hansen et al., 2018), each have been affected in some way by differences between modelled occupancy behaviours and the actual lived behaviours of occupants once using the buildings (c.f. Mallory-Hill & Gorgolewski, 2017; Cozza et al., 2020; Mitchell & Natarajan, 2019). Furthermore, increased digitalisation of buildings and their energy services requires understanding of how building occupants interact with these technologies, which is especially important for designs that hope to alter occupant energy behaviours (Balta-Ozkan et al., 2014; Hansen & Hauge, 2017; Tirado Herrero et al., 2018). Existing research highlights that while people can appreciate and engage with their energy systems using 'smart' technologies, the impacts are largely minimal and short-lived (Hargreaves et al., 2013; Shirani et al., 2020) and people can be concerned for data security and privacy (Balta-Ozkan et al., 2014). Thus, for Active Homes to achieve their multiple ambitions while allowing occupants to achieve a 'life worth living' (Henwood et al., 2016b), it is essential that insights are gained as to how home as a place in space and time is both understood and lived in by occupants, now and as Active Homes are developed into the future.

2.1.2 Energy and Homes in Transition

The development and realisation of Active Homes in the UK has the potential to alter multiple existing socio-technical assemblages at different scales. Considering proposed decarbonisation pathways for the UK (c.f. Committee on Climate Change [CCC], 2019; National Grid ESO, 2020; Ofgem, 2020; Regen, 2020) and predecessor sustainable and low carbon building designs (see Sect. 2.1.1), Active Homes will alter how energy is produced, distributed and consumed (Hansen & Hauge, 2017; Hargreaves & Middlemiss, 2020) and how homes are designed and constructed (c.f. Rossiter & Smith, 2018) and then lived in. In the UK, this change is taking place in multiple locations of new housing developments. This change is also taking place through the time it takes to develop novel building ideas, and then their construction, through to the initial and ongoing occupation of the buildings, as their residents carry out and experience every day life. As we have noted, Active Homes may also change many of the elements that comprise a home, including material elements such as architectural design, technologies and digitalisation (Rossiter & Smith, 2018), sensed elements such as the smells, sounds, sights and temperatures (Zielinska-Dabkowska, 2019) that influence the atmosphere of a place (Bille et al., 2015) and peoples emotional encounters with that place (Conradson, 2005; Hubbard, 2005; Pile, 2010). In addition, Active Homes may alter some intrinsic values of 'home' such as privacy, control and autonomy (Després, 1991).

In numerous ways, the above elements are intertwined and in constant interplay, affecting the decisions people make and how they go about their everyday life (Conradson, 2005; Hubbard, 2005; Roberts & Henwood, 2019) including their social relationships both within and outside the home (Hargreaves & Middlemiss, 2020; Shirani et al., 2017). Furthermore, there is potential for Active Homes to influence existing sociocultural norms and expectations, not just for the households involved but their neighbourhoods (through the accumulation of Active Homes in a place, in addition to their interplay with non-Active Homes in that place). Influence may become visible at wider societal scales as this change interplays with growing concerns for the environment and climate change (Climate Emergency Declaration, 2019; Parliament.UK, 2019), sustainable development (Hopwood et al., 2005; United Nations, 2015) and human health and wellbeing (Welsh Government, 2015; Robinson & Breed, 2019; The Environment Agency, 2020).

2.1.3 People: Lived Experiences, Narratives and Subjectivity

Vitally important to the achievement of the Active Home ambition, and indeed, the reason for their development are people – the end users, occupants, or household. Households are made up of different compositions of people, in different energy geography contexts, and living within different material structures, and thus each have variable requirements and expectations of their home and how they use energy (Roberts & Henwood, 2019). Indeed, how the various elements of an Active Home combine to be encountered or lived in by occupants is individual and complex. This is because 'lived experience' is personal and subjective and for each occupant will be influenced by their past experiences and histories and their present contexts (sociocultural and socioeconomic) (Jones, 2005; Henwood et al., 2016b; Middlemiss et al., 2019), their life course stage (Shirani & Henwood, 2011; Groves et al., 2016) and the lives linked to theirs (Shirani et al., 2017), in addition to their anticipated futures and ambitions (Shirani & Henwood, 2011; Ozaki, 2018). Together, these multiple technological, architectural, human and societal elements influence how, why and what meaning and value are attached to the homes and emotional encounters with the home (Conradson, 2005; Hubbard, 2005; Pile, 2010), which then affect how and what decisions are made by the home occupants in living their everyday lives (Conradson, 2005; Hubbard, 2005).

2.1.4 An Evolving Contextual Landscape

Currently, Active Homes are developing as a part of the solution to the decarbonisation of national energy grids and household energy consumption (Ofgem, 2020). Through reduced CO_2, they contribute to finding solutions to climate change (CCC, 2019). By increasing energy efficiencies that reduce household energy demand, Active Homes can also work towards reducing energy vulnerabilities and fuel poverty (c.f. Pobl, Undated; Scotcher, 2019). However, as grids decarbonise in other ways and as energy may become more expensive, in the longer term, will they realise their ambition or become less relevant? These are key elements that create the contextual landscape within which Active Buildings and Homes are being realised and affect practicalities such as development timelines and funding, through to the more conceptual developer visions and designs and finally to how they may be lived in by their occupants.

2.2 Research Framework

Our Living Well in Low Carbon Homes research aims to explore and enhance insight into the lived experience of Active Homes, understanding how and why people live within the homes in different ways. Fig. 2.2 below represents key project

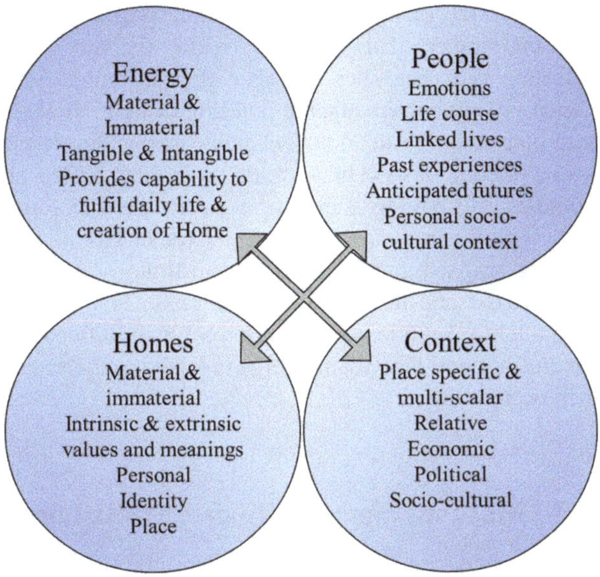

Fig. 2.2 Key research areas

research areas. We adopt an interpretive social scientific framework that considers multiple occupants' lives within different Active Homes as varying experiences of the same phenomenon. As such, this framework allows exploration of different experiences of everyday life within Active Homes and the possible comparison of experiences across Active Homes with different material, technological and imagined compositions. Interpretive phenomenology understands that object realities are subjective to individual interpretation (Creswell, 2013) that individual experience is informed in part by their past histories and to uncover the reasons behind why people do what they do requires pulling away 'layers of forgetfulness or hiddenness that are present in our everyday existence' (Frechette et al., 2020, p. 2). Within this frame, we are also able to draw on our past experience of conducting qualitative longitudinal research (Henwood & Shirani, 2012). This includes a focus on life course and relationships, temporal sensitivity (particularly to anticipated futures) and attentiveness to place dynamics within and outside of the home (Shirani et al., 2017). This focus offers insight into how people's daily lives including their energy behaviours and routines are affected by their life course (Shirani & Henwood, 2011; Groves et al., 2016), linked lives (Shirani et al., 2017) and anticipated futures (Henwood et al., 2016a; Shirani et al., 2017). Building on this, we also plan to elucidate the interplay between people and their subjective perceptions with the different material, technological and intrinsic elements that comprise their homes. This is important as how homes are subjectively perceived and valued and affects how people live within them and also how they may perceive changes to the different elements of 'home' (Després, 1991; Moore, 2000; Roberts & Henwood, 2019).

During the first year of the research project (2019–2020), a conceptual review was undertaken. This review provided the theoretical underpinning in not only selecting our case sites but also in positioning Active Buildings generally in a context of global sustainable building design. However, throughout the project we also plan to carry out document analysis (e.g. development sales literature, websites, social media posts), in addition to noting wider changing political and social contexts and paying attention to government announcements and policy changes (c.f. O'Sullivan et al., 2020). Our Active Home case sites are situated spatially and are realised and lived in through time; in multiple ways, they are affected by multi-scalar and interconnected contexts. This includes the policy, economic and socio-cultural contexts that in different ways influence the component organisations, institutions and individual people involved in their development and the people who will occupy them.

2.3 Methodology: Multi-site, Multistage and Multi-method

2.3.1 Multi-site: Locations

As our research interest is in the lived experience of Active Homes, in selecting our case sites we first had to consider what constituted an Active Home. Initial definitions of Active Buildings focused on reduced energy consumption and carbon emissions, renewable energy production and capacity to interconnect with other Active Buildings and with national grids. In addition, Active Buildings present new and alternative ways to carry out these functions (Elliott et al., 2020; Strbac et al., 2020). This understanding of Active Buildings influenced our case site selection, directing choice to novel developments that in different ways encompassed energy efficiency, renewable energy production and capacity for intelligent communication between users, the buildings and national grids.

However, the Active Building ambition can be realised in different ways, for example with the use of different energy sources, technologies and services; the use of different building materials, designs and layouts and the positioning of the building in different orientations and locations. For some Active Home developments, reduced CO_2 emissions, renewable energy production or smart intra-building and national grid communication singularly or combined may not be a key objective, or primary motivation. Indeed, for some Active Home developments, this defined role can be a by-product of their initial aim. For example, we have found that for some Active Homes a key concern is the alleviation of energy vulnerability and fuel poverty, while for others their key concern is occupant health and wellbeing. Being near or zero carbon, producing renewable energy or providing national grid flexibility in such instances is a secondary beneficial outcome in the achievement of this initial ambition. Within our selection of Active Home developments, we noted four dominant themes that to varying extents shaped the realisation of each site.

These include ambitions towards 'system-change', the improvement of occupant 'energy sufficiency', to minimise 'environmental' harm or to improve occupant's 'health and wellbeing'.

2.3.1.1 System Change

Active Homes with a focus on system change represent ambitions towards the disruption of existing socio-technical systems. We draw on Geels' (2018, p. 225) Multi-Level Perspective (MLP) to understand socio-technical systems as complex systems consisting of human society and infrastructure systems whereby a 'semi-coherent set of rules and institutions' guide the actions of actors. MLP suggests that the system is made up of three levels: Niche-innovation (small networks of actors with novel innovations), Socio-technical regime (existing, stable mix of supply chains, markets, user preferences and processes) and Socio-technical landscape (wider political, economic and social context). Change to the regime is difficult due to various 'lock-ins' or 'path dependences' but can occur when Niche-innovations emerge when pressure is placed on the regime to 'open up' by the Socio-technical Landscape. For our System Change Active Home developments, ambition to alter existing regimes includes influencing not only the energy system (at multiple scales) but also the building construction industry and housing markets. System Change developments plan to demonstrate that energy can be produced, distributed and consumed differently and more sustainably than existing systems allow, while maintaining occupant satisfaction, and that buildings can be designed and constructed to low-carbon criteria while being affordable to develop and live within.

2.3.1.2 Energy Sufficient

Active Homes with a focus on energy sufficiency represent ambitions towards the mitigation of energy vulnerability and fuel poverty for the home occupants. Energy vulnerability and fuel poverty are formed through complex interplay between energy geographies, housing conditions and personal socio-cultural and socio-economic contexts (Golubchikov & O'Sullivan, 2020) resulting in situations whereby accessing sufficient energy to participate fully in society is not possible. Fuel poverty can hold multiple other negative impacts for households, affecting not just social inclusion but morbidity, health and overall life chances (Day & Walker, 2013). In addressing these issues, policy and practical interventions have placed focus on the improvement of housing energy efficiencies (Hills, 2012) as by doing so, household energy demand will be reduced, and thus less household income is needed to purchase energy. Energy-sufficient developments plan to demonstrate that by utilising low carbon energy production and energy storage along with highly energy-efficient building design, household energy vulnerability will be reduced as households will be able to consume sufficient energy to meet their particular energy needs, without incurring high energy bills.

2.3.1.3 Environmental

Active Homes with a focus on socio-ecology represent the implementation of building design that recognises all life systems are inherently interdependent and, in particular, that humans are a part of nature, with the social and natural system intrinsically interlinked, and in constant flow and flux (Berkes et al., 1998; Zielinska-Dabkowska, 2019). Socio-ecological developments are also closely linked with concepts of sustainable development, recognising the connections at multiple scales between environmental issues, socio-economic inequality and human health (Hopwood et al., 2005). They are also linked with concepts of 'Biophilia' (Wilson, 1984) which stresses the innate connection between humans and the natural world, and that incorporating nature into built design can hold multiple benefits for the building occupants. Thus, these developments aim to demonstrate buildings with minimal impact to existing ecological systems that employ resource sustainability, maintain ecological integrity, emulate natural environments and eliminate natural debt, while meeting the needs of human inhabitants (Udomiaye et al., 2018).

2.3.1.4 Health and Wellbeing

Active Homes with a focus on health and wellbeing represent ambitions to create indoor environments that do not harm and that can benefit their occupants' physical and mental health and wellbeing. Since the 1980s when 'Sick Building Syndrome' was first recognised (Murphy, 2006), there has been growing awareness that elements contained and confined within human built indoor environments can be harmful to people's health. Indeed, as recent estimates suggest many people now spend over 90% of their time indoors (Lowther et al., 2019; Bluyssen, 2020), as pollutants can emanate from even the most mundane of our household objects and as our indoor environments become increasingly air-tight, indoor environmental quality is an increasing concern. Much of this concern focuses on indoor air quality – air ventilation and airborne particulate matter and volatile organic compounds (VOCs) (Lowther et al., 2019; Ministry of Housing, Communities, & Local Government, 2019). However, it is also recognised that lighting, acoustics and odours – as well as temperatures – play an important role in achieving (or not) occupant good health and wellbeing (Bluyssen, 2020). Thus, our Active Homes that focus on health and wellbeing have adapted such concerns into their building design.

While we have identified a key focus for each of our case sites, we should note (as Table 2.2 demonstrates) that some of our sites hold multiple foci. Thus, for all of our sites, we have identified their primary focus and a secondary focus highlighting their multifaceted ambitions and the interconnection between possible outcomes existent in all the case sites chosen. The key foci for our case sites are reflective of the political, economic and social context at the time of their conceptualisation in addition to the aims and ambitions of developing organisations, often in negotiation or partnership with other stakeholders (such as architects, building management organisations, registered social landlords and local authorities). The key focus of each site holds a strong influence on other characteristics of the development such as its location, material and technological composition and envisaged role of the household in a building's energy operations. Our Active Home case sites, while sharing some ambitions, characteristics and even stakeholders, are different in their materiality and in how it is expected that occupants will live within them. As with our previous Energy Biographies research (Henwood et al., 2016b) where we investigated the perceptions of different energy demand initiatives across different locations, in selecting a number of different Active Home developments as case sites, we can explore variances and similarities in the lived experiences of Active Households across time and space.

As we have noted, the primary focus of each development holds influence on many of the other elements comprising the home. These include the following:

- Energy source, technology mix, energy services (e.g. available technology, use of time of use tariffs)
- Expected occupant role (e.g. levels and mode of autonomy occupants have over energy consumption)
- Location (e.g. rural/urban location has implication for access to services and proximity to natural environment)
- Architectural aesthetic (e.g. aesthetics, layout and design features of the home likely to resonate with people in different ways)

2.3.2 Multistage: Qualitative Longitudinal

Active Buildings are being conceptualised, designed, developed and lived in now, and it is anticipated this will continue into the future. Further, Active Homes built now will be lived in for many years to come. Capturing this temporal experience requires a longitudinal research methodology that will allow the investigation of this multistage process as it progresses.

Qualitative longitudinal (hereafter QLL) research explores dynamic processes through an in-depth qualitative lens, giving insights into how people narrate, understand and shape their unfolding lives and the evolving world of which they are a part (Neale, 2019, p. 1). Longitudinal methodologies have variously been described as embodying the notion of time, centring time and change (Holland, 2007) and recognising that participants' thoughts, actions, emotions, attitudes and beliefs are all dynamic through time (Saldaña, 2003). This dynamic, temporal focus makes it well-suited for exploring processes of change, rather than simply outcomes (Thomson, 2007). In practical terms, a QLL study is likely to involve several occasions of data collection. QLL may be intensive (multiple research encounters

Table 2.2 Chosen active housing developments

Case site No.	Dominant theme	Secondary theme	Defining characteristics	Closest building concept	Energy specifics	Expected occupant behaviour	Rural-urban classification[b]	Architectural aesthetic	Tenure
1	System change	Energy Sufficiency	- 3 phase power supply - Grid flexibility - Aggregated energy & energy service	NZEB	- Highly insulated - Electric vehicle charge point - Ground source heat pump (GSHP) - Solar Photovoltaic (PV) - Intelligent battery storage - Thermal water storage - Underfloor heating - Radiant wall heating - Smart appliances (optional) - Energy service - Aggregated energy demand and export	- Energy service to manage occupant comfort requires communication between occupant and service technology - Occupant management of energy use via in-home controls and downloaded energy service software application. - No modification to occupant behaviour is required[a]	Rural town and fringe	Traditional UK new build	Owner-occupy

2	Health and wellbeing	Energy Sufficiency	- Modular wood construction - Non-Volatile Organic Compound (VOC) paints - Air quality sensors - Local supply chain	Solar Passivhaus	- Highly insulated - Electric vehicle charge point - Intelligent battery storage - Solar PV - Thermal & air circulation enabled building layout - Thermal water storage - Radiant wall heating - Indoor air quality sensors	- To benefit from cheapest TOU, some modification of occupant behaviour may be required[a] (limited as battery should achieve cheapest import and highest export) - Occupant management of energy use via in-home controls	Rural village in a sparse setting	Eco-minimalist	Social rent

(continued)

Table 2.2 (continued)

Case site No.	Dominant theme	Secondary theme	Defining characteristics	Closest building concept	Energy specifics	Expected occupant behaviour	Rural-urban classification[b]	Architectural aesthetic	Tenure
3	Energy Sufficiency	Health and wellbeing	- Transpired solar collectors - Manual Ventilation Heat Recovery (MVHR) - Solar roofs	Homes as power stations/ active/SOLCER	- Highly insulated - Electric vehicle charge point - Intelligent battery storage - Transpired solar collectors - Solar roofs - Thermal water storage & Powered by Air Source Heat Pump (ASHP) - Manual Ventilation Heat Recovery (MVHR) - Radiant wall heating	- To benefit from cheapest TOU, some modification of occupant behaviour may be required[a] - Occupant management of energy use via in-home controls	Urban city and town	Eco-technical/active	Social rent

4	System change	Environmental	- Ambition to alter occupant behaviour to benefit environment - Active travel links - Ecologically conscious / sustainable features - Energy aggregation and energy service	Zero carbon	- Highly insulated - Electric vehicle charge point - Intelligent battery storage - Solar PV - Thermal water storage - MVHR - GSHP - Smart appliances	- Energy Service to manage occupant comfort - Occupant management of energy use via in-home controls - No modification to occupant behaviour is required[a], however the design hopes to encourage occupant sustainable living	Urban city and town	Modern-eco	Private rent

(continued)

Table 2.2 (continued)

Case site No.	Dominant theme	Secondary theme	Defining characteristics	Closest building concept	Energy specifics	Expected occupant behaviour	Rural-urban classification[b]	Architectural aesthetic	Tenure
5	Environmental	Health and wellbeing	- WELL Certification - Urban Farm & Community Interest Company - Natural & biologically flowing interior & exterior layout and aesthetic - Sustainable features - Place-making	Biophilic building	- Highly insulated - Electric bicycle charging points - Aggregated energy demand and export - Intelligent battery storage - Solar PV - Air source heat pump (ASHP) - Thermal water storage - MVHR	- Exact energy management of building is under development. - No modification to occupant behaviour is required[a]	Urban city and town	Eco mixed-use/biophilic	Owner-occupy and social rent

[a]Occupant behaviour here is from the perspective of the developer and in relation to whether the buildings achieve their energy performance goals. This is not related to whether occupants will have to alter behaviour to manage new energy sources and technologies that may work differently to those they have lived with in previous homes

[b]Classification based on Middle Layer Super Output Areas, Wales 2011 (Office for National Statistics, 2017)

over a relatively short space of time, e.g. 1 year), extensive (research intervals many years apart) or a combination of the two. Both Neale (2019) and Thomson and McLeod (2015), who have written extensively on QLL methods, resist a prescriptive approach, instead of advocating QLL as a sensibility which involves attending to temporality and is open to unfolding futures.

The benefits of qualitative research in general, such as rich understanding of subjective experiences, are expanded by QLL, providing in-depth and contextually sensitive data with good explanatory value (Holland, 2007). In addition, larger processes of social change can be considered through how they are played out through individual trajectories (Gabb, 2008). This is a crucial insight which can only come when the life story evidence from hidden spheres can be put alongside public economic and political documentation. Understanding the intersection of biographical, historical and generational time through a detailed view of individual lives is made possible through QLL and can help to elucidate wider social processes. Thus, in QLL, time is both vehicle and topic. A particular strength of QLL research is the way in which it can shed light on the processual, on 'how' something happens over time (Thomson, 2007).

Existing studies of low-carbon homes and technologies have identified a space for longitudinal research, for example in determining whether behaviours and views of building users change over time as expectations of thermal comfort and other energy services evolve (Berry et al., 2014) and for longer-term insights into the relative success of sustainability initiatives (Collins, 2015). Some studies into sustainable technologies have used a QLL-influenced methodological approach without always explicitly identifying it as such. For example, Ozaki and Shaw (2014) describe undertaking 'repeat interviews' with residents of new properties fitted with energy saving technologies over the first 18 months of living there. Their study yielded important insights, identifying clear tensions between the assumed normative ways of using technology and actual use and illustrating a variety of reasons why residents may not use sustainable technologies installed in their homes in the way developers had intended.

The ability to show how relationships to technology change over time is a particular advantage of QLL. For example, reporting on QLL insights from a field trial of smart meters, Hargreaves et al. (2013) were able to show that while devices were initially a conspicuous and nagging presence, they quickly became embedded in the background of everyday life, with some participants no longer using the meter at all. Our own research supports these findings, indicating that interest in smart devices appears to wane over time (Shirani et al., 2020). These insights have particular importance for wider rollout of technologies. Therefore, understanding how energy demand may change over time, particularly efforts to achieve reductions, is also of direct policy relevance (Walker, 2019). QLL has unique value in being able to elucidate shifts in perceptions, beliefs and attitudes towards homes on a functional, emotional and psychosocial level (Ambrose et al., 2017). Thus, a QLL approach can have a valuable role in understanding how people come to live in low-carbon homes, which has implications for the wider roll-out of such developments.

Table 2.3 Case site 1 research schedule

Data collection	Date	Research with building occupants	Research with industry and experts
Stage 1	1–3 months pre-occupation	1. Activity pack 1 *(photo elicitation, sentence completion, media referencing)*	Interviews across the project timeline
		2. Pre-occupancy Interview (and activity discussion)	
Stage 2	2–3 months post-occupation	3. Post-occupancy interview	
Stage 3	10 months post-occupation	4. Activity pack 2	
	11–12 months post-occupation	5. Final interview (and activity discussion)	

Our empirical research activities can be broadly split between two participant groups: Industry Stakeholders/Experts and Building Occupants. Our interviews with industry stakeholders and experts were substantively focused (e.g. Butler & Pidgeon, 2011; Cherry et al., 2017) largely on their role within the specific housing development and/or their particular area of expertise in relation to the development of Active Buildings more widely. Research with our building occupant participants requires insight into their daily lives and personal perspectives through time and as they move into their new homes. As such, more frequent interaction is planned (see Table 2.3).

For our building occupants, three interviews were scheduled over the course of approximately 12 months; pre-occupation, 3–4 months post-occupation and 12+ months post-occupation. The initial pre-occupancy interviews, while also drawing on participant self-completed activities, included discussion of the following: background on current homes, current energy use, motivations for choosing a new home, awareness of climate change and whether they see low carbon homes as part of the solution. In the second interview (3–4 months post-occupancy), similar discussion points will be used, but focus will be placed on their immediate experiences within their homes and communities. Likewise, for the final interview (12+ months post occupancy), focus will be placed on experiences within their homes and communities over the preceding 9 months.

2.3.3 Multimethod: Additional Activities

In addition to the QLL interviews outlined above, our multimethod research design involves focus groups, activity packs that include sentence completion, photo elicitation and media referencing activities and, finally, a conceptual analysis of Active Homes based on document analysis and secondary data collection. We also employed researcher observations, for example visiting our case sites and capturing images and taking detailed ethnographic-style notes.

Different methods can provide varied perspectives on participants' social worlds. By using multiple methods, we can combine these multiple perspectives to form a rich and sensitive interpretation (Henwood, 2019). The interview is viewed as co-constructed with the participant, while activity packs are completed by participants alone, providing different 'contexts' for data collection (Mik-Meyer, 2021). Multimodal data provides opportunity to combine varied perspectives and form a more nuanced overall understanding, but it also allows us to use the data in complementary ways, giving 'depth' to the data (Mik-Meyer, 2021). Hence, we do not use each method in isolation, for example completed activity packs are used as discussion points and prompts during interviews (Henwood et al., 2011; Henwood et al., 2018). This allows us to 'ground' our interview questions, whereby the photos can provide both participant and researcher with a context for the discussion (Bryman, 2016; Henwood et al., 2018). Combining the activity packs with the interviews can encourage participants to think differently about objects, or elements of their life they may take for granted or have forgotten (Henwood et al., 2018), enabling greater insight to be gained than through interview alone (Henwood, 2019; Mik-Meyer, 2021). This is especially important when discussing intangible 'things' such as energy, which for many is only visible through the services it provides and the daily lives it supports (Henwood et al., 2016a; Henwood et al., 2018). As QLL requires participant engagement over an extended time period, we can use different methods to sustain interest in the project (Weller, 2012).

2.4 Status Quo, Challenges and Outlook

A number of building certifications, labels and conceptualisations have developed across the globe over the past 30 years, in particular, aiming to reduce the environmental impact of buildings, in their construction and occupation. These existing building certificates, standards and conceptualisations aiming to increase energy efficiency and reduce emissions from buildings have achieved varying results. In the UK, building regulations have incrementally increased energy efficiency requirements, for example requiring insulation, double-glazed windows and efficient energy systems (Department for Business, Energy and Industrial Strategy [BEIS], 2017). However, this forms only part of the picture, and while improved energy efficiency has likely contributed to reduced energy consumption in the UK homes since the 1990s (BEIS, 2017), building regulations themselves do not go far enough to enable net zero carbon by 2050.

Other, more purposeful interventions have been developed, but many have demonstrated a 'performance gap' (Mallory-Hill & Gorgolewski, 2017; Cozza et al., 2020). It is now increasingly recognised that much of this performance gap can be attributed to the design out of building occupants within the various concepts. In some instances, new technologies and building designs are considered too complicated for people to understand, and any requirement on them to intervene in the management of said technologies can induce anxiety and underperformance of the design. In this instance, buildings have been designed that do not require, and work better, without occupant intervention. That being said, research has demonstrated that even when people know they should not intervene with the technology, they still do (Mitchell & Natarajan, 2019). Alternatively, designs have become increasingly smart and require, at least initially, occupant input into the management of their energy system, including the ability to act on digital prompts as to when and when not to use energy in their home (Balta-Ozkan et al., 2014; Hansen & Hauge, 2017; Tirado Herrero et al., 2018), but this can be limited (Hargreaves et al., 2013).

These dichotomous perspectives are unified by the need to both understand and integrate more fully into models or conceptualisations, those who will be living within such buildings. Through the explicit integration of building occupants into an Active Building concept, their energy needs and associated practices can form a part of the solution to achieving whole system decarbonisation, as opposed to a barrier (see also, Nikolaidou et al., 2020). Furthermore, while much of the literature has pointed to a gap in how buildings perform against their energy/emissions targets, perhaps another performance gap exists in how buildings are expected to perform by the occupants and how they actually do. This alternative perspective may illuminate why people intervene in new technologies when they should not, or why they disengage with or circumvent other technologies that need their cooperation.

To understand such questions, first it must be recognised that an Active Home is a *home*. More than a material building, or an extension of energy infrastructure, homes are also places of intrinsic meaning and value, representing to some security or safe haven, privacy, control, reflection of values, relationships and emotional experience (Després, 1991). Such meanings and values are subjectively perceived (Després, 1991; Moore, 2000; Roberts & Henwood, 2019) and thus will affect daily life in different ways for different people. Active Homes will likely alter these meanings, through changing existing materialities, by making public existing private space, through adding a new, or at least perceived, security threat and through altering communication and contestations. Thus, Active Homes will also likely alter daily life, but to understand how this will occur or what form such change will take, it is essential that how home as a place in space and time is both understood and lived in by occupants is elucidated, which we aim to do in our research.

Drawing on the point above regarding the temporality of home; the way people live in their homes is related to where they are in their life course and the lives linked to the household at different points in time (Shirani et al., 2017; Hargreaves & Middlemiss, 2020). For example, a family with small children, a single person household, a household with infirm or chronically ill occupant(s), or a retired person is likely to have very different daily practices and energy needs. These are informed by sociocultural contexts combined with psychosocial norms and expectations around caring, security, socialisation and health (among others) that are more or less prominent with different households at different stages of their life course (Groves et al., 2016). These personal contexts and linkages within a household also interplay with linked lives or relationships with others outside of the household (Hargreaves & Middlemiss, 2020; Shirani et al., 2017), as well as place-specific contexts and energy geographies (Golubchikov & O'Sullivan, 2020), which are also changeable through time.

Finally, the materiality of the home, the objects within it and energy technologies will interplay with personal subjectivities to influence how comfort (thermal and otherwise) is perceived and achieved (Roberts & Henwood, 2019). Together, these multiple links shape how and why people live in certain ways within their home, including how tangible energy and other daily practices fulfil multiple other intangible and somewhat invisible psychosocial functions (Henwood et al., 2016a). Thus, Active Homes need to understand building occupants' expectations of their homes, in addition to how and why they carry out everyday practices including those which impact energy demand, in order that Active Homes can be flexible through time to adapt to people's ever-changing contexts as well as those of the wider energy system.

References

Active House. (Undated). *Active House Alliance formulates strategy 3.0.* [Online]. Available at: https://www.activehouse.info/active-house-alliance-formulates-strategy-3-0/. Accessed 12 Aug 2020.

Ade, R., & Rehm, M. (2019). Buying limes but getting lemons: Cost-benefit analysis of residential green buildings – A New Zealand case study. *Energy and Buildings, 186*, 284–296.

Ambrose, A., Goodchild, B., & O'Flaherty, F. (2017). Understanding the user in low energy housing: A comparison of positivist and phenomenological approaches. *Energy Research & Social Science, 34*, 163–171.

Balta-Ozkan, N., Watson, T., Connor, P., Axon, C., Whitmarsh , L., Davidson, R., & Taylor, G. (2014). *UKERC scenarios for the development of smart grids in the UK synthesis report scenarios for the development of smart grids in the UK – Synthesis report.*

Berkes, F., Folke, C., & Colding, J. (1998). *Linking Social and ecological systems: Management practices and social mechanisms for building resilience*. Cambridge University Press.

Berry, S., Whaley, D., Davidson, K., & Saman, W. (2014). Near zero energy homes – What do users think? *Energy Policy, 73*, 127–137.

Bille, M., Bjerregaard, P., & Flohr Sorensen, T. (2015). Staging atmospheres: Materiality, culture, and the texture of the in-between. *Emotion, Space and Society, 15*, 31–38.

Bluyssen, P. (2020). Towards an integrated analysis of the indoor environmental factors and its effects on occupants. *Intelligent Buildings International, 12(3)*, 199–207. https://doi.org/10.1080/17508975.2019.1599318

Bryman, A. (2016). *Social research methods* (5th ed.). Oxford University Press.

Buildings Performance Institute Europe. (2014). *Energy performance certificates across the EU, A mapping of national approaches*. [Online]. Available at: http://bpie.eu/publication/energy-performance-certificates-across-the-eu/. Accessed 5 Mar 2020.

Butler, C., & Pidgeon, N. F. (2011). From 'flood defence' to 'flood risk management': Exploring governance, responsibility and blame. *Environment and Planning: C, 29*(3), 533–547.

Cherry, C. E., Hopfe, C., MacGillivray, B., & Pidgeon, N. F. (2017). Homes as machines: Exploring expert and public imaginaries of low carbon housing futures in the United Kingdom. *Energy Research and Social Science, 23*, 36–45.

Collins, R. (2015). Keeping it in the family? Re-focusing household sustainability. *Geoforum, 60*, 22–32.

Committee on Climate Change [CCC]. (2019). *Net Zero – The UK's contribution to stopping global warming*. [Online]. Available at: https://www.theccc.org.uk/publication/net-zero-the-uks-contribution-to-stopping-global-warming/. Accessed 15 July 2020.

Conradson, D. (2005). Freedom, space and perspective: Moving encounters with other ecologies. In L. Bondi, J. Davidson, & M. Smith (Eds.), *Emotional geographies* (pp. 103–116). Ashgate Publishing Ltd.

Cozza, S., Chambers, J., & Patel, M. K. (2020). Measuring the thermal energy performance gap of labelled residential buildings in Switzerland. *Energy Policy, 137*, 1–14.

Creswell, J.W. (2013). Qualitative Inquiry and Research Design. SAGE Publications Ltd.

D'Agostino, D., & Mazzarella, L. (2019). What is a nearly zero energy building? Overview, implementation and comparison of definitions. *Journal of Building Engineering, 21*, 200–212.

Day, R., & Walker, G. (2013). Household energy vulnerability as 'assemblage'. In K. Bickerstaff, G. Walker, & H. Bulkeley (Eds.), *Energy justice in a changing climate* (pp. 14–29). Zed Books Ltd.

Department for Business, Energy and Industrial Strategy [BEIS]. (2017). *The clean growth strategy.* [Online]. Available at: https://www.gov.uk/government/publications/clean-growth-strategy. Accessed 15 July 2020.

Després, C. (1991). The meaning of home: Literature review and directions for future research and theoretical development. *Journal of Architectural and Planning Research, 8*(2), 96–115.

Elliott, T., Geske, J., & Green, R. (2020). *Can you make money from active buildings? Challenges facing business models*. Active Building Centre Research Programme (Energy Systems Catapult and Imperial College Business School).

European Technology Innovation Plan. (2020). *Vision 2050, integrating smart networks for the energy transition: Serving and protecting the environment*. [Online]. Available at: https://www.etip-snet.eu/wp-content/uploads/2018/06/VISION2050-DIGITALupdated.pdf. Accessed 16 July 2020.

Frechette, J., Bitzas, V., Aubry, M., Kilpatrick, K., & Lavoie-Tremblay, M. (2020). Capturing lived experience: Methodological considerations for interpretive phenomenological inquiry. *International Journal of Qualitative Methods, 19*, 1–12.

Gabb, J. (2008). *Researching intimacy in families*. Palgrave Macmillan.

Geels, F. W. (2018). Disruption and low-carbon system transformation: Progress and new challenges in socio-technical transitions research and the multi-level perspective. *Energy Research & Social Science, 37*, 224–231.

Golubchikov, O., & O'Sullivan, K. (2020). Energy periphery: Uneven development and the precarious geographies of low-carbon transition. *Energy and Buildings, 211*. https://doi.org/10.1016/j.enbuild.2020.109818

Groves, C., Henwood, K., Shirani, F., Butler, C., Parkhill, K., & Pidgeon, N. (2016). Energy biographies: Narrative genres, lifecourse transitions, and practice change. *Science, Technology, & Human Values, 41(3)*, 483–508.

Hansen, M., & Hauge, B. (2017). Scripting, control, and privacy in domestic smart grid technologies: Insights from a Danish pilot study. *Energy Research & Social Science, 25*, 112–123.

Hansen, A. R., Gram-Hanssen, K., & Knudsen, H. N. (2018). How building design and technologies influence heat-related habits. *Building Research & Information, 46(1)*, 83–98.

Hargreaves, T., & Middlemiss, L. (2020). The importance of social relations in shaping energy demand. *Nature Energy, 5*, 195–201.

Hargreaves, T., Nye, M., & Burgess, J. (2013). Keeping energy visible? Exploring how householders interact with feedback from smart energy monitors in the longer term. *Energy Policy, 52*, 126–134.

Henwood, K. (2019). Investigating risk – Methodological insights from interpretive social science and sustainable energy transitions research. In A. Olofsson & J. Zinn (Eds.), *Researching risk and uncertainty: Methodologies, methods and research strategies* (pp. 129–152). Palgrave MacMillan.

Henwood, K., & Shirani, F. (2012). Researching the temporal, Chapter 13. In H. Cooper (Editor in Chief) *Handbook of research methods in psychology* (Vol. 2). APA Publications.

Henwood, K., Shirani, & Finn, M. (2011). "So you think we've moved, changed, the representation got more what?" Methodological and analytical reflections on visual (photo-elicitation) methods used in the men-as fathers-study, Chapter 22. In P. Reavey (Ed.), *Visual methods in psychology: Using and interpreting images in qualitative research* (pp. 330–345).

Henwood, K., Groves, C., & Shirani, F. (2016a). Relationality, entangled practices, and psychosocial exploration of intergenerational dynamics in sustainable energy studies. *Families, Relationships and Societies, 5*(3), 393–410. https://doi.org/10.1332/204674316X147584383416945

Henwood, K., Pidgeon, N., Groves, C., Shirani, F., Butler, C., & Parkhill, K. (2016b). *Energy biographies research report*. Available via: http://orca.cf.ac.uk/87333/

Henwood, K., Shirani, F., & Groves, C. (2018). Using photographs in interviews: When we lack the words to say what practice means. Chapter 38. In U. Flick (Ed.), *The Sage handbook of qualitative data collection* (pp. 599–614). Sage.

Hills, J. (2012). *Getting the measure of fuel poverty: Final report of the fuel poverty review* (CASE report 72). Department of Energy and Climate Change.

Holland, J. (2007). Qualitative longitudinal research: Exploring ways of researching lives through time. Real life methods node of the ESRC National Centre for Research Methods. *Workshop held at LSBU.*

Hopwood, B., Mellor, M., & O'Brien, G. (2005). Sustainable development: Mapping different approaches. *Sustainable Development, 13*, 38–52.

Hubbard, P. (2005). The geographies of 'going out': Emotion and embodiment in the evening economy. In L. Bondi, J. Davidson, & M. Smith (Eds.), *Emotional geographies* (pp. 117–134). Ashgate Publishing Ltd.

Jones, O. (2005). An ecology of emotion, memory, self and landscape. In L. Bondi, J. Davidson, & M. Smith (Eds.), *Emotional geographies* (pp. 117–134). Ashgate Publishing Ltd.

Liu, Z., Zhou, Q., Tian, Z., He, B., & Jin, G. (2019). A comprehensive analysis on definitions, development, and policies of nearly zero energy buildings in China. *Renewable and Sustainable Energy Reviews, 114*, 1–13.

Lowther, S. D., Jones, K. C., Wang, X., Whyatt, J. D., Wild, O., & Booker, D. (2019). Particulate matter measurement indoors: A review of metrics, sensors, needs, and applications. *Environment, Science and Technology, 53*(20), 11644–11656.

Mallory-Hill, S., & Gorgolewski, M. (2017). Mind the gap: Studying actual versus predicted performance of green buildings in Canada. *Building Performance Evaluation*, 261–274.

Mattoni, B., Guattari, C., Evangelisti, L., Bisegna, F., Gori, P., & Asdrubali, F. (2018). Critical review and methodological approach to evaluate the differences among international green building rating tools. *Renewable and Sustainable Energy Reviews, 82*(1), 950–960.

Middlemiss, L., Isenhour, C., & Martiskainen, M. (2019). Introduction: Power, politics and (un)sustainable consumption. In C. Isenhour, M. Martiskainen, & L. Middlemiss (Eds.), *Power and politics in sustainable consumption research and practice* (pp. 1–9). Routledge.

Mik-Meyer, N. (2021). Multimethod qualitative research. In D. Silverman (Ed.), *Qualitative research* (5th ed.). Sage.

Ministry of Housing, Communities & Local Government. (2019). *Ventilation and indoor air quality in new homes*. Aecom Ltd. [Online]. Available at: https://www.gov.uk/government/publications/ventilation-and-indoor-air-quality-in-new-homes. Accessed 1 Dec 2020.

Mitchell, R., & Natarajan, S. (2019). Overheating risk in Passivhaus dwellings. *Building Services Engineering Research & Technology, 40(4)*, 446–469.

Moore, J. (2000). Placing home in context. *Journal of Environmental Psychology, 20*, 207–217.

Murphy, M. (2006). *Sick building syndrome and the problem of uncertainty*. Duke University Press.

National Grid ESO. (2020). *Future energy scenarios*. [Online]. Available at: https://www.nationalgrideso.com/document/173821/download. Accessed 12 Aug 2020.

Neale, B. (2019). *What is qualitative longitudinal research?* Bloomsbury.

Nikolaidou, E., Fosas, D., Roberts, M., Allen, S., Walker, I., & Coley, D. (2020). *Buildings as energy infrastructure, not passive consumers*. Active Building Centre Research Programme (University of Bath).

Nilsson, A., Wester, M., Lazarevic, D., & Brandt, N. (2018). Smart homes, home energy management systems and real-time feedback: Lessons for influencing household energy consumption from a Swedish field study. *Energy and Buildings, 179*, 15–25.

O'Sullivan, K., Henwood, K., & Pidgeon, N. (2020). *Active buildings in the changing policy landscape: Conceptual challenges and social scientific perspectives*. Active Building Centre Research Programme (Cardiff University).

Office for National Statistics. (2017). *Rural Urban Classification (2011) map of the MSOAs in Wales*. [Online]. Available at: https://geoportal.statistics.gov.uk/datasets/fed098274cdf4bc18a9bbfe896ccf086. Accessed 7 Dec 2020.

Ofgem. (2020). Ofgem decarbonisation programme action plan. [Online]. Available at: https://www.ofgem.gov.uk/publications-and-updates/ofgem-s-decarbonisation-action-plan. Accessed 15 July 2020.

Ozaki, R. (2018). Follow the price signal: People's willingness to shift household practices in a dynamic time-of-use tariff trial in the United Kingdom. *Energy Research & Social Science, 46*, 10–18.

Ozaki, R., & Shaw, I. (2014). Entangled practices: Governance, sustainable technologies, and energy consumption. *Sociology, 48*(*3*), 590–605.

Passivhaus Trust. (Undated). *What is Passivhaus?* [Online]. Available at: https://scholar.google.com/scholar?as_ylo=2019&q=passivhaus+standard&hl=en&as_sdt=0,5. Accessed 7 Dec 2020.

Pile, S. (2010). Emotions and affect in recent human geography. *Transactions of the Institute of British Geographers, 35*(*1*), 5–20.

Pobl. (Undated). *Construction underway on homes that will generate, store & release their own energy*. [Online]. Available at: https://www.poblgroup.co.uk/about-us/pobl-story/pobl-story-oct-2018/construction-underway-on-homes-that-will-generate-store-release-their-own-energy/. Accessed 1 Dec 2020.

Regen. (2020). *Net Zero South Wales*. [Online]. Available at: https://www.regen.co.uk/wp-content/uploads/NZSW-Learning-Report-FINAL.pdf. Accessed 15 July 2020.

Roberts, E., & Henwood, K. (2019). "Its an old house and that's how it works": Living sufficiently well in insufficient homes. *Housing, Theory and Society, 36*(*4*), 469–488.

Robinson, J. M., & Breed, M. F. (2019). Green prescriptions and their co-benefits: integrative strategies for public and environmental health. *Challenges, 10*(*9*). https://doi.org/10.3390/challe10010009

Rossiter, W., & Smith, D. J. (2018). Green innovation and the development of sustainable communities: The case of Blueprint Regeneration's Trent Basin development. *The International Journal of Entrepreneurship and Innovation, 19*(*1*), 21–32.

Saldaña, J. (2003). *Longitudinal qualitative research: Analyzing change through time*. AltaMira.

Scotcher, S. (2019). *Tackling fuel poverty through low carbon homes*. [Online]. Available at: https://www.iwa.wales/agenda/2019/03/tackling-fuel-poverty-through-low-carbon-homes/. Accessed 1 Dec 2020.

Semple, S., & Jenkins, D. (2020). Variation of energy performance certificate assessments in the European Union. *Energy Policy, 137*, 1–11.

Shirani, F., & Henwood, K. (2011). Taking one day at a time: Temporal experiences in the context of unexpected life course transitions. *Time and Society, 20*(*1*), 49–68.

Shirani, F., Groves, C., Parkhill, K., Butler, C., Henwood, K., & Pidgeon, N. (2017). Critical moments? Life transitions and energy biographies. *Geoforum, 86*, 86–92.

Shirani, F. Groves, C. Henwood, K. Pidgeon, N., & Roberts, E. (2020). 'I'm the smart meter': Perceptions and experiences of smart technology amongst vulnerable consumers. *Energy Policy, 144*, article number: 111637.

Strbac, G., Woolf, M., Pudjianto, D., Zhang, X., Walker, S., & Vahidinasab, V. (2020). *The role of active buildings in the transition to a net zero energy system.* Active Building Centre Research Programme (Imperial College London and Newcastle University).

The Environment Agency. (2020). *State of the environment: health, people and the environment.* [Online]. Available at: https://www.gov.uk/government/publications/state-of-the-environment/state-of-the-environment-health-people-and-the-environment. Accessed 1 Dec 2020.

Thomson, R. (2007). The qualitative longitudinal case history: practical, methodological and ethical reflections. *Social Policy and Society, 6(4)*, 571–582.

Thomson, R., & McLeod, J. (2015). New frontiers in qualitative longitudinal research: An agenda for research. *International Journal of Social Research Methodology, 18(3)*, 243–250.

Tirado Herrero, S., Nicholls, L., & Strengers, Y. (2018). Smart home technologies in everyday life: do they address key energy challenges in households? *Current Opinion in Environmental Sustainability, 31*, 65–70.

Udomiaye, E., Okon, I. U., Uzodimma, O. C., & Patrick, N. (2018). Eco-friendly buildings: The architect's perspectives. *International Journal of Civil Engineering, Construction and Estate Management, 6(2)*, 14–26.

United Nations [UN]. (2015). *Transforming our world: The 2030 agenda for sustainable development.* [Online]. Available at: https://sustainabledevelopment.un.org/post2015/transformingourworld/publication. Accessed 1 Dec 2020.

Walker, G. (2019). The dynamics of energy demand: Change, rhythm and synchronicity. *Energy Research and Social Science, 1*, 49–55.

Weller, S. (2012). Evolving creativity in qualitative longitudinal research with children and teenagers. *International Journal of Social Research Methodology, 15(2)*, 119–133.

Welsh Government. (2015). *The wellbeing of future generations act.* Welsh Government.

Wilson, E. O. (1984). *Biophilia.* Harvard University Press.

Zielinska-Dabkowska, K. (2019). The green, green grass... of home (or the office) is becoming ever more important in lighting design. *Arc Lighting in Architecture, 110*, 49–52. [Online]. Available at: https://depot.ceon.pl/handle/123456789/18515. Accessed 25 Sept 2020.

Zuo, J., & Zhao, Z. (2014). Green building research–current status and future agenda: A review. *Renewable and Sustainable Energy Reviews, 3*, 271–281.

Chapter 3
Electrical Energy Storage Devices for Active Buildings

Mohamad Amin Rajabinezhad, Hooman Firoozi, Hosna Khajeh, and Hannu Laaksonen

Status Quo, Challenges, and Outlook
The future electricity networks would require backup power as stored energy in order to respond to different kinds of variation and uncertainty. The electrical storages, as one potential type, could be utilized in all scales including the end-user scale. However, utilizing different types of electrical storage on the demand side, e.g., in active buildings, could be challenging due to their operational requirements and their applications. Therefore, the first step is to answer the question that what kind of electrical storage could be utilized in active buildings? What are their characteristics and technology requirements? Which one could be more applicable compared to the others?

3.1 Overview on Energy Storage Systems (ESS)

Several types of energy storage for various applications in electrical networks have been introduced during the last three decades. Their application varies from the power generation system to end-users. Based on the desired application, different energy storage can be deployed. There have been introduced several types of energy storage as follows: thermal energy storage (TES), electrical energy storage (EES), mechanical energy storage (MES), etc. In this chapter, different types of ESS systems in terms of their technologies and applications have been studied.

M. A. Rajabinezhad
School of Electrical Engineering, Iran University of Science and Technology, Tehran, Iran

H. Firoozi (✉) · H. Khajeh · H. Laaksonen
Flexible Energy Resources, School of Technology and Innovations, University of Vaasa, Vaasa, Finland
e-mail: hooman.firoozi@uwasa.fi

V. Vahidinasab, B. Mohammadi-Ivatloo (eds.), *Active Building Energy Systems*, Green Energy and Technology, https://doi.org/10.1007/978-3-030-79742-3_3

There are several advantages to utilize ESS in the active distribution networks and microgrids. ESS has a prominent role in increasing the system's efficiency since they can manage the surplus of electricity generated by renewable energy sources (RESs) and reduce energy loss. Therefore, ESSs help to maximize the hosted capacity of the active networks and increase the share of RESs in grid's demand supply as well (Aleem et al., 2020). Further, the ESSs can eliminate the fluctuations of consumption/production by providing the required flexibility (Khajeh et al., 2020). Therefore, they can improve the stability and reliability of the active distribution network. Besides, using ESS can help increase supply security and power quality by maintaining frequency/voltage at the required levels. For instance, power storage options can ensure a reliable power supply with voltage drop occurring in the event of a power outage (Nadeem et al., 2018; Ibrahim et al., 2008).

ESSs can be classified in two groups. One is based on their manufacturing technology and the other is based on their operating time. In general, ESSs have short-term and long-term applications in terms of time. This means that in some cases we need ESSs to act quickly and inject power (Andrijanovits et al., 2012). However, in some cases, we need the ESS to stay longer in the grid, and its momentary payback does not matter. The first way of operation is called the short-term function, while the other one, which is performed on a minute to hour scale, is called the long-term function. Short-term applications include the use of ESSs in frequency power converters, transient stability, FACTS devices, and damping of low-frequency fluctuations. The utilization of ESSs as an emergency energy source, alternative energy source, voltage regulation, peak load adjustment, and power management is among its long-term applications (Li et al., 2010; Suvire et al., 2010). The two important factors distinguish the performance of energy storage technologies including the amount of energy that the device can store as well as the speed at which energy can be injected into the grid (Hossain et al., 2020). A graphical overview on different types of energy storage in active buildings (ABs) can be found in Fig. 3.1.

In this chapter, the electrical ESSs in microgrids and ABs are investigated. Firstly, different types of EES and their manufacturing technologies, their benefits, and drawbacks are discussed. In the following, the application of each one of the energy storage is illustrated.

3.2 Static Storages

3.2.1 Electrical Storage

Electrical energy can be stored in electric and magnetic fields using supercapacitors (SCs) and superconducting magnets, respectively. They have high power and medium energy density, which means they can be used to smooth power fluctuations

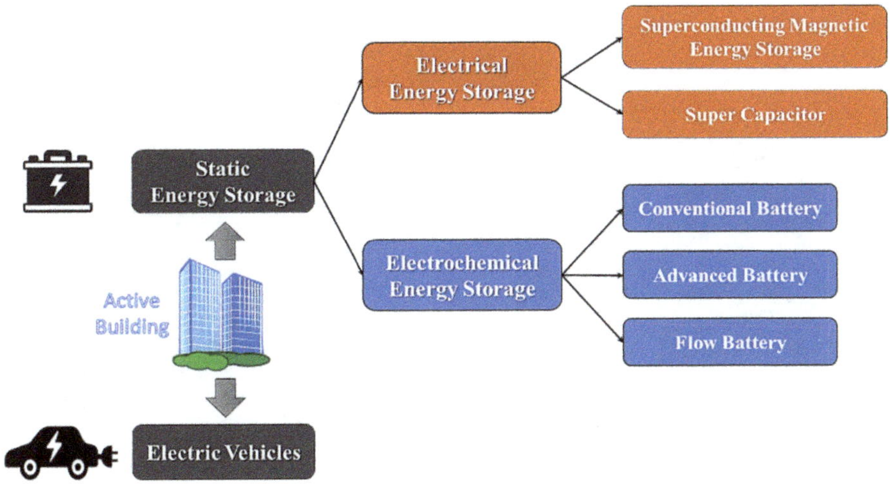

Fig. 3.1 Overview on different types of energy storage in ABs

and meet maximum power requirements and energy recovery in transportation devices (Nadeem et al., 2018). In this section, the technology of electrostatic and magnetic energy storage is investigated.

3.2.1.1 Superconducting Magnetic Energy Storage (SMES)

Although the concept of superconductivity was discovered in 1911, its application as a storage for power grids was not considered until the 1980s. SMES is a direct current (DC) storage that stores the energy in the magnetic field generated by DC flows via superconducting coils. The technology of making superconducting magnetic energy storage consists of an inductive coil made of superconducting material (Ali et al., 2010). The temperature of the superconducting coil should keep under its critical value. This temperature is retained by a crystal or rotary containing helium or nitrogen fluid vessels.

Since the superconductor has a small internal resistance (almost zero), by passing the DC current through it, low magnetic losses will be created. Energy could be stored as a circulating current meaning that the energy can be received from a SMES with an almost immediate response from a second to a few hours (Guney & Tepe, 2017; Zakeri & Syri, 2015). The SMESs are used in two different shapes: toroid and solenoid. In the toroid SMES, the external field is low, but the cost of its superconductor and another component is higher than the solenoid form. The stored energy and power in the SMES unit can be obtained as in (3.1) and (3.2):

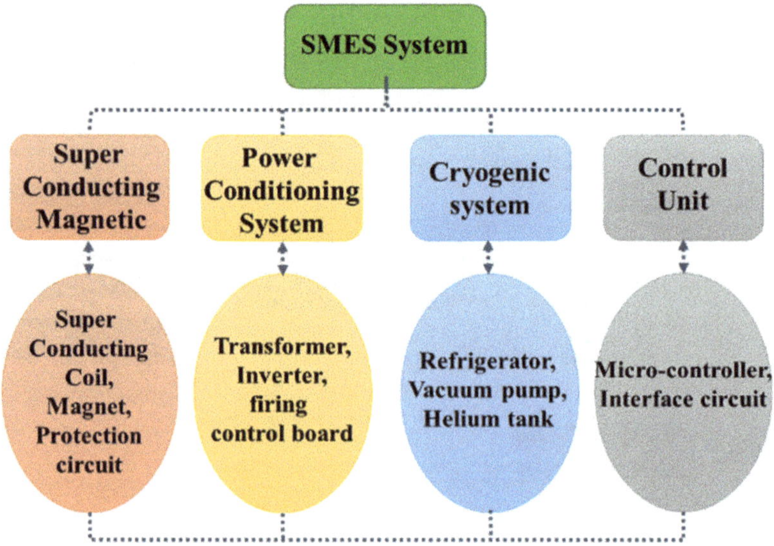

Fig. 3.2 Component of a SMES unit

$$E = \frac{1}{2} L I^2 \tag{3.1}$$

$$P = \frac{de}{dt} = LI\frac{dI}{dt} = vI \tag{3.2}$$

In the above equations, I, L, and V represent the DC current, inductance of the superconductor system, and voltage of the coil, respectively. According to the above formulas, the stored energy in the magnetic field is directly related to the inductance of the coil and the DC current passing through it. Hence, by increasing each one, the energy stored in the coil can be increased. SMES unit is connected to the AC grid in system or grid-connected ABs through a DC/AC power conditioning system. There are two common methods to connect the SMES to the network. In the first one, the current source converter (CSC) could be utilized. The voltage source converter (VSC) could also be utilized as the second method. However, it should be considered that in the VSC manner, the usage of DC-DC chopper is also necessary. The price of SMES depends on two independent factors. One is the cost of the stored energy capacity, while the other is the cost of output power management. Costs associated with stored energy include conductor construction cost, coil building elements, refrigeration tube, cooling system, protection, and control equipment. The component and structure of a SMES unit are depicted in Figs. 3.2 and 3.3, respectively. The costs of output power management include the cost of design/build a power conditioning system.

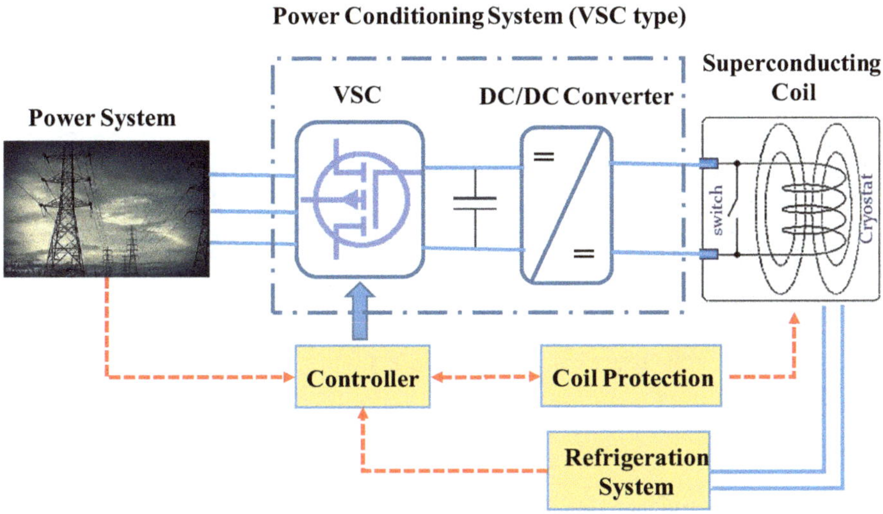

Fig. 3.3 Structure of SMES system

It has high accessibility and can provide a lot of energy in milliseconds with an efficiency about 95–98% (Gallo et al., 2016). The low energy density of the SMES unit is one important drawback of this storage type. Due to its structure requirements, it could be somewhat difficult to use it for large-scale applications. Therefore, a combination of SMES to another storage such as the battery was proposed as an option.

3.2.1.2 Supercapacitor

Capacitors are the well-known devices to store electrical energy. They have fast-response characteristics, long life cycles up to thousands of times, and high efficiency. Due to the low energy density, high self-discharge rate of conventional capacitors, and filling the gap between these equipment and conventional batteries, the concept of SCs has arisen. SC is also known as electric double-layer capacitors (ELDC) or ultra-capacitors (UC). The principles of operation of SC and ordinary capacitors are similar. However, in SCs instead of conventional dielectrics such as ceramics, polymer films, etc., an electrolytic physical barrier consisting of activated carbon is used. This means SC including two porous electrodes is usually made up of activated carbon plunge in an electrolyte dilution. The introduced structure effectively makes two capacitors, one at each carbon electrode, connected in series. This ion conduction increases the energy density of the SC (Kalaiselvam & Parameshwaran, 2014). Therefore, the SCs can have the characteristics of conventional capacitors as well as electrochemical batteries at the same time. Due to the low energy density of the capacitors, they can store or inject high currents only

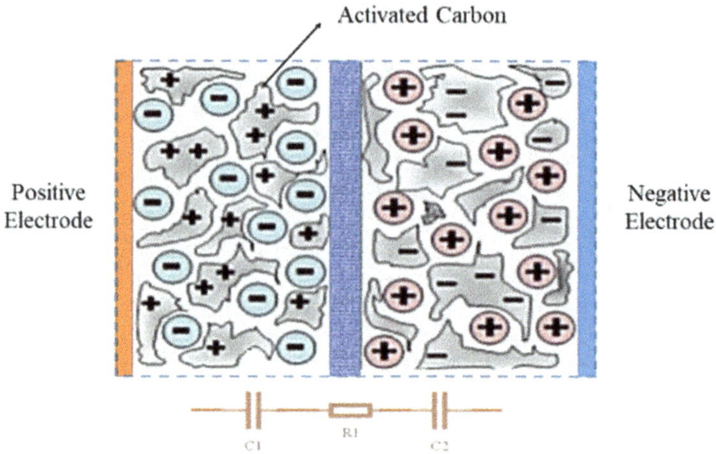

Fig. 3.4 Construction of individual cell of a SC

for very short time intervals (Luo et al., 2015). The maximum operating voltage of the capacitor depends on the insulation endurance of its dielectric. The construction of the individual cell of a SC is shown in Fig. 3.4. The physical principles of the capacitors are illustrated as follows:

$$C = \varepsilon \frac{A}{d} \tag{3.3}$$

$$E = \frac{1}{2} c v^2 \tag{3.4}$$

where C, A, d, E, and ε represent capacity, the area of the plates, the distance between two plates, energy, and dielectric constant of capacitor, respectively. As in the previous formulas, the amount of stored energy is proportional to the capacitance C and the square of the voltage V across capacitor terminals. The area of the plates and the thickness of electrodes are higher, while the distance between electrodes is lower in SCs than traditional capacitors (Yao et al., 2006).

The main application of traditional DC capacitors in distribution systems, in form of large-scale storage, is in dynamic voltage restorer (DVR). DVR is a FACTS device that is used for temporary voltage sag mitigation. SCs are mostly deployed for a specific situation that several charge/ discharge cycles and instantaneous high-power output are needed, e.g., power smoothing and power quality related services (Chatzivasileiadi et al., 2013).

3.2.2 *Electrochemical Storage Systems*

Electrochemical energy storages transmute chemical energy to electrical energy. At least two reaction parts participate in a chemical procedure during this conversion. The obtained energy from this reaction is accessible as an electric current at a specified voltage. Electrochemical storages can generally be categorized into primary and secondary classification according to their principle of operation. The primary battery is usually not rechargeable. The second type of battery energy storage categorization reclassifies different kinds of rechargeable batteries. Each cell of this storage has three main components including negative and positive electrodes along with an electrolyte. The main factors for the application of battery energy storage systems (BESS) are high energy density, high-energy capability, consecutive charge/discharge ability, longevity, and primary cost. Electrochemical storages can be classified as conventional, advanced, and flow batteries.

3.2.2.1 Conventional Batteries

Lead-Acid Battery (Pb-Acid)

Lead-acid batteries were invented by Gaston Planet in 1859, and they have been used for 150 years (Tan et al., 2013). They are the most inexpensive energy storage of all the available battery technologies. However, the limited cycling capability of them is an undesirable factor in the economic view of the power network. Lead-acid battery includes lead metal and lead oxide electrodes and sulfuric acid dilution.

In discharging state, both the positive and negative plates are converted to lead sulfate. The electrolyte then loses most of the dissolved sulfur and converts mainly to water. At full charge mode, the negative plate contains lead, and the positive plate contains lead dioxide. In this case, the electrolyte has a high concentration of sulfuric acid, which stored chemical energy. It should be noted that, since the electrolysis of water will produce oxygen and hydrogen, overcharging with high voltage causes electrolyte water loss. The following chemical reactions occur at the anode/cathode during the charge/discharge process. These reactions are exactly opposite in charge and discharge mode and are shown in (3.5) and (3.6), respectively.

$$\text{Charge mode}: 2PbSO_4 + 2H_2O \rightarrow PbO_2 + Pb + 2H_2SO_4 \tag{3.5}$$

$$\text{Discharge mode}: PbO_2 + Pb + 2H_2SO_4 \rightarrow 2PbSO_4 + 2H_2O \tag{3.6}$$

The valve-regulated lead-acid (VRLA) battery is a special type of lead-acid battery that has special advantages over ordinary acid batteries, which make them easier to use. These batteries do not need to add acid or distilled water, are completely closed, and do not require any vents. It can also be placed in any position

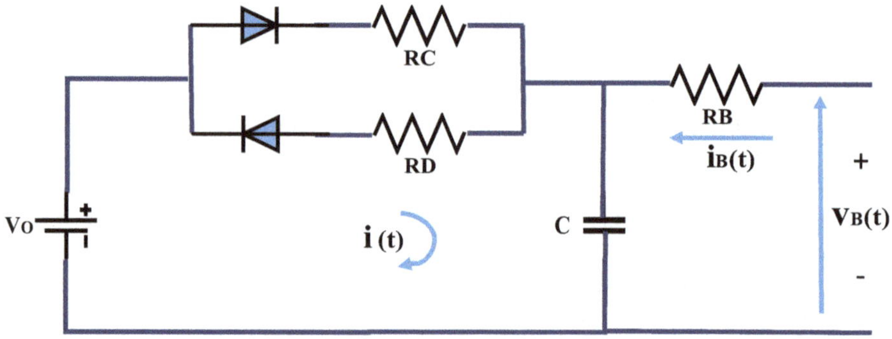

Fig. 3.5 Equivalent nonlinear Thevenin-based circuit of lead-acid battery

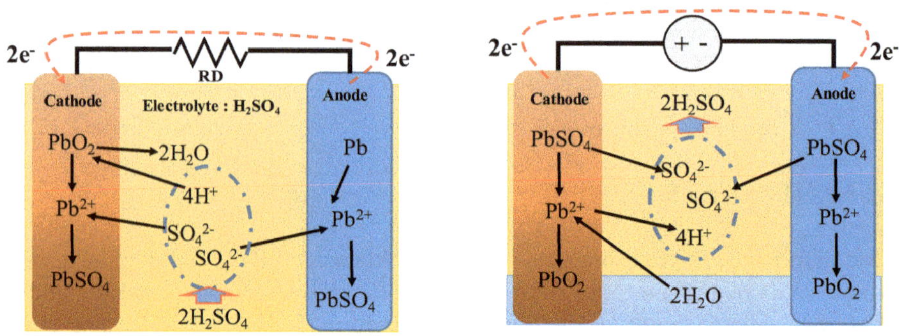

Fig. 3.6 Schematic diagram of lead-acid battery's principle operation

(i.e., vertical, horizontal, and inverted). They are also commonly known as sealed batteries or maintenance-free batteries. In order to mathematically model lead-acid batteries, an equivalent nonlinear Thevenin-based circuit is proposed. This model is shown in Fig. 3.5 (Hossain et al., 2020). In this model, R_C and R_D represent the charge and discharge resistance that the battery current encounters them, and C illustrates the capacitive effect of the battery.

The schematic diagram of lead-acid battery principle operation could be seen in Fig. 3.6.

One of the best features of this battery is longer storing capability due to its low self-discharge rate. Lead-acid batteries are used for starter device in motor vehicles and usually applied for emergency power supply, renewables in energy communities such as wind turbine units and photovoltaic systems, and grid-scale integration. This is caused by their safety, temperature endurance, and low cost. The VRLA batteries have low initial cost, fast charge ability, high power density, and also low maintenance mostly used for electric vehicle (EV) applications (Noack et al., 2015).

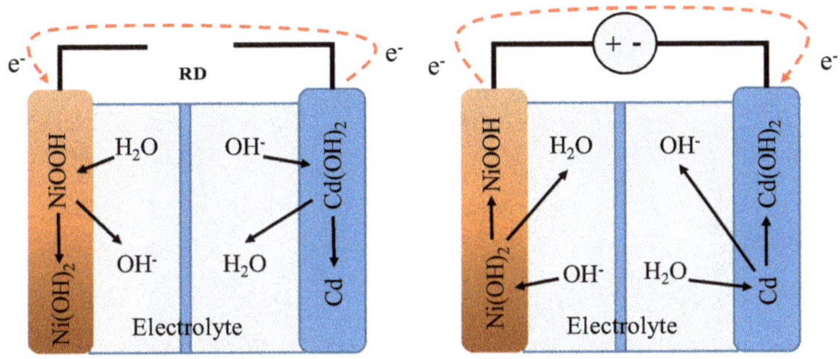

Fig. 3.7 Schematic diagram of NiCd battery principle operation

Nickel Cadmium (NiCd)

In the nickel-based storages, nickel hydroxide is used for positive plate and various negative plate materials, in which according to the negative electrode, there are several types of nickel-based batteries such as NiCd and NiMH. The electrolyte of these batteries is potassium hydroxide. Equation (3.7) illustrates the chemical reactions occur at anode/cathode during the charge/discharge process in a NiCd battery.

$$Cd + 2NiO\,(OH) + 2H_2O \Longleftrightarrow 2Ni(OH)_2 + Cd(OH)_2 \tag{3.7}$$

The schematic diagram of NiCd battery principle operation is also depicted in Fig. 3.7.

In this battery type, the alkaline electrolyte unlike lead-acid battery is not consumed in its chemical reaction. Therefore, its electrolyte gravity is almost constant and cannot be used to determine the state of charge. The advantages of NiCd batteries such as high energy density, reliability, and less maintenance make them more popular for power system applications, portable instrument, industrial UPS, telecommunication, etc. The disadvantages of NiCd batteries are environmental issues due to the toxic feature of cadmium and nickel, memory effect, and capacity reduction during repeatedly recharging after several insignificant discharges (Linden & Reddy, 2002).

Nickel Metal Hydride (NiMH)

NiMH batteries are a type of NiCd battery that has been developed to improve its characteristics. The NiMH battery was patented in 1986 by Stanford Ovshinsky while researching hydrogen storage materials. In order to become more environmentally friendly, cadmium was eliminated from their electrodes. As mentioned earlier, in the nickel-metal hydride (NiMH) battery, a hydrogen-absorbing alloy is used in place of cadmium. Contrary to NiCd storages, NiMH batteries have lower

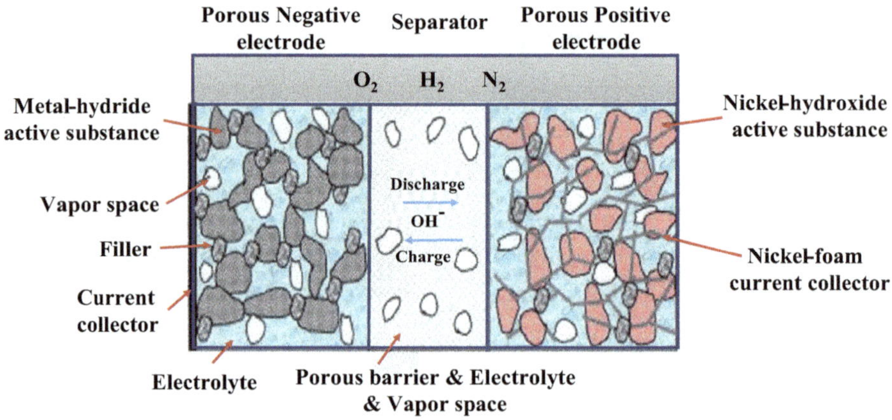

Fig. 3.8 Schematic of NiMH battery's principle operation

memory and environmental effect. NiMH batteries are maintenance-free. Therefore, these benefits show that the NiMH batteries are more applicable than NiCd storages to use in EVs (Ferreira et al., 2013; Poullikkas, 2013). The schematic diagram of NiMH battery principle operation is shown in Fig. 3.8. The overall reaction of NiMH batteries is as in (3.8):

$$M + Ni(OH)_2 \Longleftrightarrow MH + NiO\,(OH) \tag{3.8}$$

where M is an intermetallic alloy that forms a metal hydride phase.

3.2.2.2 Advanced Batteries

Lithium-Ion (Li-ion)

In Li-ion batteries, the positive electrode anode is made of lithiated metal oxide, and the negative electrode is made of graphitic carbon with a layering structure. Since the lithium element is extremely reactive, the Li-ion batteries do not utilize elemental lithium. Instead, they typically contain lithium-metal oxide, such as lithium cobalt oxide ($LiCoO_2$), which means Li-ion batteries provide power through the movement of ions.

The performed electrochemical reaction in this storage is as in (3.9):

$$LiC_6 + CoO_2 \Longleftrightarrow C_6 + LiCoO_2 \tag{3.9}$$

As depicted in Fig. 3.9, in discharge mode, Li^+ transfer from the cathode electrolyte to anode ones, and during its processes, the lithium atoms formed. In the charge mode, a reverse reaction is executed. This type of batteries has high

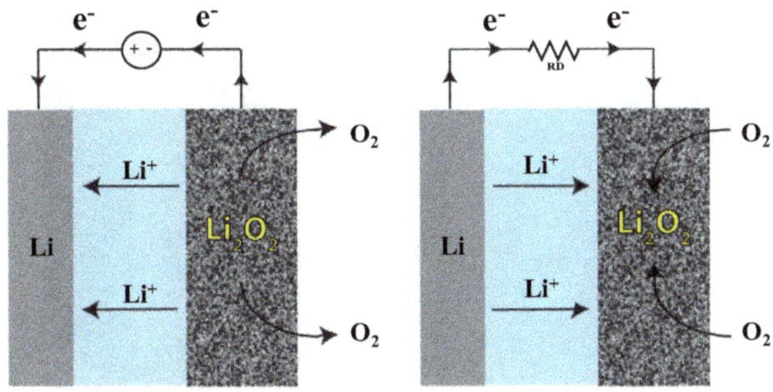

Fig. 3.9 Schematic diagram of Li-ion battery's principle operation

energy/power density as well. Hence, Li-ion batteries are widely used in mobile electronic equipment, such as cell phones and EVs.

Sodium Sulphur (NaS)

Molten salt technology is used in NaS batteries. The negative and positive plate consists of molten sodium and molten sulfur, respectively. Moreover, solid ceramic sodium alumina is used as the electrolyte. The performed reaction inside these batteries is as in (3.10):

$$2Na + 4S \Longleftrightarrow Na_2S_4 \tag{3.10}$$

To ensure the molten status of the negative and positive plate of NaS, a temperature around 574–624 °K is required. The acceptable features of NaS batteries can be considered as follows:

- It has partly high energy densities and higher capacity.
- The daily self-discharge rate of this battery is nil.
- The utilized material is not toxic and expensive.
- The main disadvantage of this type of secondary battery is its operating temperature and the requirement of an extra-sensing system to ensure its operating temperature.

3.2.2.3 Flow Batteries

Redox-Flow Battery (RFB)

In RFBs, similar to other types of batteries, materials react in reversible chemical reactions during energy conversion. Flow batteries (FB) store energy in active electricity species. Electrically active species are dissolved in the fluid electrolytes in tanks. Converting chemical energy to electrical energy is performed by pumping the liquid by the electrochemical cell. RFB is one type of these batteries. The most

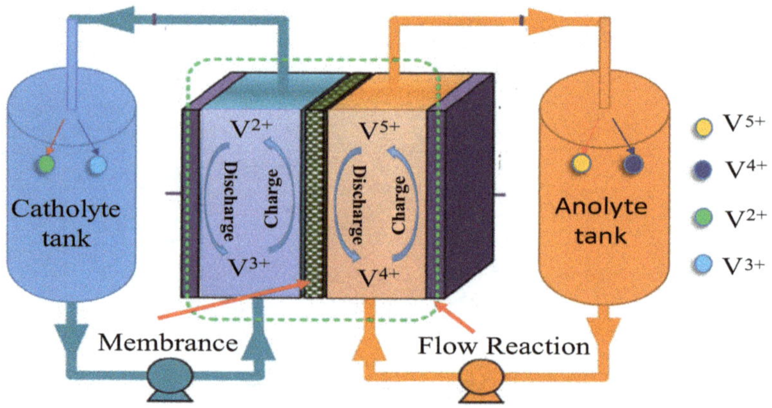

Fig. 3.10 Schematic diagram of vanadium redox flow battery principle operation

important feature of FB is that its power rating and capacity are not dependent to each other (Gallo et al., 2016). This means that the power rating is determined by the active region of the cell stack, while the capacity is determined by the volume of dilution of electrolytes. RFBs demonstrate a long life cycle, stability, high-output efficiency, and energy flexibility that all lead RFBs to become more popular to apply in any standalone networks such as microgrids and energy communities (Firoozi et al., 2020). As shown in Fig. 3.10, the two liquid electrolytes are pumped toward the reverse sides of the cell with dissolved metal ions. This storage has a lifetime of around 15–20 years, and it can discharge for 4–10 h. The efficiency of FBs is about 60–70% (Hannan et al., 2017).

3.3 Electric Vehicles

Totally, there are several types of EVs utilized by occupants of buildings. These vehicles include battery-based EVs, internal combustion engine/electric hybrid vehicle, solar-powered vehicles, flywheels, or SC-based EV and fuel cell-based EVs (Larminie & Lowry, 2012). Some EV type is depicted in Fig. 3.11.

In an active distribution network, distributed energy resources play important roles in providing the required energy and flexibility to the grid. Solar panels, wind turbines, and grid-connected EVs of the building are some examples of DERs. In recent decades, due to environmental benefits as well as advances in energy storage technology, a tendency to the high utilization of EVs has increased. However, an increase in the number of EVs and their uncertain charging behavior can also result in some issues in the electrical grid. For this reason, the use of these vehicles has an enormous impact on the load curve, performance, and design of the power system (Lopes et al., 2010; Omrani & Jannesari, 2019; Khooban et al., 2017).

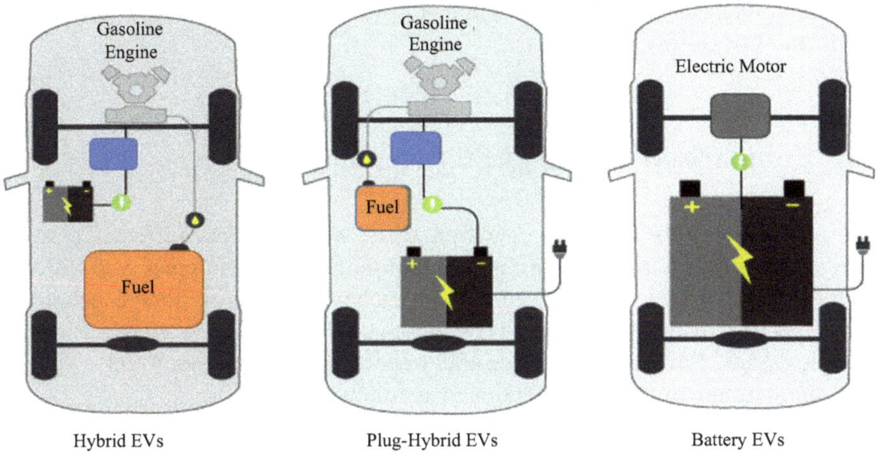

Fig. 3.11 Different types of EV

Recent studies demonstrate that drivers use their car for only around 2 h during the day and for the rest of it, the car is in the building's parking. Thus, the long presence of EVs in the building can help the energy management system to schedule charging/discharging the EVs in a more efficient way. For example, if these cars have the ability to inject energy back to the grid (V2G), the energy stored in their batteries can be used to reduce peak power and provide other auxiliary services to the grid (Almeida et al., 2017; Peng et al., 2017; Denholm & Short, 2006). In this way, not only the grid flexibility requirements can be satisfied. However, the building can also enjoy the monetary benefits of providing services. It is worth mentioning that considering the parameters of reducing battery life, the use of EVs in frequency control will be more profitable for car owners than using them to reduce the peak (Liu et al., 2018).

Regarding active network management, the main aim of the contribution of EVs to the grid service provision is minimizing costs and improving network reliability. In this regard, the use of EVs with network connection ability in smart grids will reduce network dependency on large-scale expensive units. Since the number of EVs as small-scale generation units is great and they are located at different nodes of distribution networks, they can highly participate in providing energy and flexibility. However, due to a huge number of these vehicles, the problem of optimizing their application will be more complicated (Tan et al., 2014; Corchero et al., 2014; Aliasghari et al., 2018). Therefore, the management of these resources in distribution networks is of vital necessity. The optimal (charging/discharging) scheduling of these vehicles can also assist islanded microgrid with its required energy and flexibility. In addition, adaptive charging/discharging EVs is one of the important factors facilitating active distribution management of distribution networks. Nevertheless, the preference of the owner needs to be taken into account when determining the EV's charging/discharging schedules. In addition, the owner must be compensated sufficiently if the EV was supporting the grid through the provision of flexibility services.

3.4 Comparison of Energy Storages in ABs

It is obvious that the efficient operation of network requires different types of storage-based resources. In addition, each building can be equipped with different kinds of storage-based resources. At present, each type of energy storage technology has its specific advantages and disadvantages to use in an active distribution network and microgrids. Several criteria can affect choosing energy storage for a specific application. Some of them are mentioned as follows:

- Power rating
- Lifetime
- Response time
- Cost
- Memory effects
- Self-discharge
- Environmental impact

A comparison of power and energy density of various types of ESS is also depicted in Fig. 3.12. According to this figure, in case of high power and high energy density for a specific amount of energy, the total required storage volume is reduced.

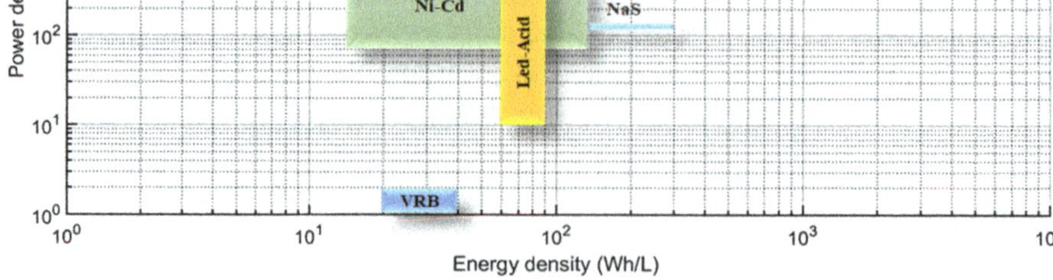

Fig. 3.12 Comparison of energy storages in terms of power and energy density

Although the application of energy storages in grid scale, microgrids, and AB scale is almost the same, there are still apparent differences between these applications. The application of ESS in the grid scale aims to optimize energy consumption, provide flexibility through time-shifting, peak shaving, and renewable energy resources integration in a low- and medium-voltage scale (Castillo & Gayme, 2014). Besides, the energy storage type used for the grid-scale applications can be classified based on their operational power rate. For instance, mechanical storages such as pumped hydro and compressed air storages are mostly used in large-scale cases. On the other hand, regarding medium- and small-scale power rating, flow battery and conventional battery (in large-scale) are generally used. Despite the recent development of energy storage technology such as SMES and supercapacitor, they have not received much attention in grid-scale application yet due to their high costs (Luo et al., 2015).

The main goal of energy storage integration into the microgrids and active distribution networks (through ABs) can consist of power quality (PQ) improvement, stability issues, enhancing the hosting capacity of RES (Laaksonen et al., 2020), etc. A wide range of applications in grid-scale applications has led to the possibility of utilizing various types of energy storage technologies. For instance, regarding maintaining PQ, the ESS technology with very fast response (in *ms* time intervals) is needed. As can be seen in Fig. 3.11, SMES, capacitors, supercapacitors, and also conventional batteries are suitable for power quality applications (Abbey & Joos, 2007).

Due to the stochastic feature of renewable energy resources such as wind and solar, they have fluctuations in their output power that can cause several problems for the grid. Application of EES, in this case, can overcome some of those problems and can increase the maximum hosting capacity of the network. For instance, the device could store power at off-peak hours and inject it back to the grid during peak moments. In addition, the intermittent production of RES can be smoothed and stabilized by using EES (Hodson, 2013). In this case, several ESS technologies can be deployed including NiCd and lead-acid batteries and supercapacitors. However, it seems that the use of hybrid energy storage is more suitable especially for power fluctuation compensation. Using hybrid energy storage can increase the lifetime of each energy storage device. It is possible to utilize a combination of ESS disregarding different types of storages. For instance, by combining battery storage and SC, relatively high capacity and high energy density can be obtained.

Another important proposed application for energy storage is for balancing between production and load demand in real time. In this regard, ESS can help the system stay reliable and stable. This application of EES is highly helpful especially for microgrids with high penetration of intermittent RESs and uncertain loads. EES also can also be used for black-start applications. It means they can help the system to startup from a black-shutdown without catching high power from the main grid. FBs and conventional batteries can be more suitable for this application (Lobato et al., 2013; Kloess & Zach, 2014).

On the other hand, by increasing the installation of highly sensitive electric equipment in the distribution system that is very sensitive to the power outage, the necessity of EES application becomes more important. In this way, power bridging with ESS maintains supplying sensitive equipment or grid during the transition period. Therefore, the deployed energy storage devices must have a moderate power rating and response time ranges, between 100 kW to 10 MW and around 1 s, respectively. Conventional and FBs are proper for this type of application because they have an approximately fast response and almost long discharge time (up to several hours) (Chen et al., 2009; Koohi-Fayegh & Rosen, 2020).

Regarding building-scale applications, consider the case in which the building is not responsible for providing any flexibility services to the grid. In this situation, the building mostly utilizes the ESS to increase its self-sufficiency. However, a cost-benefit analysis should be performed in order to assess the profitability of the ESS. In other words, the ESS should bring monetary profits for the building in order to cover its capital and installation cost. A comprehensive technical comparison of different types of EES from several aspects is gathered in Table 3.1.

3.5 Summary and Conclusion

By changing the structure of traditional electrical networks to active networks, the role of ESSs has been highlighted. There are several types of currently accessible energy storage including TES, EES, MES, etc. This chapter provides a comprehensive overview of the main features of EES and EV, their technologies, and their various applications regarding the active electrical network. Electrical ESS is regarded as a promising solution to deal with variable/uncertain renewables. The ESSs act as key elements of the grid that contribute to creating a more reliable and flexible power system. By having a rapid bi-directional flow capability, ESSs can rapidly smooth the fluctuations in both over-frequency and under-frequency situations. Moreover, ESSs have a rapid response to unexpected variations, which is crucial for the system's stability. Different types of EESs presented in this chapter with their characteristics have several value functions for the grid as well as demand-side ABs such as (1) helping to meet required peak load, (2) balancing between generation and demand, (3) power quality/reliability improvement, (4) supplying isolated loads, and (5) reducing the total cost. Furthermore, deploying EVs can create many potential opportunities for active distribution management such as vehicle-to-grid technology, which can also support RESs integration.

Table 3.1 Electrical energy storages' technologies

Type	Year	Power rating (MW)	Power density (kW/m^3)	Energy density (kWh/m^3)	Life time (cycle)	Response time	Self discharge (%/ day)	Operating Temp. (°C)	Discharge time
Pb-acid	1859	<70	90–700	75	2000	ms	0.1–0.3	+25	s – 3 h
Ni-Cd	1899	<40	75–700	<200	1500	ms	0.2–0.6	−40 to +45	s – h
NaS	1960	0.5–50	120–160	<400	2000–45000	ms	20	+300	s – h
Ni-MH	1967	10-6–0.2	500–3000	<350	300–500	ms	0.4–1.2	−20 to +45	hour
SMES	1969	0.01–10	2600	6	10000	ms	10–15	−270 to −140	ms – 5 min
SC	1978	0.01–1	$(4–12) \times 10^4$	10–20	$>5 \times 10^5$	ms	2–40	−40 to +85	ms – 1 h
V-redox	1984	0.03–7	250–270	20–35	>13000	<1ms	0–10	0 to +40	s – 10 h
Li-ion	1985	0.1–5	$1300–10^4$	250–620	>4000	ms – s	0.1–0.3	−10 to +50	min – h

References

Abbey, C., & Joos, G. (2007). Supercapacitor energy storage for wind energy applications. *IEEE Transactions on Industry applications, 43*(3), 769–776.

Aleem, S. A., Hussain, S., & Ustun, T. S. (2020). A review of strategies to increase PV penetration level in smart grids. *Energies, 13*(3), 636.

Ali, M. H., Wu, B., & Dougal, R. A. (2010). An overview of SMES applications in power and energy systems. *IEEE Transactions on Sustainable Energy, 1*(1), 38–47.

Aliasghari, P., Mohammadi-Ivatloo, B., Alipour, M., Abapour, M., & Zare, K. (2018). Optimal scheduling of plug-in electric vehicles and renewable micro-grid in energy and reserve markets considering demand response program. *Journal of Cleaner Production, 186*, 293–303.

Almeida, P. R., Iria, J., Soares, F., & Lopes, J. P. (2017). Electric vehicles in automatic generation control for systems with large integration of renewables. In *2017 IEEE power & energy society general meeting*. IEEE, pp. 1–5.

Andrijanovits, A., Hoimoja, H., & Vinnikov, D. (2012). Comparative review of long-term energy storage technologies for renewable energy systems. *Elektronika ir Elektrotechnika, 118*(2), 21–26.

Castillo, A., & Gayme, D. F. (2014). Grid-scale energy storage applications in renewable energy integration: A survey. *Energy Conversion and Management, 87*, 885–894.

Chatzivasileiadi, A., Ampatzi, E., & Knight, I. (2013). Characteristics of electrical energy storage technologies and their applications in buildings. *Renewable and Sustainable Energy Reviews, 25*, 814–830.

Chen, H., Cong, T. N., Yang, W., Tan, C., Li, Y., & Ding, Y. (2009). Progress in electrical energy storage system: A critical review. *Progress in Natural Science, 19*(3), 291–312.

Corchero, C., Cruz-Zambrano, M., & Heredia, F.-J. (2014). Optimal energy management for a residential microgrid including a vehicle-to-grid system. *IEEE Transactions on Smart Grid, 5*(4), 2163–2172.

Denholm, P., & Short, W. (2006). *Evaluation of utility system impacts and benefits of optimally dispatched plug-in hybrid electric vehicles* (Revised). National Renewable Energy Lab.(NREL).

Ferreira, H. L., Garde, R., Fulli, G., Kling, W., & Lopes, J. P. (2013). Characterisation of electrical energy storage technologies. *Energy, 53*, 288–298.

Firoozi, H., Khajeh, H., & Laaksonen, H. (2020). Optimized operation of local energy community providing frequency restoration reserve. *IEEE Access*.

Gallo, A., Simões-Moreira, J., Costa, H., Santos, M., & dos Santos, E. M. (2016). Energy storage in the energy transition context: A technology review. *Renewable and Sustainable Energy Reviews, 65*, 800–822.

Guney, M. S., & Tepe, Y. (2017). Classification and assessment of energy storage systems. *Renewable and Sustainable Energy Reviews, 75*, 1187–1197.

Hannan, M., Hoque, M., Mohamed, A., & Ayob, A. (2017). Review of energy storage systems for electric vehicle applications: Issues and challenges. *Renewable and Sustainable Energy Reviews, 69*, 771–789.

Hodson, H. (2013). *Texas-sized battery aims to green up the grid*. Elsevier.

Hossain, E., Faruque, H. M. R., Sunny, M., Haque, S., Mohammad, N., & Nawar, N. (2020). A comprehensive review on energy storage systems: Types, comparison, current scenario, applications, barriers, and potential solutions, policies, and future prospects. *Energies, 13*(14), 3651.

Ibrahim, H., Ilinca, A., & Perron, J. (2008). Energy storage systems – Characteristics and comparisons. *Renewable and Sustainable Energy Reviews, 12*(5), 1221–1250.

Kalaiselvam, S., & Parameshwaran, R. (2014). *Thermal energy storage technologies for sustainability: Systems design, assessment and applications*. Elsevier.

Khajeh, H., Laaksonen, H., Gazafroudi, A. S., & Shafie-khah, M. (2020). Towards flexibility trading at TSO-DSO-customer levels: A review. *Energies, 13*(1), 165.

Khooban, M. H., Niknam, T., Blaabjerg, F., & Dragičević, T. (2017). A new load frequency control strategy for micro-grids with considering electrical vehicles. *Electric Power Systems Research, 143*, 585–598.

Kloess, M., & Zach, K. (2014). Bulk electricity storage technologies for load-leveling operation–an economic assessment for the Austrian and German power market. *International Journal of Electrical Power & Energy Systems, 59*, 111–122.

Koohi-Fayegh, S., & Rosen, M. (2020). A review of energy storage types, applications and recent developments. *Journal of Energy Storage, 27*, 101047.

Laaksonen, H., Parthasarathy, C., Hafezi, H., Shafie-khah, M., Khajeh, H., & Hatziargyriou, N. (2020). Solutions to increase PV hosting capacity and provision of services from flexible energy resources. *Applied Sciences, 10*(15), 5146.

Larminie, J., & Lowry, J. (2012). *Electric vehicle technology explained*. Wiley.

Li, Q., Choi, S. S., Yuan, Y., & Yao, D. (2010). On the determination of battery energy storage capacity and short-term power dispatch of a wind farm. *IEEE Transactions on Sustainable Energy, 2*(2), 148–158.

Linden, D., & Reddy, T. (2002). *Handbook of batteries*. McGraw-Hill.

Liu, H., Huang, K., Yang, Y., Wei, H., & Ma, S. (2018). Real-time vehicle-to-grid control for frequency regulation with high frequency regulating signal. *Protection and Control of Modern Power Systems, 3*(1), 13.

Lobato, E., Sigrist, L., & Rouco, L. (2013). Use of energy storage systems for peak shaving in the Spanish Canary Islands. In *2013 IEEE power & energy society general meeting*. IEEE, pp 1–5.

Lopes, J. A. P., Soares, F. J., & Almeida, P. M. R. (2010). Integration of electric vehicles in the electric power system. *Proceedings of the IEEE, 99*(1), 168–183.

Luo, X., Wang, J., Dooner, M., & Clarke, J. (2015). Overview of current development in electrical energy storage technologies and the application potential in power system operation. *Applied Energy, 137*, 511–536.

Nadeem, F., Hussain, S. S., Tiwari, P. K., Goswami, A. K., & Ustun, T. S. (2018). Comparative review of energy storage systems, their roles, and impacts on future power systems. *IEEE Access, 7*, 4555–4585.

Noack, J., Roznyatovskaya, N., Herr, T., & Fischer, P. (2015). The chemistry of redox-flow batteries. *Angewandte Chemie International Edition, 54*(34), 9776–9809.

Omrani, M. M., & Jannesari, H. (2019). Economic and environmental assessment of reusing electric vehicle lithium-ion batteries for load leveling in the residential, industrial and photovoltaic power plants sectors. *Renewable and Sustainable Energy Reviews, 116*, 109413.

Peng, C., Zou, J., & Lian, L. (2017). Dispatching strategies of electric vehicles participating in frequency regulation on power grid: A review. *Renewable and Sustainable Energy Reviews, 68*, 147–152.

Poullikkas, A. (2013). A comparative overview of large-scale battery systems for electricity storage. *Renewable and Sustainable Energy Reviews, 27*, 778–788.

Suvire, G., Mercado, P., & Ontiveros, L. (2010). Comparative analysis of energy storage technologies to compensate wind power short-term fluctuations. In *2010 IEEE/PES transmission and distribution conference and exposition: Latin America (T&D-LA)*. IEEE, pp. 522–528.

Tan, X., Li, Q., & Wang, H. (2013). Advances and trends of energy storage technology in microgrid. *International Journal of Electrical Power & Energy Systems, 44*(1), 179–191.

Tan, Z., Yang, P., & Nehorai, A. (2014). An optimal and distributed demand response strategy with electric vehicles in the smart grid. *IEEE Transactions on Smart Grid, 5*(2), 861–869.

Yao, Y., Zhang, D., & Xu, D. (2006). A study of supercapacitor parameters and characteristics. In *2006 international conference on power system technology*. IEEE, pp. 1–4.

Zakeri, B., & Syri, S. (2015). Electrical energy storage systems: A comparative life cycle cost analysis. *Renewable and Sustainable Energy Reviews, 42*, 569–596.

Chapter 4
Hybrid AC/DC Electrical Power Grids in Active Buildings: A Power Electronics Perspective

Vitor Monteiro, Catia Oliveira, Sergio Coelho, and Joao L. Afonso

4.1 Introduction

As demonstrated in the last decades, electrical power grids are in continuous progress, and nowadays, more than ever, they must be prepared to deal with the growing penetration of new technologies without neglecting policies targeting to mitigate climate change. Consequently, electrical grids are facing a revolution, from a centralized generation paradigm (characterized by a unidirectional power flow) to a decentralized and dynamic generation (characterized by a bidirectional power flow and communication). Aligned with this context, around the world, several countries are modernizing the distribution electrical grids to obtain access to affordable, reliable, and sustainable energy.

As broadly publicized and debated, the matter of sustainable energy is appropriate for coming generations, including the transition to more and more emergent smart grids as an essential pillar for this topic (Sakis Meliopoulos et al., 2011; Gungor et al., 2012; Yu et al., 2011a). Accordingly, electrical grids are facing great challenges caused by a novel paradigm, where a more intensive and large-scale integration of different technologies has been occurring. In addition to the exponential growth of renewable energy sources (RES), the recent preponderance of energy storage systems (ESS), electric vehicles (EV), and new controlled electrical appliances has also been contributing to aggravate this situation. With the growth of smart grids, it is predictable the change of traditional consumers to active consumers, acting as active players in the energy market and management. Consequently, it is expected a shift toward active buildings, where the possibility of bidirectional flow of power and information is a key feature to offer a higher

V. Monteiro (✉) · C. Oliveira · S. Coelho · J. L. Afonso
University of Minho, Braga, Portugal
e-mail: vmonteiro@dei.uminho.pt

V. Vahidinasab, B. Mohammadi-Ivatloo (eds.), *Active Building Energy Systems*, Green Energy and Technology, https://doi.org/10.1007/978-3-030-79742-3_4

degree of flexibility in terms of defining adjustable profiles of power consumption or injection into the power grid, as explored in refs. (Penya et al., 2013; Bulut & Wallin, 2014; Bulut et al., 2016). However, the shift of paradigm toward active buildings is emerging and involves different features related with power.

Based on ref. (Isaksson, 2011), the concept of active building encompasses the control of the indoor temperatures according to the electrical appliances, as well as the household actions. On the other hand, based on ref. (Calbureanu et al., 2018), the concept of active building is more related with the self-power production and consumption not necessarily covering the interface with the power grid. Therefore, converging on this concept of active building, all of the aforementioned technologies are identified as priority technologies and imposing new requirements and more flexibility and innovative challenges targeting to control the electrical grid and reduce the pollution caused by the electricity production, as investigated in refs. (Abdullah et al., 2014; Bloemink & Green, 2013; Bollen & Hassan, 2011; Ai et al., 2016). Therefore, power electronics play a central role for such purpose, supporting the evolution and modernization targeting smart grids and contributing to establishing future directions to achieve the goal of sustainability, as demonstrated in ref. (Yu et al., 2011b).

In fact, power electronics is already present in some stages of the electrical grids, where particular and exemplificative cases of application include the transmission on HVDC, solutions for power quality in high-level and low-level applications, as well as the integration of renewables, EV and ESS. An investigation concerning forthcoming power electronics grids is introduced in ref. (Boroyevich et al., 2013). Additionally, power electronics has also penetrated in other applications, as is the case of electrical substations, mainly in the scope of the transformer that it is considered the heart of the latter.

Regarding the diverse engagements and control strategies of DC substations, important investigations are introduced in ref. (Ma et al., 2019a) and ref. (Ma et al., 2019b), where different configurations and control strategies are highlighted. Nevertheless, despite the many advantages demonstrated by these technologies in terms of innovative and cooperative operation modes, many individual power electronics converters are required for interfacing the power grid, as demonstrated in refs. (Monteiro et al., 2018a; Bhattacharjee et al., 2019; Muttaqi et al., 2019), representing high-cost and global low-efficiency solutions.

4.1.1 Status Quo, Challenges, and Outlook

As a contribution to overcoming the mentioned aspects, unified power electronics systems for active buildings (e.g., smart homes) are assuming new preponderance in the context of smart grids. In fact, smart homes are identified as active buildings with the integration of innovative characteristics, like self-production from RES, local ESS, and participation in the power management due to the offered flexibility. In this sense, several unified systems are approached, namely, a unified three-port structure for linking an EV and a RES is proposed in ref. (Monteiro et al., 2018b); dedicated unified systems for interfacing EV, ESS, and RES are proposed in ref. (Monteiro et al., 2019); and multiobjective off-board EV chargers as smart home enablers offering the possibility to provide ancillary services for the electrical grid are proposed in refs. (Monteiro et al., 2018c; Aldik et al., 2018; Monteiro et al., 2020). Besides, aligned with the new paradigm of distributed energy resources in low voltage, new challenges concerning solid-state power electronics technologies are also emerging. The goal is to provide precise and fast controllability of the power grid voltages, as well as to prevent power quality issues, as demonstrated in refs. (Huber & Kolar, 2019; Saleh et al., 2019a; Saleh et al., 2019b; Agrawal et al., 2019).

The aforementioned innovative features are also considered important assets for the concept of active buildings since they can be seen as microgrids within the smart grid (Zhong et al., 2019). In this context, the control methodologies proposed in refs. (Li et al., 2018a; Vuyyuru et al., 2019), besides being used for the flexible interlinking and harmonized power control of several DC microgrids, can also be adopted for various smart homes when operating as individual microgrids. However, some pertinent key features are not contemplated in these prior works, opening new challenges and opportunities. Thus, considering the numerous electrical appliances and technologies natively operating in DC, the discussion of AC versus DC is, nowadays, being revived for different contexts, including smart homes as part of boosting smart grids.

The debate of AC versus DC is not new, and it can be outlined back to the battle between Edison and Tesla/Westinghouse (Fairley, 2012), where the advantages of AC are contributed to the actual worldwide electrical grids (mainly due to the transformers, allowing to change the voltage levels). However, due to the nature of some technologies available nowadays, this battle has been reactivated, and it is gaining preponderance compared to the current situation of electrical grids, as investigated in refs. (Lotfi & Khodaei, 2017). Specifically, electrical grids are facing an intense revolution targeting the presence of DC grids, since the aforementioned technologies (e.g., EV and RES) have contributed definitively to the opportunity of shifting for hybrid AC/DC electrical grids requiring high-speed controllability of the power flow. Figure 4.1 shows the architecture of a hybrid AC/DC electrical grid, where is shown the interface of EV, RES, ESS, and loads. As it can be seen, both AC grid and DC grid can be interfaced by AC-DC converters.

4.1.2 Contribution and Organization

Along with this book chapter, as a core contribution, the hybrid AC/DC electrical grids are presented from the power electronics point of view, showing that it is a promising and emergent technology with an abundant prospect toward sustainable smart grids. More specifically, the distinct possibilities in terms of structures for hybrid AC/DC electrical grids for active buildings are investigated in Sect. 4.2, while a broad explanation and presentation of the main power electronics topologies for

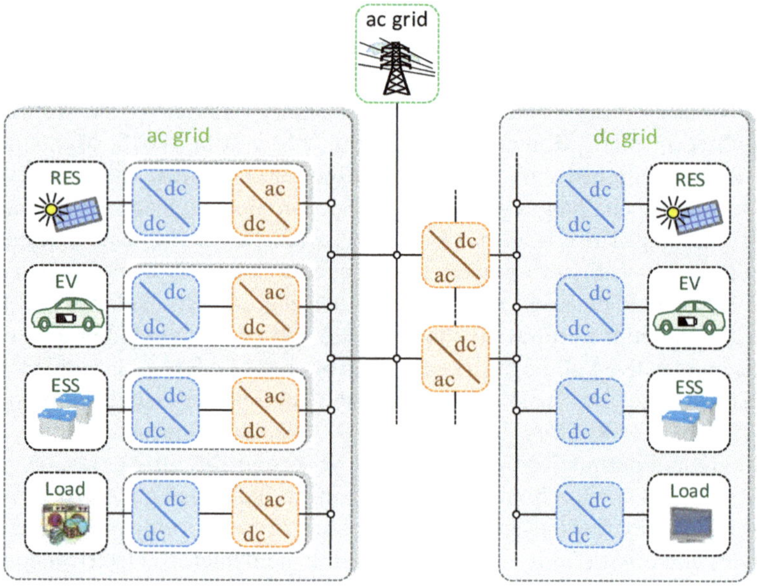

Fig. 4.1 Hybrid AC/DC electrical power grid

hybrid AC/DC structures are presented in Sect. 4.3. Regarding the experimental validation, different power electronics prototypes and results are presented in Sect. 4.4. Innovative perspectives for integrating a hybrid AC/DC electrical grid of a smart home, as an example of active building, within smart grids, are presented in Sect. 4.5. The active building studied includes a developed prototype and experimental results, while the main conclusions that can be retained are highlighted in Sect. 4.6.

4.2 Hybrid AC/DC Electrical Power Grids for Active Buildings

As introduced, electrical grids are facing a decisive challenge since DC grids are gaining greater preponderance due to the available technologies nowadays, natively operating in DC. This is mainly influenced by modern solid-state power electronics technologies, where DC systems offer pertinent advantages in terms of simplicity, cost, and efficiency. As exemplificative cases, in ref. (Pei et al., 2015) is presented a multi-terminal DC grid for integrating EV, in ref. (Blaabjerg et al., 2017) are presented recent advances for DC grids, in ref. (Chuangpishit et al., 2014) is presented DC grid in offshore wind systems, and in ref. (Suryanarayana & Sudhoff, 2017) is presented the role of power electronics for DC grids in the distribution perspective. Moreover, futuristic perspectives of hybrid AC/DC electrical grids are presented in ref. (Wang et al., 2013), whereas in ref. (Bracale et al., 2012) are presented the hybrid AC/DC electrical grids in the perspective of power quality, in refs. (Paez et al., 2019; Khan et al., 2020) are presented reviews of power electronics converters for DC grids, and in refs. (Unamuno & Barrena, 2015a; Unamuno & Barrena, 2015b) are presented extensive and detailed reviews regarding classifications in terms of control and topologies.

4.2.1 Advantages of Hybrid AC/DC Electrical Power Grids

Although the employability of AC grids remains interesting and necessary, several factors have contributed to make DC systems a viable possibility for low-voltage applications at home level (i.e., as a possibility of active building): (i) the RES production at home level is mainly in DC (since PV panels are the most common option); (ii) the energy storage is mainly in DC (typically, batteries are used as ESS); (iii) the growing number of electrical appliances that are used nowadays is natively DC (e.g., TV and LED lighting systems); (iv) the spreading introduction of plug-in EV (i.e., the internal ESS can be directly supplied in DC); (v) the DC grids do not present voltage drop in the inductive reactance of the line impedance; (vi) the global efficiency is improved since the power stages are reduced (e.g., using a common DC-link for native DC grid) and also because there is no skin effect; (vii) there are not power quality issues such as harmonics and low power factor; (viii) the controllability of DC electrical appliances is simpler; and (ix) during power outages in the AC grid, the DC grid can operate guaranteeing that the loads are always supplied with the required voltage with high levels of power quality.

Summarizing, the interest and potential advantages of hybrid AC/DC electrical grids have emerged and gained attention as a focus of research efforts, which is seen as a natural next step in active building (e.g., at the home level), also contributing to the widespread of smart homes, as demonstrated in refs. (Rodriguez-Diaz et al., 2016a; Vossos et al., 2014; Patterson, 2012; Huang et al., 2015; Nasir et al., 2018). In this context, previous studies can be identified: an economic perspective of the electrical appliances for the viability of DC homes is presented in ref. (Sousa et al., 2019a); a comparative analysis of power electronics topologies to interface DC homes with the main AC grid is presented in ref. (Sousa et al., 2019b); and a performance comparison of a typical nonlinear load in AC and DC power grids is presented in refs. (Sousa et al., 2018a, 2019c). Nevertheless, it is worth emphasizing that there are some challenges to overcome (e.g., standardization and grid codes, and protection systems); however, the market of smart homes with hybrid AC/DC electrical grids is forecasted to gain more attraction in the next decades, as demonstrated by the worldwide preliminary projects aligned with this perspective (Hirose et al., 2013; Diaz et al., 2015). However, this context of smart homes does not include the possibility of providing ancillary services for the power grid (e.g., as offered by the selective harmonic measurement and compensation in distributed microgrids, investigated in ref. (Sousa et al., 2018b), and the control of interlinking converter in hybrid AC/DC electrical grids, investigated in refs. (Shen et al., 2019; Li et al., 2018b), since the focus is only for the sustainable operation from the smart home point of view. The controllability of power flow in hybrid AC/DC electrical grids can be found in refs. (Renedo et al., 2017; Zhang et al., 2016; Miao et al., 2016), where two main models are identified, namely, the unified and the sequential.

Specifically focusing on the distinctive applications of DC grids, advanced designs as an innovative generation of distribution grids are examined in ref. (Dragicevic et al., 2014). Analyzing more in detail the structure of DC grids in detriment of the AC grids, or combined with them, it is necessary to select proper voltage levels as a conciliation, e.g., between compatibility and efficiency (Rodriguez-Diaz et al., 2016b). A distribution system constituted by a DC grid and considering the interface of some specific and complex electrical appliances is planned in ref. (Salomonsson & Sannino, 2007). An innovative hierarchical control, especially dedicated to DC grids, is formulated in ref. (Li et al., 2018c), where it was considered as a specific requirement in a significant dispersion of RES into the grid. Considering DC grids in the perspective of urban scenarios, in ref. (Sun et al., 2018) are presented the fundamental operation modes. A flexible control, targeting to manage the power flow in DC grids, is proposed in ref. (Rouzbehi et al., 2016). The regulation and management of the power flow considering different DC grids is accessible in ref. (Vuyyuru et al., 2019).

By considering all of the aforementioned technologies for smart homes, including unified systems and solid-state power electronics technologies with multiport and multifunctionalities, it is possible to identify some similarities and convergence to disruptive technology for future innovative smart homes introducing different and diverse features as analyzed in refs. (Fan et al., 2017; Sahoo & Ned, 2014; López et al., 2006; Wang et al., 2016; Gu et al., 2016a). Notwithstanding the abovementioned benefits, the spread of DC grids is fronting two core facts: the protection systems and the legislation/standards as scrutinized in refs. (Dragicevic et al., 2016a; Dragicevic et al., 2016b; Mohammadi et al., 2019). However, it is important to note that the European Union introduced some documentation, establishing the voltage values between 75 V and 1500 V (European Parliament, 2014).

4.2.2 Unipolar and Bipolar DC Electrical Power Grids

Currently, concerning the possible structures for the implementation of DC grids, two leading structures are recognized, specifically the unipolar and the bipolar as studied in refs. (Shen et al., 2019; Li et al., 2018b; Renedo et al., 2017; Zhang et al., 2016). These two possibilities are shown in Fig. 4.2. The unipolar structure is characterized by imposing a configuration based on two wires, consequently, presenting a single DC voltage level. In different circumstances, the bipolar structure is characterized by imposing a configuration based on three wires, consequently,

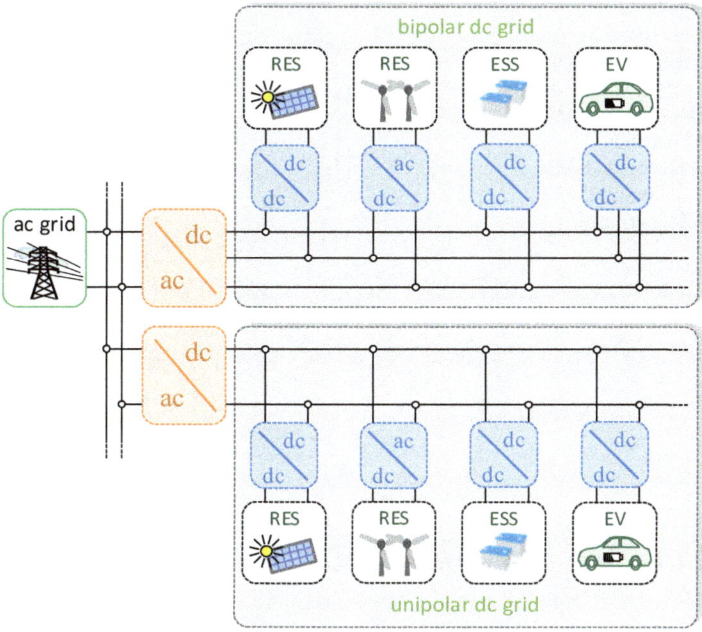

Fig. 4.2 Electrical DC grid based on unipolar and bipolar structures

presenting two DC voltage levels (Gu et al., 2016b; Kakigano et al., 2010). By analyzing the two structures, it is easy to verify that the unipolar structure has less complexity, both from the perspective of the structure itself (since it requires fewer wires) and from the control algorithms implementation difficulty point of view. Nonetheless, regardless of the additional wire and the consequent control complexity, the bipolar structure offers much more flexibility to deal with technologies supplied by different voltage levels. In this sense, this type of structure also allows higher efficient power stages and, since it presents an additional wire, more controllability in conditions of a line fault. An embracing assessment of emergent technologies directly related to bipolar DC grids is introduced in ref. (Rivera et al., 2020).

Notwithstanding the benefits of bipolar DC grids, problems of unbalances can occur, conditional to the technologies interfaced, as well as their operation in terms of bidirectional power flow. In this perspective, it is opportune the application of specific power electronics converters aiming to guarantee balanced currents in the bipolar DC grids, representing important support targeting stability. As shown, some of the technologies are linked to the bipolar DC grid through a two-wire DC-DC power electronics converter, while other technologies are linked through a three-wire DC-DC power converter.

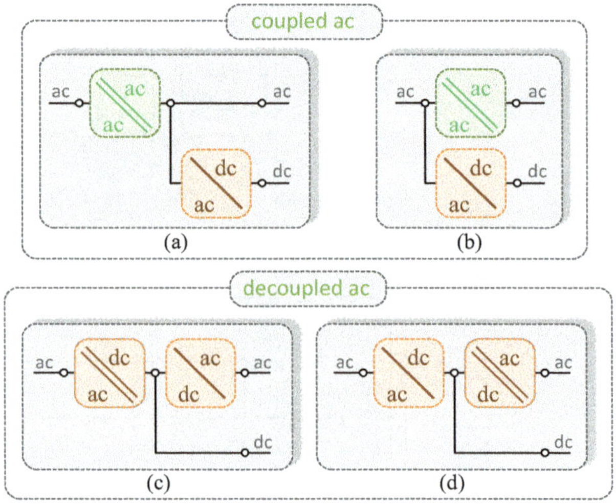

Fig. 4.3 Hybrid AC/DC electrical grids based on coupled AC and decoupled AC structures: (**a**) Coupled AC completely isolated; (**b**) Coupled AC partially isolated; (**c**) Decoupled AC completely isolated; (**d**) Decoupled AC partially isolated

4.2.3 Structures of Hybrid AC/DC Electrical Power Grids

The design of a hybrid AC/DC electrical grid requires the analysis of some characteristics such as the principle of hierarchy, the power quality assurance, the compensation of reactive power, the use of resources, and the configuration to be adopted. Concerning the structure of a hybrid AC/DC electrical power grid, two main groups can be identified, namely, coupled and decoupled AC configurations, as it can be seen in Fig. 4.3. In the case of the coupled AC configurations, the AC grid is directly attached to the main electrical grid through a transformer, and an AC-DC converter is required for the DC grid. In this case, the galvanic isolation is guaranteed by conventional low-frequency (power grid frequency) power transformers. As shown in Fig. 4.3a, the DC grid is established after the galvanic isolation, guaranteeing that the hybrid AC/DC is completely isolated from the main electrical grid. On the other hand, as shown in Fig. 4.3b, the DC grid is established with the AC-DC converter directly linked to the main electrical grid, meaning that the hybrid AC/DC electrical grid is only partially isolated (i.e., in this case, only the AC grid is isolated).

Analyzing the decoupled AC configuration, it is identified that the AC grid is not directly connected to the main electrical grid through a low-frequency transformer. In this type of configuration, it is introduced, at least, one AC-DC and one DC-DC stage. The associated costs of the coupled AC grid are lower when compared with the decoupled one due to the reduced size of the AC-DC converter required for the management of the power flow between the main electrical grid and the DC grid. However, in the decoupled grid configuration, the AC grid is connected to the main electrical grid by a DC stage, providing isolation, and independent control algorithms are applied for both grids. For this type of configuration, solid-state transformers (SSTs) are introduced to guarantee galvanic isolation as an alternative for the low-frequency transformers as shown in refs. (Yan et al., 2015; Bignucolo et al., 2015; Hannan et al., 2020).

Regarding the SST, the direct AC-AC only consists of power conversion from a high-voltage AC to a low-voltage AC. This type of architecture does not require complex control, but the absence of a DC-link offers few capabilities. On the other hand, an SST with more power stages presents more complexity in relation to the control system and power system, since several stages are introduced (e.g., AC-DC interfacing the main electrical grid, DC-DC isolated in high frequency, and the final DC-AC stage). In the case of the decoupled AC configurations, as shown in Fig. 4.3c, the AC grid and the DC grid are defined after the galvanic isolation offered by the front-end AC-DC power stage, establishing a hybrid AC/DC completely isolated from the main electrical grid. On the other hand, as shown in Fig. 4.3d, the galvanic isolation is no longer guaranteed by the front-end DC-AC converter; therefore, the hybrid AC/DC is only partially isolated (i.e., in this case, only the AC grid is isolated).

4.2.4 Load-Shift Systems for Hybrid AC/DC Electrical Power Grids

Numerous studies are being carried out to obtain efficient solutions for demand-side response in hybrid AC/DC electrical grids. In order to optimize the power transfer between local grids and central grids and within hybrid AC/DC electrical grids, ref. (Zhao et al., 2018) proposes a new problem formulation that allows the reduction of the number of variables and restrictions for more efficient power conversion. Part of the demand-side response is achieved through load-shift systems, allowing the management of electrical loads in customer's installations, taking advantage of the energy price difference throughout the day, as analyzed in ref. (Kinhekar et al., 2016). This type of systems interfaces ESS to store the energy during the off-peak periods and to discharge it during the on-peak periods. The power transfer and storage during these two different periods can also minimize peak demand as considered in ref. (Dusonchet et al., 2013). Load-shift systems are also often integrated with RES, namely, solar PV panels. In this case, the solar PV panels supply the DC loads in certain periods, also allowing the peak shaving and the reduction of the electricity expenses as explored in ref. (Jahromi & Seifi, 2016). On the other hand, the integration of RES in load-shift systems allows the reduction of greenhouse gas emissions. Furthermore, as investigated in refs. (Yu et al., 2011c; Golovanov et al., 2013), the consumer can participate in the energy market, selling power for other users or providers. In ref. (Chauhan et al., 2017) is proposed a new demand-side management system for DC grids applied in buildings. In order to improve the system efficiency, it is proposed a control algorithm that shifts the load consumption from periods where there is not power production from RES to periods where there is power production from these types of sources.

4.3 Power Electronics for Hybrid AC/DC Electrical Power Grids

This section introduces the main power electronics topologies for hybrid AC/DC electrical grids in active buildings, namely, single-phase and three-phase front-end AC-DC converters and back-end DC-DC converters. In this sense, the main advantages of each topology that can be applied in hybrid AC/DC electrical grids are presented. A comparison between the different power electronic topologies is described in order to understand which topology is more suitable for each situation.

4.3.1 Power Converters: Single-Phase AC-DC

Regarding single-phase front-end AC-DC converters, when focusing on the essential bidirectional operation, the full-bridge converter, presented in Fig. 4.4a, is the most used. This converter allows bidirectional power operation without ignoring a sinusoidal waveform of the consumed current and the operation with unitary power factor. Additionally, operates as an active rectifier when transferring power from the AC grid to the DC grid, and as a grid-tied inverter when transferring power from the DC grid to the AC grid. In terms of control, the most used algorithm is based on two stages: an outer control for regulating the DC-link voltage and an inner control for controlling the AC-side current. Therefore, the AC grid current is controlled, in terms of amplitude, according to the required power for the DC grid and the power for regulating the DC-link voltage. In this converter, each switching device must deal with the nominal current and the nominal DC-link voltage, which is recognized as the main drawback of this topology.

Nevertheless, with the foremost objective of reducing the size of the coupling passive filters, as well as minimizing the current in each switching device, when compared with the full-bridge converter presented in Fig. 4.1a, the main choice lies in interleaved topologies (i.e., the association of additional legs with the respective coupling filter). Figure 4.4b shows a bidirectional interleaved full bridge (i.e., consisting of two full-bridge converters with a common DC-link and individual coupling filters), representing a dual-stage interleaved structure, since are used two legs for the phase and two for the neutral. Therefore, each switching device must deal with half of the nominal current, which is an interesting option to reduce the switching losses. In terms of control, it can be implemented the same structure as for the full bridge, changing the modulation where two carriers are used.

On the other hand, when the focus is on reducing the maximum voltage applied to each switching device, also guaranteeing the reduction of the coupling passive filters, the main choice lies in multilevel topologies. Figure 4.4c, d shows two multilevel topologies, each one capable of operating with five voltage levels. In both of them, a split DC-link is considered, permitting to obtain the intermediary voltage levels. In both cases, an intermediary circuit between the full-bridge structure and

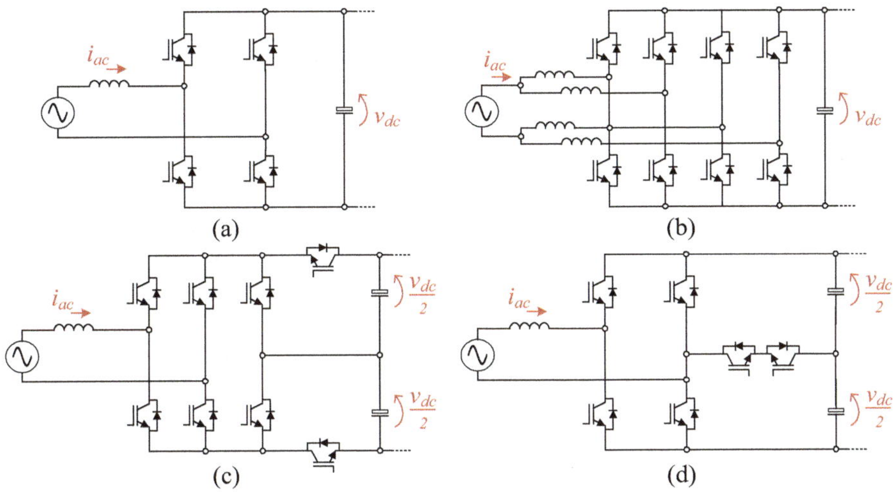

Fig. 4.4 Single-phase front-end AC-DC converters for interfacing an AC grid and a DC grid: (**a**) Full bridge; (**b**) interleaved; (**c**) multilevel; (**d**) multilevel T type

the split DC-link is required. Regarding the switching devices of the full bridge, each one must deal with the nominal voltage of the DC-link and the nominal current, while the switching devices of the intermediary circuit must deal with half of the nominal voltage of the DC-link and the nominal current. Comparing both multilevel topologies, it is possible to identify that the topology presented in Fig. 4.4d requires a lower number of switching devices, but it is more complex in terms of control, mainly due to the required dead times.

4.3.2 Power Converters: Three-Phase AC-DC

Regarding three-phase front-end AC-DC converters, the most commonly used bidirectional converter is the full bridge, represented in Fig. 4.5a. This converter is composed of three legs. As in the single-phase topologies, this converter also operates as an active rectifier when transferring power from the AC grid to the DC grid and as a grid-tied inverter when transferring power from the DC grid to the AC grid. The DC-link is divided, allowing to connect its middle point to the neutral wire of the AC grid.

Besides the traditional full bridge presented in Fig. 4.5a, bidirectional VIENNA topologies have gained more attention due to their characteristics. As shown in Fig. 4.5b, it consists of a three-phase full bridge and an organization of switching devices to link each phase to a common point connected in the midpoint of the split DC-link. In this topology, each semiconductor must deal with half of the nominal DC-link voltage and the maximum current present in each phase.

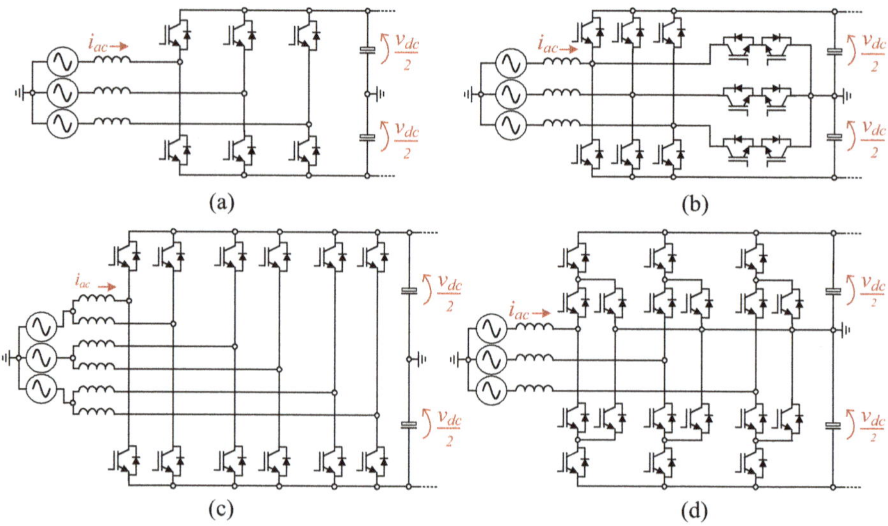

Fig. 4.5 Three-phase front-end AC-DC converters for interfacing an AC grid and a DC grid: (**a**) Full bridge; (**b**) Vienna type; (**c**) interleaved; (**d**) multilevel

Consequently, when it is required to reduce the current on each semiconductor, the main possibility is the use of an interleaved topology. Figure 4.5c shows a double-stage interleaved converter, which consists in combining two full bridges in parallel. In this way, the current is divided by two, permitting to diminish the current in each switching device, at the expense of increasing the number of switching devices. By applying a suitable modulation based on two carriers, the ripple of the resulting current is significantly reduced, and it can be zero when the duty cycle is 50%. This distinctive feature is extremely important targeting to minimize the requirements of coupling passive filters.

On the other hand, when the objective is to diminish the voltage that each switching device must deal, the alternatives are the multilevel topologies, as occurs with the single-phase converters. Figure 4.5d presents the regularly used bidirectional multilevel topology, which is the active neutral point clamped. This converter presents, as main characteristics, the ones described for the multilevel single-phase converters.

4.3.3 Power Converters: DC-DC

Regarding back-end DC-DC converters, when focusing on bidirectional converters, the leading converter is the half-bridge DC-DC converter, as shown in Fig. 4.6a, offering the possibility to operate in buck mode or in boost mode (the operation in each mode is guaranteed by controlling one switching device in each mode). This

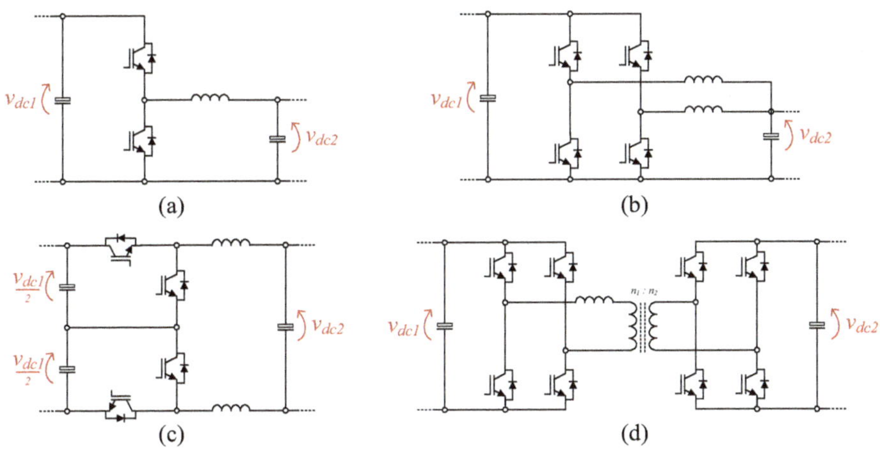

Fig. 4.6 Back-end DC-DC converters for DC grids: (**a**) Half-bridge; (**b**) interleaved half-bridge; (**c**) multilevel three level; (**d**) dual active bridge

converter interfaces the DC grid (established by the front-end AC-DC converter) and an energy storage device, capable of receiving or providing active power. The main disadvantage of this converter is related to the voltage and current in each switching device, since each one of them must deal with the nominal values of the converter.

In this context, with the objective of reducing the current in each switching device, interleaved converters can also be considered, as the one presented in Fig. 4.6b, which is a double-stage interleaved converter. As shown, this converter is constituted by two half-bridge converters, i.e., two legs in parallel, allowing to reduce in half the current that each switching device operates, both in buck mode and boost mode.

Also, for back end DC-DC converters, multilevel structures can be used, as shown in Fig. 4.6c. This converter permits the operation in buck mode or boost mode, being controlled by two semiconductors in each mode. Depending on the value of the voltages on each DC interface, the converter operates with three distinct voltage levels, which is a very interesting feature, since the value of the coupling passive filters can be reduced, as well as the voltage that each switching device must deal. Moreover, this converter also presents as innovative feature the possibility of operating similarly to an interleaved converter, obtaining a current ripple with a frequency that is the double of the switching frequency. Additionally, since this converter has a split DC-link in one of the interfaces, it is seen an interesting topology to be considered for bipolar DC grids.

As a complement to the converters presented above, when it is absolutely necessary to guarantee galvanic isolation with the DC grid, or when the voltage levels at the DC interfaces are very different, the main converter topology to be used is the dual active bridge. This converter, presented in Fig. 4.6d, is constituted by two full-bridge converters and by a high-frequency transformer. However, by replacing one of the full-bridge converters for a diode rectifier, a single active bridge topology is obtained. This converter has the same characteristics as the dual active bridge, but only allows unidirectional operation.

4.4 Experimental Validation of Power Electronics and Operation Modes: Future Perspectives for Smart Grids

This section demonstrates some illustrative laboratory prototypes of power electronics topologies and respective experimental validation, including front-end AC-DC three-phase and single-phase topologies for interfacing the AC grid and creating the DC grid, as well as back-end DC-DC topologies for interfacing the different technologies with the DC grid (e.g., EV and RES). The presented converters and respective laboratory prototypes were selected with the objective of allowing a direct interface between the topologies, i.e., the front-end AC-DC converters can be directly coupled to the DC-DC converters. This configuration is considered a relevant aspect since it offers a complete solution for hybrid AC/DC electrical grids. Additionally, in this section, it is also presented a developed solution for a hybrid AC/DC electrical grid, adding, as innovative aspect, the operation modes offered by the front-end AC-DC converter framed with future perspectives of smart grids.

4.4.1 Power Electronics Validation: Three-phase AC-DC Converter

This item presents an example of a front-end AC-DC converter that allows the interface with the AC grid, being a three-phase converter. As an innovative feature, in addition to allow the operation in bidirectional mode (as an active rectifier or as a grid-tied inverter), this converter is a double-stage interleaved, in which, in the interface of each phase of the AC grid, two legs of the converter are used. Moreover, this converter also permits the operation in the other quadrants of the PQ plane, allowing, if required by the application scenario, the operation with reactive power for the AC grid regardless of the operating power of the DC grid. In terms of hardware, the requirements of this converter are double compared to the traditional solution based on the full-bridge converter. However, the current ripple on the AC grid is halved, representing the main added value of this converter. In addition, as the current in each phase on the AC grid is divided by the two legs of the converter, switching and conduction losses are lower. At the interface with the DC grid, this converter has a capacitive filter and a medium connection point, which can be a very important feature for implementing bipolar DC grids.

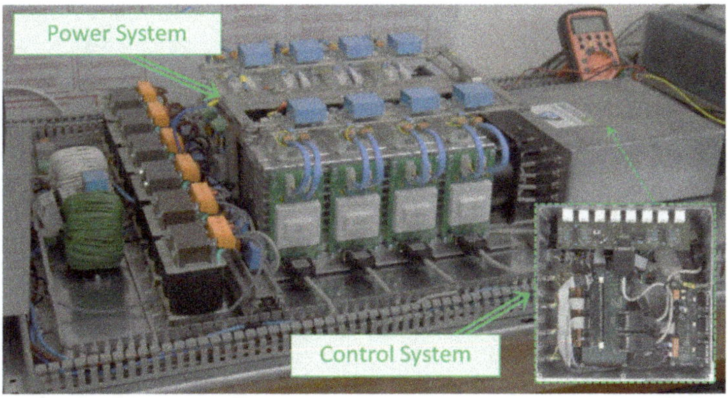

Fig. 4.7 Laboratory prototype of a three-phase front-end AC-DC converter based on an interleaved structure

In this respect, Fig. 4.7 shows the three-phase double-stage interleaved power converter, as well as the developed converter framed in the laboratory workbench, where the power system and the control system stand out. Summarizing, the power system consists of IGBT modules (model SKM100GB12T4) and drivers (model SKHI21AR), both from Semikron and voltage and current sensors (model LA25NP and LA100P, respectively, both from LEM), while the control consists of a DSP F28335 from Texas Instruments and a set of additional boards (e.g., command and signal conditioning boards).

Thus, Fig. 4.8 shows the main illustrative experimental results of this converter. Specifically, Fig. 4.8a shows the three voltages on the AC grid and the respective currents, allowing to verify that the converter operates with sinusoidal currents, presenting a very low value of THD (1.7%). Since the converter was validated when directly connected to the AC grid, with a nominal phase-to-phase voltage of 400 V, the voltages present a THD of 3.5%. At the interface with the DC grid (i.e., the interface provided for the back-end DC-DC converters), a DC voltage of 800 V was made available, with access to the midpoint. It is important to note that such DC-link voltage can be controlled to another reference value according to the DC grid requirements. More particularly and as an example, in Fig. 4.8b it is presented a detail of the current in each leg of phase a. As it can be seen, the current ripples of the legs are in phase opposition (i.e., 180 degrees), ensuring the interleaved operation in each phase. A particular and important case occurs when the converter is operating with a duty cycle of 50%, resulting in the absence of ripple in the resulting current at the AC grid interface (since the currents are added and the ripple is cancelled). It is also possible to verify that the voltage produced by the converter has two voltage levels since this case is analyzed for the positive semi-cycle.

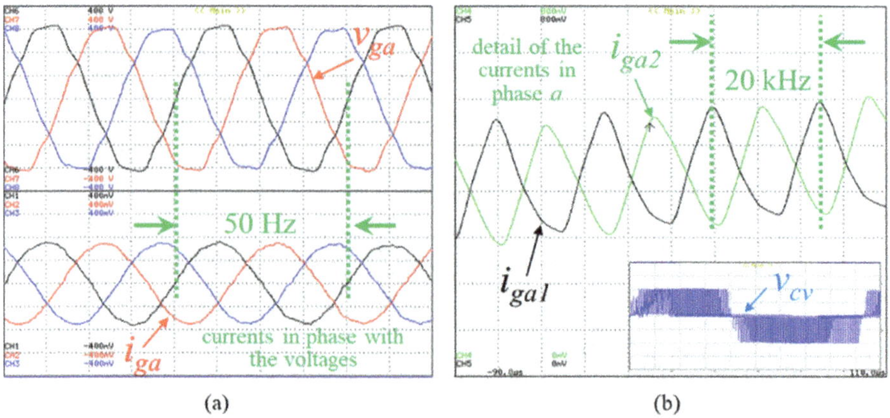

Fig. 4.8 Experimental results of the three-phase front-end AC-DC converter based on an interleaved structure, showing the AC grid currents (i_g{a,b,c}), the AC grid voltages (v_g{a,b,c}), the currents on each leg of phase a (i_{ga1}, i_{ga2}), and the voltage produced by the converter (v_{cv}): (**a**) With the AC-DC converter providing power to the DC grid (**b**) Detail of the currents on each leg of phase a the resultant current

4.4.2 Power Electronics Validation: Single-Phase AC-DC Converter

In this item, it is presented a multilevel front-end AC-DC converter for interfacing the AC grid and the DC grid. As a multilevel topology, it is possible to synthesize a voltage with more voltage levels, i.e., a signal closer to a sinusoidal signal, permitting to reduce the needs and the capacity/size of the coupling passive filters. Figure 4.9 shows the laboratory workbench, where it is possible to verify the developed multilevel converter, consisting of the power system (discrete IGBTs, model FGA25N120, discrete drivers with HCPL3120 and LEM sensors, as mentioned in the previous item) and the control circuit (Texas Instruments DSP F28335 and external signal conditioning and control circuit).

The developed prototype is a five-level topology that consists of a full-bridge structure with a split DC-link, where an auxiliary circuit is connected between the midpoint of the split DC-link and the neutral of the AC grid (cf. Fig. 4.4d). This auxiliary circuit, which allows to obtain the intermediate voltage levels in the positive and negative half cycles ($+v_{dc}/2$ and $-v_{dc}/2$), has the particularity to operate in bidirectional and bipolar modes.

Figure 4.10 shows some experimental results that validate this multilevel AC-DC converter. Figure 4.10a shows the voltage and current on the AC grid, and it can be verified that the current is sinusoidal and that the power factor is unitary. In this figure, it is also possible to verify the voltage produced by the converter in a prominent way, highlighting the five levels (two voltage levels for each half cycle and the zero level). As it can be seen, this figure also shows the voltage on the DC-

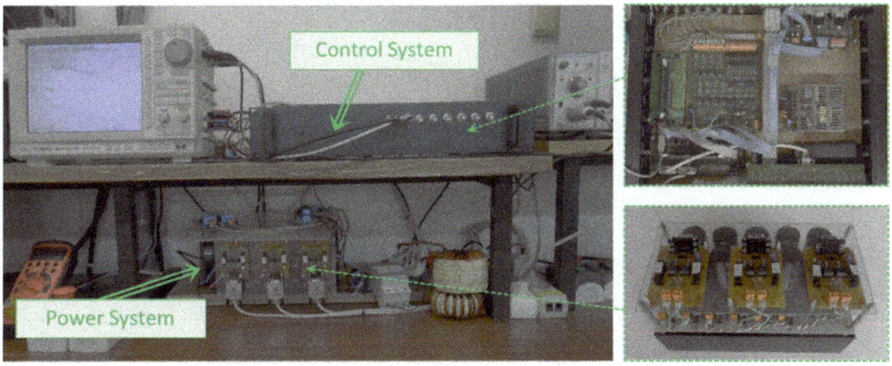

Fig. 4.9 Laboratory prototype of a single-phase front-end AC-DC converter based on a multilevel T-type structure

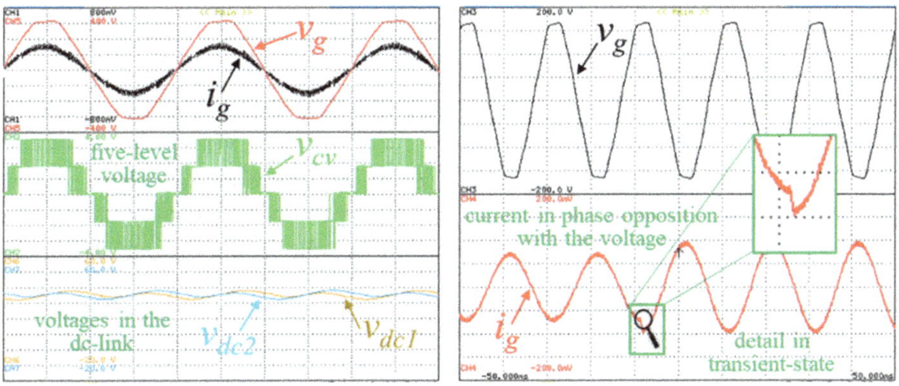

(a) (b)

Fig. 4.10 Experimental results of the single-phase front-end AC-DC converter based on a multilevel T-type structure, showing the AC grid current (i_g), the AC grid voltage (v_g), the voltages on the DC grid side (v_{dc1}, v_{dc2}), and the voltage produced by the converter (v_{cv}): (**a**) With the AC-DC converter providing power to the DC grid; (**b**) With the AC-DC converter receiving power from the DC grid and operating in transient-state.

link, which, as it is a split DC-link, presents two voltages (e.g., indicated for bipolar DC grids). These voltages are controlled according to the half cycle of the voltage on the AC grid.

On the other hand, in Fig. 4.10b the voltage and current are shown once again, but considering an operation in transient state when the DC grid provides power to the AC grid, a situation that can occur, e.g., when the production from renewable sources is higher than the consumption in the DC grid or when required by the AC grid. This result allows to verify that the converter also operates correctly in this operation mode. As aspects to highlight, it is important to verify that, regardless of

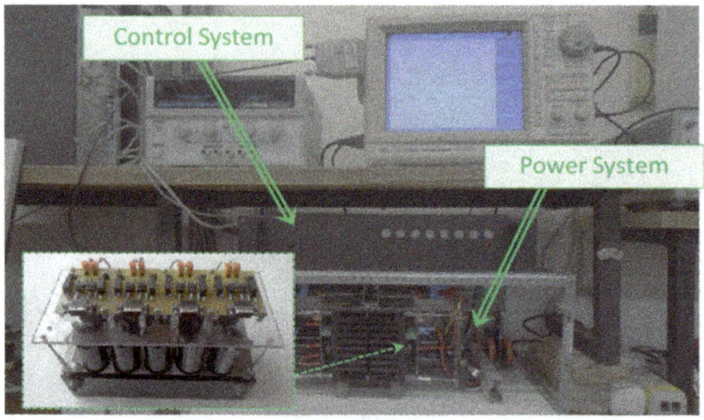

Fig. 4.11 Laboratory prototype of the back-end DC-DC converter based on a multilevel structure

the operation mode and the operation in transient or steady state, the current on the AC grid is always sinusoidal.

4.4.3 Power Electronics Validation: DC-DC Converter

This item presents an example of a multilevel DC-DC converter, which can be used for two purposes: (a) as interface with the front-end AC-DC converter with the objective to create a bipolar DC grid and (b) as interface between the bipolar DC grid and specific technology, such as EV or ESS. Thus, this converter has one of the interfaces with a split DC-link, which is fundamental to create or to interface a bipolar DC grid. With this structure, it is possible to operate in bidirectional mode (e.g., exchanging power with the DC grid), with three voltage levels and also with voltage or current control. As mentioned, this topology is ideal when it is intended to implement the interface with a bipolar DC grid.

Moreover, the voltage levels are dependent on the voltage in each DC interface, i.e., if the voltage in the unipolar DC interface is greater or lower than half of the bipolar DC interface. This feature is also particularly important during the operation in transient state, where the DC-DC converter can assume the three voltage levels in order to guarantee that the controlled variables follow the references in few sampling periods, reducing the error in transient state.

Thus, Fig. 4.11 shows the laboratory workbench and, in detail, the developed prototype, consisting of the power circuit and the control circuit, where the same components as those mentioned for the previous converters were used (e.g., the IGBTs FGA25N120, LEM voltage and current sensors, and the DSP F28335).

In terms of experimental validation, Fig. 4.12 shows the main results that validate this DC-DC converter. Figure 4.12a shows the current in the unipolar DC interface and the voltage produced by the converter (i.e., the measured voltage before the coupling inductive filters). This result was obtained with an EV connected to the unipolar DC interface and considering that it is receiving power from the DC grid, where an algorithm based on current control was imposed. In this specific case,

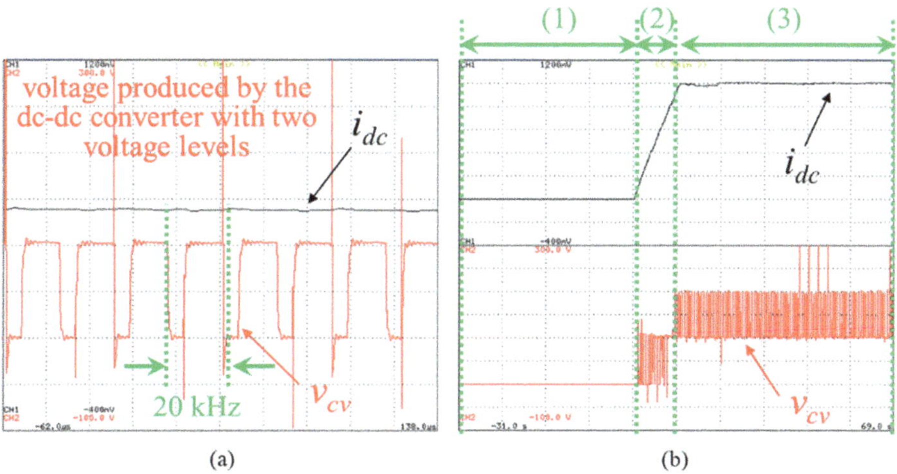

Fig. 4.12 Experimental results of the back-end DC-DC converter based on a multilevel structure, showing the DC grid current (i_{dc}) and the voltage produced by the converter (v_{cv}): (**a**) Detail of the voltage produced by the converter operating with two voltage levels; (**b**) with the DC-DC converter operating with three voltage levels

since the input voltage of the converter (i.e., in the unipolar DC interface) is lower than half of the output voltage (i.e., in the bipolar DC interface), the converter operates with two voltage levels. This figure shows the detail of such voltage, where is also highlighted the fixed switching frequency of 20 kHz. On the other hand, Fig. 4.12b shows a result in the same operating conditions of the previous one, but considering a transient state to verify the relationship between the voltage levels and the operating current of the converter in the unipolar DC interface. As expected, the converter operates with three voltage levels, which are dependent on the voltage levels on each side of the converter. In this case, the three voltage levels were achieved since the voltage on the unipolar DC interface (where is connected the EV) exceeded half of the voltage on the bipolar DC interface.

4.5 Smart Home with Hybrid AC/DC Electrical Power Grid: Innovative Operation Modes Framed with Perspectives for Smart Grids

This section presents a structure of power electronics converters for a hybrid AC/DC electrical grid, incorporated within a smart home. Such a framework allows innovative features in terms of operating modes for the main AC grid, contributing to the dynamic participation and influence of smart homes in smart grids. Figure 4.13 shows the structure considered as an illustrative case example of a hybrid AC/DC electrical grid in a smart home and that was considered for the experimental

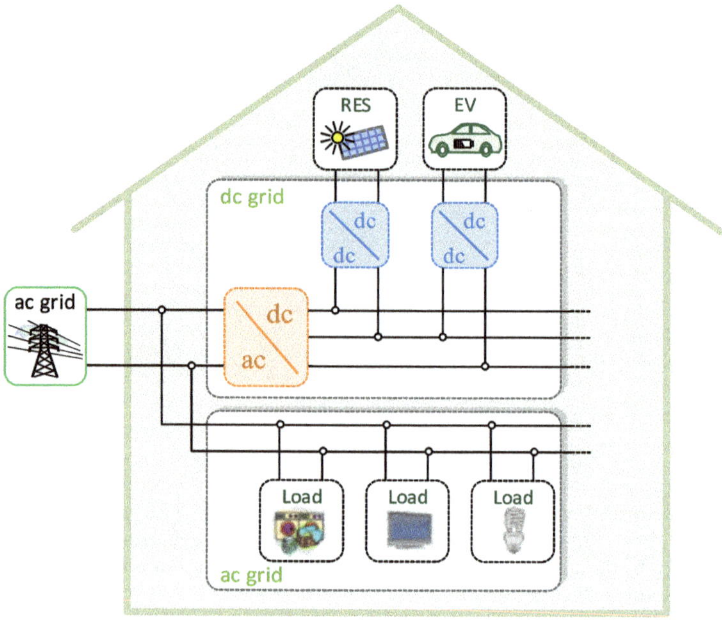

Fig. 4.13 Structure of the hybrid AC/DC electrical grid for a smart home, which was considered for the experimental validation

validation. As expected, there is a front-end AC-DC converter that makes the interface between the main AC grid and the created DC grid where back-end DC-DC converters are coupled.

In this specific demonstration case, it is considered the DC interface of an EV and a conventional set of solar photovoltaic panels as a case of RES in a DC grid. Obviously, the converter that interfaces the RES is unidirectional, and in a realistic future perspective of electric mobility in smart grids, the converter that interfaces the EV operates in bidirectional mode. Thus, the converter that interfaces the EV can operate in grid-to-vehicle mode (G2V) (the EV receives power from the AC grid), in renewable-to-EV mode (R2V) (the EV receives power from the RES), or in vehicle-to-grid mode (V2G) (where the EV supplies power to the AC grid or to the DC grid). As shown, as examples of loads linked to the AC grid, lighting systems and general electrical appliances, as TV or washing machines, were considered. In terms of power quality, these loads are characterized by the nonlinear behavior and by the operation with low power factor, which are undesirable characteristics for the AC grid, but realistic loads within a smart home. The smart home is presented as an example in the context of active buildings. The smart home presented plays an active role in the interaction with the power grid and in its own operation since it can integrate several technologies for its autonomous operation and the interaction between active buildings and active integration with the power grid, enable a flexible operation in terms of demand-side power consumption and specific patterns of

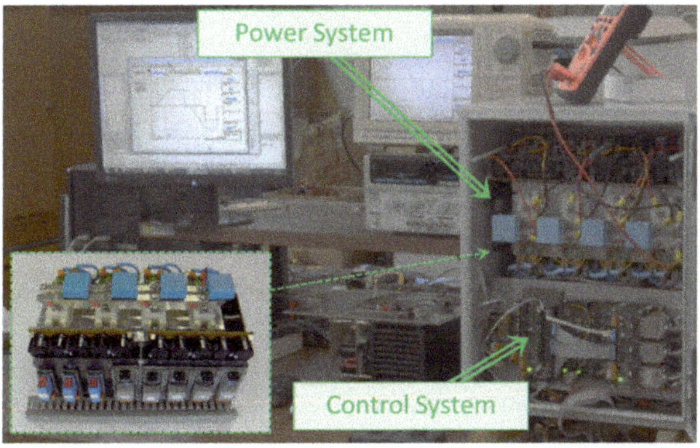

Fig. 4.14 Laboratory prototype for a smart home framed in smart grids, showing a single-phase front-end AC-DC converter based on a full bridge, which was used to form a DC grid where are linked two back-end DC-DC converters based on half-bridge structure

power injection due to the bidirectional operation, and contribute to decentralized power solutions in smart grids and a goal of zero emissions.

Despite the advantages that this structure offers in comparison with the conventional approach (with individual front-end AC-DC converters), where only one converter is used for the interface with the AC grid, the main advantages are related to the operation of the front-end converter for the AC grid regardless the operation of the back-end DC-DC converters. This technology is very important for smart homes, since from the power grid point of view, the smart home is seen as a linear load, operating with sinusoidal current as well as with unitary power factor. Consequently, by replicating this technology for multiple smart homes, the effect on the power grid is much more advantageous, especially with regard to power quality issues, as they practically cease to exist (i.e., in terms of current harmonics and power factor), since all smart homes start to operate as linear loads. It is important to mention that, nowadays, due to the various electrical appliances, the traditional homes present a nonlinear behavior, which results in several power quality problems for the power grid. In addition, the bidirectional power operation is possible, allowing, if necessary, the power exchange between smart homes according to the power management between smart homes and smart grid. Moreover, with a single front-end AC-DC converter in each home, it is possible to improve efficiency, and the requirements in terms of power converters are drastically reduced.

In this sense, Fig. 4.14 shows the laboratory workbench and a detail of the developed prototype, which was also implemented with the components used in the previous cases, both for the power system and for the control system (IGBTs and drivers from Semikron, LEM sensors, and the DSP F28335). The experimental evaluation was carried out for the key operation modes and considering the most important characteristics that the front-end AC-DC converter can offer. Therefore, Fig. 4.15a shows a case where the main AC grid is providing power to the conventional nonlinear loads (cf. Fig. 4.13) of the smart home, linked in the AC grid, as well as for the DC grid through the AC-DC converter. As shown, the current consumption due to the nonlinear loads presents a distorted waveform, while the current consumption from the DC grid (i.e., through the front-end AC-DC converter)

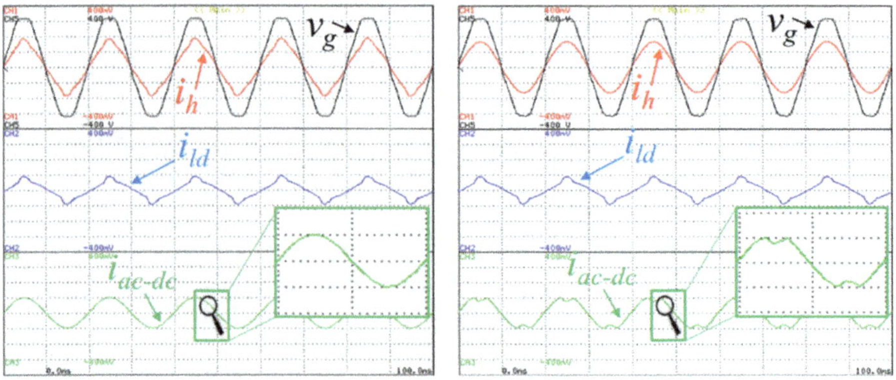

(a) (b)

Fig. 4.15 Experimental results of the single-phase front-end AC-DC converter combined with two back-end DC-DC converters for a smart home framed in smart grids, showing the AC grid voltage (v_g), the total current of the smart home (i_h), the current of the loads (i_{ld}), and the current of the AC-DC converter ($i_{ac\text{-}dc}$): (**a**) with the AC-DC converter providing power to the DC grid; (**b**) with the AC-DC converter providing power to the DC grid and compensating the harmonics and power factor

presents a sinusoidal waveform. Consequently, the total current of the smart home presents a non-sinusoidal waveform, which is characteristic of a hybrid AC/DC electrical grid in smart homes. In this case, it was considered that the EV is operating in the charging mode (G2V) and that the RES is injecting power into the DC grid, but the power provided by the RES for the EV charging is not sufficient, and therefore, it is necessary that the AC grid provides part of the power.

On the other hand, Fig. 4.15b shows the same situation of the previous case, with the EV in charging mode and regarding the operating power of the hybrid AC/DC electrical grid, but considering the operation of the front-end AC-DC for compensating the harmonic distortion and the low power factor caused by the connected nonlinear loads in the AC grid. Consequently, as shown, the total current of the smart home presents a sinusoidal waveform in phase with the voltage, which is an extremely important characteristic of a future hybrid AC/DC electrical grid in smart homes aiming to perverse power quality. As a consequence, the current consumed by the front-end AC-DC converter is no longer sinusoidal, but presents harmonic distortion.

4.6 Conclusions

This paper presents hybrid AC/DC electrical power grids for active buildings from the perspective of power electronics. For this purpose, the main structures of hybrid AC/DC electrical power grids are presented, as well as the main topologies of power electronics converters that can be used for interfacing the AC grid and the DC grid, whether single-phase or three-phase front-end AC-DC converters, as well as back-end DC-DC converters. In the scope of this book chapter, a smart home scenario was considered as an example of active building; however, it can be easily replicated in a residential context with multiple smart homes, in a commercial scenario, in an industrial application scenario, or in another type of scenario. The smart home scenario was selected taking into account the added value that it represents for the continuous development of smart grids, knowing that smart homes are a major driver of smart grids. Besides, smart homes are the ideal scenario to support the concept of hybrid AC/DC electrical grids, since the various technologies that are emerging, as well as those that are already implemented, are of DC nature, such as the energy storage systems, most of the renewables, and the electric vehicles. In the perspective of power electronics, several laboratory prototypes, both single phase and three phase, are presented, targeting to be used in the interface with the AC grid or the DC grid. For each laboratory prototype, key experimental results are presented and described in detail. Additionally, are also presented innovative contributions in terms of operation modes, in particular, when the control is performed in a smart grid context, where the front-end AC-DC converter can have multiple functionalities, such as the combined compensation of current harmonics and reactive power.

References

Abdullah, M. A., Agalgaonkar, A. P., & Muttaqi, K. M. (2014). Assessment of energy supply and continuity of service in distribution network with renewable distributed generation. *Applied Energy, 113*, 1015–1026.

Agrawal, A., Nalamati, C. S., & Gupta, R. (2019). Hybrid DC–AC Zonal microgrid enabled by solid-state transformer and centralized ESD integration. *IEEE Transactions on Industrial Electronics, 66*(11), 9097–9107.

Ai, Q., Fan, S., & Piao, L. (2016). Optimal scheduling strategy for virtual power plants based on credibility theory. *Protection and Control of Modern Power Systems, 1*(1), 1–8.

Aldik, A., Al-Awami, A. T., Sortomme, E., Muqbel, A. M., & Shahidehpour, M. (2018). A planning model for electric vehicle aggregators providing ancillary services. *IEEE Access, 6*, 70685–70697.

Bhattacharjee, A. K., Kutkut, N., & Batarseh, I. (2019). Review of multiport converters for solar and energy storage integration. *IEEE Transactions on Power Electronics, 34*(2), 1431–1445.

Bignucolo, F., Bertoluzzo, M., & Fontana, C. (2015). Applications of the solid state transformer concept in the electrical power system. In *International annual conference*.

Blaabjerg, F., Dragicevic, T., Doolla, S., & Kang, N. (2017). Recent advances in control, analysis and design of DC distribution systems and microgrids. *Electric Power Components & Systems, 45*(10), 1031.

Bloemink, J. M., & Green, T. C. (2013). Benefits of distribution-level power electronics for supporting distributed generation growth. *IEEE Transactions on Power Delivery, 28*(2), 911–919.

Bollen, M. H., & Hassan, F. (2011). *Integration of distributed generation in the power system*. Wiley.

Boroyevich, D., Cvetkovic, I., Burgos, R., & Dong, D. (2013). Intergrid: A future electronic energy network? *IEEE Journal of Emerging and Selected Topics in Power Electronics, 1*(3), 127–138.

Bracale, A., Caramiaa, P., Carpinelli, G., Mottola, F., & Proto, D. (2012). A hybrid AC/DC smart grid to improve power quality and reliability. In *Proceedings of the IEEE International Energy Conference Exhibition (ENERGYCON), Florence, Italy*, pp. 507–514.

Bulut, M. B., & Wallin, F. (2014). Buildings as components of smart grids – Perspectives of different stakeholders. *ELSEVIER Energy Procedia, 61*, 1630–1633.

Bulut, M. B., Odlare, M., Stigson, P., Wallin, F., & Vassileva, I. (2016). Active buildings in smart grids – Exploring the views of the Swedish energy and buildings sectors. *ELSEVIER Energy and Buildings, 117*, 185–198.

Calbureanu, M., Malciu, R., & Calbureanu, D. (2018). Active house – A building that offers more than consumes. *Advanced Engineering Forum, 27*, 193–201.

Chauhan, R. K., Phurailatpam, C., Rajpurohit, B. S., Gonzalez-Longatt, F. M., & Singh, S. N. (2017). Demand-side management system for autonomous DC microgrid for building. *Technology and Economics of Smart Grids and Sustainable Energy, 2*(1).

Chuangpishit, S., Tabesh, A., & Moradi-Shahrbabak, Z. (2014). Topology design for collector systems of offshore wind farms with pure DC power systems. *IEEE Transactions on Industrial Electronics, 61*(1), 320–328.

Diaz, E. R., Su, X., Savaghebi, M., Vasquez, J. C., Han, M., & Guerrero, J. M. (2015). Intelligent DC microgrid living laboratories – A Chinese-Danish Cooperation Project. In *IEEE 1st international conference on DC microgrids*, pp. 365–370.

Dragicevic, T., Vasquez, J. C., Guerrero, J. M., & Skrlec, D. (2014). Advanced LVDC electrical power architectures and microgrids: A step toward a new generation of power distribution networks. *IEEE Electrification Magazine, 2*(1), 54–65.

Dragicevic, T., Lu, X., Vasquez, J. C., & Guerrero, J. M. (2016a). DC microgrids – Part I: A review of control strategies and stabilization techniques. *IEEE Transactions on Power Electronics, 31*(7), 4876–4891.

Dragicevic, T., Lu, X., Vasquez, J. C., & Guerrero, J. M. (2016b). DC microgrids – Part II: A review of power architectures, applications, and standardization issues. *IEEE Power Electronics, 31*(5), 3528–3549.

Dusonchet, L., Ippolito, M. G., Telaretti, E., Zizzo, G., & Graditi, G. (2013). An optimal operating strategy for combined RES-based generators and electric storage systems for load shifting applications. *International Conference on Power Engineering, Energy and Electrical Drives, 5*, 552–557.

European Parliament. (2014). Directive 2014/35/EU of the European Parliament and of the Council. *Official Journal of the European Union, L96*, 357–374.

Fairley, P. (2012). DC versus AC: The second war of currents has already begun [In my view]. *IEEE Power and Energy Magazine, 10*(6), 104–103.

Fan, B., Li, Y., Wang, K., et al. (2017). Hierarchical system design and control of an MMC-based power-electronic transformer. *IEEE Transactions on Industrial Information, 13*(1), 238–247.

Golovanov, N., Lazaroiu, G. C., Roscia, M., & Zaninelli, D. (2013). Power quality assessment in small scale renewable energy sources supplying distribution systems. *Energies, 6*(2), 634–645.

Gu, C., Zheng, Z., Xu, L., et al. (2016a). Modeling and control of a multiport power electronic transformer (PET) for electric traction applications. *IEEE Transactions on Power Electronics, 31*(2), 915–927.

Gu, Y., Li, W., & He, X. (2016b). Analysis and control of bipolar LVDC grid with DC symmetrical component method. *IEEE Transactions on Power Systems, 31*(1), 685–694.

Gungor, V. C., Sahin, D., Kocak, T., Ergut, S., Buccella, C., Cecati, C., & Hancke, G. P. (2012). Smart grid and smart homes – Key players and pilot projects. *IEEE Industrial Electronics Magazine, 6*, 18–34.

Hannan, M. A., et al. (2020). State of the art of solid-state transformers: Advanced topologies, implementation issues, recent progress and improvements. *IEEE Access, 8*, 19113–19132.

Hirose, K., Reilly, J., & Irie, H. (2013). The Sendai microgrid operational experience in the aftermath of the Tohoku earthquake: A case study. *New Energy Industrial Technology Development Organization, 308*, 1–6.

Huang, R., Pu, T., Liu, K., Yang, Z., & Chen, N. (2015). Design of hierarchy and functions of regional energy internet and its demonstration applications. *Automation of Electric Power Systems, 39*(9), 26–33.

Huber, J., & Kolar, J. W. (2019). Applicability of solid-state transformers in today's and future distribution grids. *IEEE Transactions on Smart Grid, 10*, 317–326.

Isaksson, C. (2011, May). From a passive to an active house. In *World Renewable Energy Congress, Linköping, Sweden*, pp. 1789–1796.

Jahromi, R. G., & Seifi, A. R. (2016). The demand side management program considering AC/DC hybrid distribution system concept based on the energy hub. In *Iranian conference on electrical engineering*, pp. 384–388.

Kakigano, H., Miura, Y., & Ise, T. (2010). Low-voltage bipolar-type DC microgrid for super high quality distribution. *IEEE Transactions, 25*(12), 3066–3075.

Khan, Z. W., Minxiao, H., Kai, C., Yang, L., & Rehman, A. (2020, January). State of the art DC-DC converter topologies for the multi-terminal DC grid applications: A review. In *IEEE PESGRE international conference on power electronics, smart grid and renewable energy*, pp. 1–7.

Kinhekar, N., Padhy, N. P., Li, F., & Gupta, H. O. (2016). Utility oriented demand side management using smart AC and micro DC grid cooperative. *IEEE Transactions on Power Systems, 31*(2), 1151–1160.

Li, X., et al. (2018a). Flexible interlinking and coordinated power control of multiple DC microgrids clusters. *IEEE Transactions on Sustainable Energy, 9*(2), 904–915.

Li, X., et al. (2018b). A unified control for the DC-AC interlinking converters in hybrid AC/DC microgrids. *IEEE Transactions on Smart Grid, 9*(6), 6540–6553.

Li, X., Guo, L., Hong, C., Zhang, Y., Li, Y. W., & Wang, C. (2018c). Hierarchical control of multiterminal DC grids for large-scale renewable energy integration. *IEEE Transactions on Sustainable Energy, 9*(3), 1448–1457.

López, M., Briz, F., Saeed, M., et al. (2006, September). Comparative analysis of modular multiport power electronic transformer topologies. In *International conference on energy conversion congress and exposition, Milwaukee, USA*, pp. 1–8.

Lotfi, H., & Khodaei, A. (2017). AC versus DC microgrid planning. *IEEE Transactions on Smart Grid, 8*(1), 296–304.

Ma, J., Zhu, M., Xu, C., & Li, Y. W. (2019a). DC substation for DC grid – Part I: Comparative evaluation of DC substation configuration. *IEEE Transactions on Power Electronics, 34*(10), 9719–9731.

Ma, J., Zhu, M., Xu, C., & Li, Y. W. (2019b). DC substation for DC grid – Part II: Hierarchical control strategy and verification. *IEEE Transactions on Power Electronics, 34*(9), 8682–8696.

Miao, J., Ning, Z., & Chongqing, K. (2016, July). Generalized steady-state model for energy router with applications in power flow calculation. In *International conference on power and energy society general meeting, Boston, USA*, pp. 1–5.

Mohammadi, J., Ajaei, F. B., & Stevens, G. (2019). Grounding the DC microgrid. *IEEE Transactions on Industrial Applications, 55*(5), 4490–4499.

Monteiro, V., Ferreira, J., et al. (2018a). Experimental validation of a novel architecture based on a dual-stage converter for off-board fast battery chargers of electric vehicles. *IEEE Transactions on Vehicular Technology, 67*, 1000–1011.

Monteiro, V., Pinto, J. G., & Afonso, J. L. (2018b). Experimental validation of a three-port integrated topology to interface electric vehicles and renewables with the electrical grid. *IEEE Transactions on Industrial Informatics, 14*, 2364–2374.

Monteiro, V., Sousa, T. J. C, Couto, C., Martins, J. S., Melendez, A. A. N, & Afonso, J. L. (2018c, October). A novel multi-objective off-board EV charging station for smart homes. In *IEEE IECON industrial electronics conference*, pp. 1893–1988.

Monteiro, V., Afonso, J., Sousa, T., & Afonso, J. L. (2019). The role of off-board EV battery chargers in smart homes and smart grids: operation with renewables and energy storage systems. In A. Ahmadian, B. Mohammadi-ivatloo, & A. Elkamel (Eds.), *Electric vehicles in energy systems* (1st ed., pp. 47–72). Springer.

Monteiro, V., Lima, P., Sousa, T. J. C., Martins, J. S., & Afonso, J. L. (2020). An off-board multi-functional electric vehicle charging station for smart homes: Analysis and experimental validation. *MDPI Energies, 13*(8), 1–23.

Muttaqi, K. M., Islam, M. R., & Sutanto, D. (2019). Future power distribution grids: Integration of renewable energy, energy storage, electric vehicles, superconductor, and magnetic bus. *IEEE Transactions on Applied Superconductivity, 29*(2), 1–5.

Nasir, M., Khan, H. A., Zaffar, N. A., Vasquez, J. C., & Guerrero, J. M. (2018). Scalable solar dc microgrids: On the path to revolutionizing the electrification architecture of developing communities. *IEEE Electrification Magazine, 6*(4), 63–72.

Paez, J. D., Frey, D., Maneiro, J., Bacha, S., & Dworakowski, P. (2019). Over-view of DC–DC converters dedicated to HVdc grids. *IEEE Transactions on Power Delivery, 34*(1), 119–128.

Patterson, B. T. (2012). DC, come home: DC microgrids and the birth of the "Enernet". *IEEE Power and Energy Magazine, 10*(6), 60–69.

Pei, W., Deng, W., Zhang, X., et al. (2015). Potential of using multiterminal LVDC to Improve plug-in electric vehicle integration in an existing distribution network. *IEEE Transactions on Industrial Electronics, 62*(5), 3101–3111.

Penya, Y. K., Borges, C. E., Haase, J., & Bruckner, D. (2013). Smart buildings and the smart grid. In *IEEE IECON industrial electronics conference*, pp. 113–117.

Renedo, J., García-Cerrada, A., Rouco, L., et al. (2017, February). Development of a PSS/E tool for power-flow calculation and dynamic simulation of VSC-HVDC multi-terminal systems. In *International conference on AC and DC power transmission, Manchester, UK*, pp. 1–6.

Rivera, S., Lizana, R., Kouro, S., Dragicevic, T., & Wu, B. (2020). Bipolar DC power conversion: State-of-the-art and emerging technologies. *IEEE Journal of Emerging and Selected Topics in Power Electronics*.

Rodriguez-Diaz, E., Vasquez, J. C., & Guerrero, J. M. (2016a). Intelligent DC homes in future sustainable energy systems: When efficiency and intelligence work together. *IEEE Consumer Electronics Magazine, 5*(1), 74–80.

Rodriguez-Diaz, E., Chen, F., Vasquez, J. C., Guerrero, J. M., Burgos, R., & Boroyevich, D. (2016b). Voltage-level selection of future two-level LVdc distribution grids: A compromise between grid compatibiliy, safety, and efficiency. *IEEE Electrification Magazine, 4*(2), 20–28.

Rouzbehi, K., Candela, J. I., Luna, A., Gharehpetian, G. B., & Rodriguez, P. (2016). Flexible control of power flow in multiterminal DC grids using DC–DC converter. *IEEE Journal of Emerging and Selected Topics in Power Electronics, 4*(3), 1135–1144.

Sahoo, A. K., & Ned, M. (2014, May). A power electronic transformer with sinusoidal voltages and currents using modular multilevel converter. In *International conference on power electronics conference. ECCE-ASIA, Hiroshima, Japan*, pp. 3750–3757.

Sakis Meliopoulos, A. P., Cokkinides, G., Huang, R., Farantatos, E., Choi, S., Lee, Y., & Yu, X. (2011). Smart grid technologies for autonomous operation and control. *IEEE Transactions on Smart Grid, 2*(1), 1–10.

Saleh, S. M., Richard, C., et al. (2019a). Solid-State transformers for distribution systems–Part I: Technology and construction. *IEEE Transactions on Industry Applications, 55*, 4524–4535.

Saleh, S. M., Alsayid, B., et al. (2019b). Solid-state transformers for distribution systems–Part II: Deployment challenges. *IEEE Transactions on Industry Applications, 55*, 5708–5716.

Salomonsson, D., & Sannino, A. (2007). Low-voltage DC distribution system for commercial power systems with sensitive electronic loads. *IEEE Transactions on Power Delivery, 22*(3), 1620–1627.

Shen, X., Tan, D., Shuai, Z., & Luo, A. (2019). Control techniques for bidirectional interlinking converters in hybrid microgrids: Leveraging the advantages of both ac and dc. *IEEE Power Electronics Magazine, 6*(3), 39–47.

Sousa, T. J. C., Monteiro, V., Pinto, J. G., & Afonso, J. L. (2018a, November). Performance comparison of a typical nonlinear load connected to ac and dc power grids. In *EAI GreeNets – International conference on green energy and networking*, pp. 1–10, Guimarães, Portugal.

Sousa, T. J. C., Monteiro, V., Afonso, J. A., & Afonso, J. L. (2018b, July). Selective harmonic measurement and compensation in a distributed microgrid. In IEEE INDIN international conference on industrial informatics, pp. 439–444, Porto, Portugal.
Sousa, T. J. C., Monteiro, V., & Afonso, J. L. (2019a, June). On the viability of DC homes: An economic perspective from domestic electrical appliances. In *ICEE international conference on energy and environment: Bringing together engineering and economics*, Guimarães Portugal.
Sousa, T. J. C., Monteiro, V., Martins, J. S., Sepúlveda, M. J., Lima, A., & Afonso, J. L. (2019b, September). Comparative analysis of power electronics topologies to interface dc homes with the electrical ac power grid. In *IEEE SEST international conference on smart energy systems and technologies*, Porto, Portugal.
Sousa, T. J. C., Monteiro, V., Pinto, J. G., & Afonso, J. L. (2019c, November). Performance comparison of a typical nonlinear load connected to ac and dc power grids. In *EAI endorsed transactions on energy web*, pp. 1–8.
Sun, K., Li, K.-J., Wang, Z.-d., Sun, H., Wang, M., Liu, Z., & Wang, M. (2018). Operation modes and combination control for urban multivoltage-level DC grid. *IEEE Transactions on Power Delivery, 33*(1), 360–370.
Suryanarayana, H., & Sudhoff, S. D. (2017). Design paradigm for power electronics-based DC distribution systems. *IEEE Journal of Emerging and Selected Topics in Power Electronics, 5*(1), 51–63.
Unamuno, E., & Barrena, J. A. (2015a). Hybrid ac/dc microgrids – Part I: Review and classification of topologies. *Elsevier Renewable and Sustainable Energy Reviews, 52*, 1251–1259.
Unamuno, E., & Barrena, J. A. (2015b). Hybrid ac/dc microgrids – Part II: Review and classification of control strategies. *Elsevier Renewable and Sustainable Energy Reviews, 52*, 1123–1134.
Vossos, V., Garbesi, K., & Shen, H. (2014). Energy savings from direct-DC in U.S. residential buildings. *Energy and Buildings, 68*, 223–231.
Vuyyuru, U., Maiti, S., & Chakraborty, C. (2019). Active power flow control between DC microgrids. *IEEE Transactions on Smart Grid, 10*(5), 5712–5723.
Wang, P., Goel, L., Liu, X., & Choo, F. H. (2013). Harmonizing AC and DC: A hybrid AC/DC future grid solution. *IEEE Power and Energy Magazine, 11*(3), 76–83.
Wang, X., Liu, J., Ouyang, S., et al. (2016). Control and experiment of an H-bridge based three-phase three-stage modular power electronic transformer. *IEEE Transactions on Power Electronics, 31*(3), 2002–2011.
Yan, J., Zhu, X., & Lu, N. (2015). Smart hybrid house test systems in a solid-state transformer supplied microgrid. *IEEE Power and Energy Society General Meeting*, 1–5.
Yu, X., Cecati, C., Dillon, T., & Simoes, M. G. (2011a). The new frontier of smart grids: An industrial electronics perspective. *IEEE Industrial Electronics Magazine, 5*(3), 49–63.
Yu, X., Cecati, C., Dillon, T., & Simões, M. G. (2011b). The new frontier of smart grids. *IEEE Industrial Electronics Magazine, 5*(3), 49–63.
Yu, X., Cecati, C., Dillon, T., & Simes, M. G. (2011c). The new frontier of smart grids. *IEEE Industrial Electronics Magazine, 5*(3), 49–63.
Zhang, Y., Jayashri, R., & John, F. (2016). Power flow and transmission loss analysis of modular multi-level converter based multi-terminal high-voltage DC systems. *IET Renewable Power Generation, 10*(6), 767–775.
Zhao, C., Dong, S., Gu, C., Li, F., Song, Y., & Padhy, N. P. (2018). New problem formulation for optimal demand side response in hybrid AC/DC systems. *IEEE Transactions on Smart Grid, 9*(4), 3154–3165.
Zhong, Q., Wang, Y., & Ren, B. (2019). Connecting the home grid to the public grid: field demonstration of virtual synchronous machines. *IEEE Power Electronics Magazine, 6*(4), 41–49.

Chapter 5
A Modelling Workflow for Predictive Control in Residential Buildings

E. O'Dwyer, E. Atam, P. Falugi, E. C. Kerrigan, M. A. Zagorowska, and N. Shah

5.1 Introduction

With electrification of heating and transport, as well as an increased reliance on varying renewable generation sources, buildings will be required to act in a flexible manner, allowing for the shifting of demands and coordination of local low-carbon generation (Kathirgamanathan et al., 2021). Rather than simply acting as a rigid demand that must be satisfied by the grid, a building can act as a useful system of components by offering the opportunity for flexibility provision (Greater London Authority, 2018). By exploiting the energy storage capacity of a building's fabric and using information about occupancy and comfort requirements, the heat/cold supplied to a building can be shifted away from times of high demand (Gonzato et al., 2019). Furthermore, storage devices including batteries (Al Essa, 2019), thermal stores (O'Dwyer et al., 2020) and electric vehicles (Barone et al., 2019), along with renewable generation sources (particularly solar generation), can be managed appropriately to allow power to be absorbed from the grid and supplied back to the grid as needed. Aside from the physical storage and generation assets, communication technology and computational techniques are needed to determine optimal control decisions for the various system components in a timely fashion.

E. O'Dwyer (✉) · N. Shah
Department of Chemical Engineering, Imperial College London, London, UK
e-mail: e.odwyer@imperial.ac.uk; n.shah@imperial.ac.uk

E. Atam · P. Falugi · M. A. Zagorowska
Department of Electrical and Electronic Engineering, Imperial College London, London, UK
e-mail: p.falugi@imperial.ac.uk; m.zagorowska@imperial.ac.uk

E. C. Kerrigan
Department of Electrical and Electronic Engineering, Imperial College London, London, UK

Department of Aeronautics, Imperial College London, London, UK
e-mail: e.kerrigan@imperial.ac.uk

V. Vahidinasab, B. Mohammadi-Ivatloo (eds.), *Active Building Energy Systems*, Green Energy and Technology, https://doi.org/10.1007/978-3-030-79742-3_5

A technique commonly proposed for the task of optimal real-time decision-making in the building sector is Model Predictive Control (MPC) (Shaikh et al., 2014). An MPC-based strategy uses a predictive model of a system's dynamics and constraints to form an optimisation problem, which is solved to determine input trajectories over a receding horizon. The ability of MPC to incorporate future behaviour in the decision-making process makes it particularly suited to the problem of optimally time-shifting energy demands. Furthermore, the manner in which system constraints can be explicitly applied makes MPC a natural fit for the role of building energy management (Drgoňa et al., 2020). Despite the promising potential of MPC to enable demand-side flexibility, the uptake of the technology in the sector remains limited. This is particularly challenging for the domestic sector, as a high-level of adoption would be required to ensure sufficient *shiftable* demand is available to encourage worthwhile interaction with the grid.

A commonly cited obstacle to its wider roll-out is the modelling challenge (Killian & Kozek, 2016; Atam & Helsen, 2016). Buildings can vary greatly in their design and are inherently formed of multiple inputs and outputs. Furthermore, how a building is used (particularly in a domestic setting) is highly dependent on the individual occupants and one-size-fits-all solutions are unlikely to provide adequate performance. Modelling techniques that are not contingent on large time investments from building modelling experts would greatly facilitate the deployment of MPC in the sector, thus expediting the shift towards more flexible operation and a lower-carbon energy sector (Fig. 5.1).

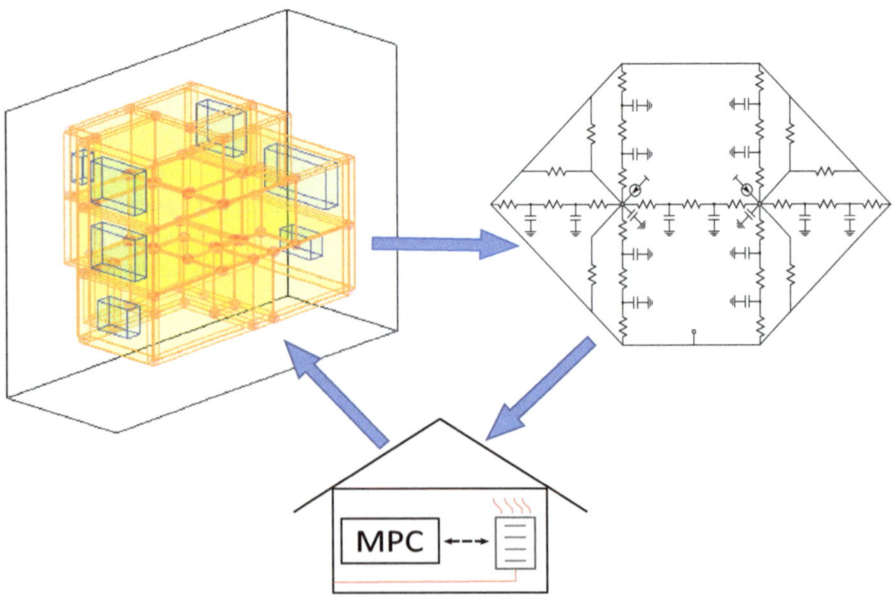

Fig. 5.1 Modelling workflow for MPC design in buildings

This chapter establishes a systematic workflow, from detailed building modelling based on standard tools, to MPC design. This workflow begins with detailed physics-based modelling followed by low-order control-oriented modelling and heating system component modelling. The workflow can be summarised as follows:

1. The building shape is drawn and building materials are defined.
2. A detailed building simulation model is created.
3. The underlying mathematical representation of the model is extracted.
4. Option A: The model complexity is reduced using truncation techniques for control implementation
5. Option B: Low-complexity models are derived from data.
6. Design and analysis of control strategy is carried out using detailed simulation modelling platform.

A case study is used to illustrate the application of the proposed methods, whereby the performance of MPC strategy for managing a heating system is analysed and the ability to interact with an external price signal is demonstrated. The contribution of the work is primarily as a guide for practitioners applying MPC to residential buildings, by describing and illustrating, in detail, modelling principles that can underpin a successful strategy implementation. Furthermore, some of the remaining open challenges and key design questions are emphasised, encouraging future research in the area.

In the second section, the development of detailed physics-based modelling is considered both from a theoretical and software perspective. By commencing with such an approach, comprehensive analyses and design evaluations can be carried out, while the use of industry-standard software reduces the need for specialist expertise. The third section examines the generation of low-order, low-complexity representations of building models to better fit the optimisation-based focus of an MPC strategy. Low-order modelling techniques are outlined from two perspectives. Firstly, a model-based approach is summarised whereby a detailed physics-based model is truncated to emphasise the dynamics most relevant to the timescales of an MPC controller. Following this, data-centric approaches based on system identification are considered. Finally, the outlined techniques are combined in the design of both a centralised and a decentralised MPC strategy, which minimises energy cost while maintaining user comfort. By outlining suitable tools, models and design decisions, the chapter is intended to provide a systematic methodology for application of MPC in the sector.

The chapter is structured as follows: Sect. 5.2 briefly describes the current state and future outlook of the field, Sect. 5.3 outlines the production of a detailed simulation environment, Sect. 5.4 describes methods for deriving low-complexity control-oriented models and Sect. 5.5 introduces suitable MPC formulations for a residential case study. The methods are illustrated through the use of a case study in which an MPC strategy for a heating system in a three-storey dwelling is developed and evaluated.

5.2 Status Quo, Challenges and Outlook

The need for new tools and methods to tackle the modelling challenge faced in the building energy domain has been recognised, leading to the development of interoperable tools that can bridge the gap between building simulation software and control design. The Building Controls Virtual Test Bed (BCVTB) (Simulationresearch, 2016), for example, is an environment that enables co-simulation of different building modelling tools and programming languages (e.g. EnergyPlus, Modelica, Matlab) for the purpose of control design, within the open-source Ptolemy II framework. The Building Resistance-Capacitance Modelling (BRCM) toolbox (Sturzenegger et al., 2014) is focussed on MPC for buildings and can interface with EnergyPlus (as done in this chapter). Aside from software tools, modelling techniques for building simulation come in different forms, from the physics-based focus of white-box techniques (Zwickel et al., 2019), to black-box or machine learning techniques that seek to model building behaviour without in-depth knowledge of the building structure (Rätz et al., 2019). Furthermore, grey-box techniques that use data to parameterise models are often proposed with the advantage of maintaining a desirable model structure without needing detailed knowledge of the building fabric, with a review of methods for matching simulation models to data presented in (Coakley et al., 2014). An overall review of building modelling methods and tools can be found in Harish and Kumar (2016).

Despite progress in these areas, challenges remain. Data-driven techniques can be used to improve model accuracy and reduce modelling effort, but care must be taken with regard to system excitation. As noted in (Lin et al., 2012), closed loop data may not provide sufficient richness for successful parameter identification. Carrying out forced-response tests on a building can be a challenge if the building is occupied and would add to the implementation cost, which is already a hindrance to wider scale-up (Sturzenegger et al., 2016). The challenge for researchers at present is not to demonstrate that MPC can lead to improved performance in a building, but to demonstrate that it can be implemented at scale without excessive cost. Nonetheless, recent development in data-driven predictive control approaches (e.g. Bünning et al., 2020; Coulson et al., 2020) may be promising as they can provide low implementation effort while maintaining certain robustness guarantees. A review on data-driven building control methods can be found in Maddalena et al. (2020). Furthermore, as simulation tools become more common for building design, it has been recognised that such tools can also be useful for operational analysis. In this manner, the digital twin concept (see for example the Gemini principles CDBB, 2018) fits with these goals. In O'Dwyer et al. (2019), the BCVTB environment is deployed as a digital twin for a distributed set of building energy and smart city assets. Using a digital representation of a system enables control strategies to be analysed in-silico prior to deployment, while running a digital twin in parallel allows for operational insights to be attained that may otherwise have been missed.

5.3 Physics-Based Modelling of the Building

There are different reasons for developing a detailed physics-based model of a building and its components. Many methods and international standards have been developed to predict the expected energy performance of a building, thus allowing the evaluation of design decisions and retrofit options. Such methods tend to be focussed on long-term steady-state performance, mostly independent of the control approach taken (the Tabula webtool (EASME, 2017), is an excellent example of a database of such models based on thermal properties and heat balances). The approaches discussed in this section are distinct from these insofar as they are specifically focussed on the impact of control techniques and, as such, require the transient behaviour over a wider range of timescales to be captured. The thermal response of the air in the building (which has a relatively low heat capacity) must be captured as well as the thermal response of larger concrete slabs (which have relatively high heat capacities). With such models, it should be possible to carry out detailed analysis and evaluation in-silico prior to implementation of a strategy. Simulation allows for control techniques to be compared without being impacted by changeable external influences that cannot otherwise be controlled (most notably the weather). The commonly used resistance-capacitance (RC) methodology is detailed here along with software packages that can be used for implementation.

5.3.1 Modelling Background

The equations and concepts underpinning the development of an appropriate simulation model are first described. Suitable methods are outlined for modelling the thermal fabric, the heating system and the occupants in the building. These components can be then brought together in a single simulation model.

5.3.1.1 Building Fabric: The Resistance-Capacitance Model

When some knowledge of the building fabric dimensions and materials are known, a thermal model of the building can be formed as a structure analogous to an electrical circuit composed of resistors and capacitors (Ma, 2012; Oliveira Panão et al., 2019). In such an RC-network, each component of the physical structure can be represented as a system of resistors and capacitors with parameters corresponding to the thermal properties of the structure, while each room or airspace can be represented by a single capacitor. Heat flows are represented in such a circuit as current, while temperature differences are represented as voltages. To illustrate the method, a model of a single wall connecting a zone of temperature T_z to the external air at temperature T_e is shown in Fig. 5.2. In this diagram, the zone capacitance is denoted C_1, while the three wall resistances are denoted R_1, R_2 and R_3. Two intermediate

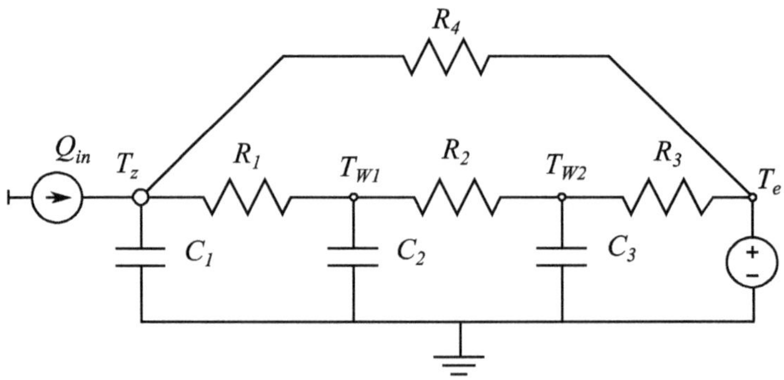

Fig. 5.2 RC-network representation of single external wall with window

wall temperatures are defined (T_{W1} and T_{W2}), with the wall capacitances given as C_2 and C_3. The window is represented by a single resistance, given as R_4. The heat supply to the zone given by Q_{in}.

A key feature of this form is that the full building can be represented as a linear dissipative state-space system. In such a system, the internal states correspond to wall and air temperatures, heat supplies to the zones can be regarded as controlled inputs while external influences, such as the ambient temperature or the solar radiation incident on the walls, can be categorised as disturbances to the model. With this approach, the wall of Fig. 5.2 can be represented in a standard form by the following set of differential equations:

$$\begin{bmatrix} \dot{T}_z(t) \\ \dot{T}_{W1}(t) \\ \dot{T}_{W2}(t) \end{bmatrix} = \begin{bmatrix} \frac{-1}{C_1}\left(\frac{1}{R_1}+\frac{1}{R_4}\right) & \frac{1}{C_1 R_1} & 0 \\ \frac{1}{C_2 R_1} & \frac{-1}{C_2}\left(\frac{1}{R_1}+\frac{1}{R_2}\right) & \frac{1}{C_2 R_2} \\ 0 & \frac{1}{C_3 R_2} & \frac{-1}{C_3}\left(\frac{1}{R_2}+\frac{1}{R_3}\right) \end{bmatrix} \begin{bmatrix} T_z(t) \\ T_{W1}(t) \\ T_{W2}(t) \end{bmatrix}$$
$$+ \begin{bmatrix} \frac{1}{C_1} \\ 0 \\ 0 \end{bmatrix} Q_{in}(t) + \begin{bmatrix} \frac{1}{C_1 R_4} \\ 0 \\ \frac{1}{C_3 R_3} \end{bmatrix} T_e(t) \tag{5.1}$$

$$T_z(t) = \begin{bmatrix} 1 & 0 & 0 \end{bmatrix} \begin{bmatrix} T_z(t) \\ T_{W1}(t) \\ T_{W2}(t) \end{bmatrix}. \tag{5.2}$$

By representing all walls, floors and ceilings in a similar manner, the full building can be given a general form, where the wall and zone air temperature states are given by the vector $\mathbf{x}(t)$ and the measured zone air temperatures are given by the vector $\mathbf{y}(t)$. Controlled inputs are given by $\mathbf{u}(t)$, disturbances (e.g. the external temperature, ground temperature, solar incident radiation and internal gains) given here as $\mathbf{d}(t)$, C is a matrix mapping states to measurements and A, B and E are parameter matrices:

$$\dot{\mathbf{x}}(t) = A\mathbf{x}(t) + B\mathbf{u}(t) + E\mathbf{d}(t) \tag{5.3}$$

$$\mathbf{y}(t) = C\mathbf{x}(t). \tag{5.4}$$

This is a standard linear state-space representation. Generally (5.4) can also contain an additional term denoted $D\mathbf{u}(t)$ representing the direct feedthrough from input to output. This is omitted as no such feedthrough is present in the model.

While (5.3) describes the flow of heat through the building fabric, the heating or cooling systems associated with the inputs can vary in form and appropriate modelling methods will depend on the specific systems used. For example, the control variables themselves may be system flow rates (in a radiator-based system) or air-flows (in an air-based system) and the heating or cooling supply in a particular zone can become a function of the product of the controlled flow rate and the air temperature of the zone, which is a model state. Thus, the model of (5.3) would become bilinear. Furthermore, the system may include boilers, chillers or heat pumps, which may introduce nonlinearities in the overall system behaviour as discussed in the next section.

5.3.1.2 Heating System: Heat Pump Modelling

As the set of potential heating or cooling system components is open-ended, an exhaustive list of different modelling techniques for different technologies is not presented here. Nonetheless, to demonstrate the use of the techniques and the modelling toolchain established here via case studies, it is assumed that a heating system is present in the building, with heat supplied by an electric heat pump. The use of predictive control for energy management of a heat pump-based system is sensible, since the lower-system temperatures preferable for heat pump utilisation can lead to longer heat-up times, thus making predictive strategies favourable. Furthermore, electrification of heating is likely to become ever more prevalent, leading to an increased use of heat pumps, and by choosing a heat pump as the heat source, the interaction between the electrical grid and the building can be better illustrated. Advanced control strategies such as MPC are needed to manage this interaction, once again making a heat pump the most suitable choice for illustrating an MPC-based modelling workflow. A suitable heat pump model is shown in this section.

Whether using an air or ground source heat pump, the relationship between electrical power consumed (given here as P_e) and the heat supplied to the heating system (given here as Q_h) is defined by its coefficient of performance (COP) as:

$$COP = \frac{Q_h}{P_e}. \tag{5.5}$$

The COP is often assumed to be a constant for modelling simplicity, however, more detailed representations consider COP to be a function of a number of operational variables, of which perhaps the most important is the *lift*. In the context of a heat pump, lift is the temperature difference between the source temperature (ground or ambient air) and the sink temperature (the heating system flow temperature). Lower lift values tend to correspond to higher COP values. This makes lower temperature heating systems favourable, while also implying that an efficiency reduction can be expected during cold weather. While manufacturers may supply spec sheets with COPs included, field tests of heat pumps tend to show a wide variation in performance depending on context. To account for this, an extensive review of domestic heat pumps was carried out by Staffell et al. (2012), whereby COP values taken from manufacturers data, field trials and experimental studies were plotted as a function of lift. The resulting empirical relationships derived for air-source (ASHP) and ground-source (GSHP) heat pumps are as follows (where lift is denoted ΔT):

$$COP_{ASHP} = 6.81 - 0.121\Delta T + 0.00063\Delta T^2,\ 15 \leq \Delta T \leq 60, \tag{5.6}$$

$$COP_{GSHP} = 8.77 - 0.15\Delta T + 0.000734\Delta T^2,\ 12 \leq \Delta T \leq 60. \tag{5.7}$$

These equations are suitable for the purpose of a simulation model that can capture the nonlinear relationship between system efficiency and lift, while not requiring detailed descriptions of the inner workings of the heat pumps themselves.

5.3.1.3 Renewable Generation

Photovoltaic (PV) generation can be included in the simulation model in the following manner. The power generated can be calculated as a nonlinear function of the solar irradiance hitting the panel and the ambient temperature, following the approach described by Pepe et al. (2018). Where the power generated is given as P_{pv}, the ambient temperature is T_a and θ_1, θ_2 and θ_3 are parameters, the model can be represented as:

$$P_{pv} = \theta_1 \left(1 + \theta_2 I_{panel} + \theta_3 T_a\right) I_{panel} \tag{5.8}$$

These parameters can be derived from data using simple regression techniques. In the absence of data, nominal values are provided in Pepe et al. (2018), with θ_2 and θ_3 falling in the following ranges:

$$\theta_2 \in [-2.5 \times 10^{-4}, -1.9 \times 10^{-5}] \tag{5.9}$$

$$\theta_3 \in [-4.8 \times 10^{-3}, -1.7 \times 10^{-3}] \tag{5.10}$$

Similarly, solar thermal generation can be readily included. Solar thermal panels transfer heat to a fluid (water for the purposes of this framework) which can be used as a heat source for a heat pump or a thermal store. The model suggested here is the widely adopted Hottel–Whillier–Bliss model (Duffie & Beckman, 2013). The useful energy removed by the panel (Q_u) can be calculated as a function of the collector area (A_{Sol}), the overall loss coefficient (U_L), the heat removal factor (F_R), the inlet flow temperature (T_i) and the ambient temperature (T_a) as follows:

$$Q_u = A_{Sol} F_R \left(I_{panel} - U_L (T_i - T_a) \right) \tag{5.11}$$

Parameter values (F_R and U_L) can be found from literature, for example, using values from the experimental set-up of Anderson et al. (2009). Once again, these values could be calibrated to a specific system with measured data.

Hybrid technologies that combine PV and solar thermal could also be considered. The efficiency of a PV panel tends to decrease with increasing temperature (note that the parameter θ_3 in Eq. (5.8) is negative). As such, PV efficiency improvements can be achieved by removing heat from the panel. This heat can be used in the manner of a solar thermal collector, resulting in a PV-thermal (PVT) hybrid.

To represent this, the Hottel–Whillier–Bliss model of Eq. (5.11) as modified by Florschuetz (1976) to account for the electrical efficiency drop due to temperature increase can be used. The modified thermal efficiency calculation can be represented as:

$$\eta_t = F_{R_{PVT}} \left[(\tau\alpha)_e (1 - \eta_e) - U_L \left(\frac{T_{i_{PVT}} - Ta}{I_{panel}} \right) \right] \tag{5.12}$$

where η_e is the electrical efficiency at ambient and $(\tau\alpha)_e$ is the effective transmittance. The full area of a PVT collector will not be covered in PV cells, and as such, a term representing packing factor is included to relate the electrical cell efficiency to an efficiency per panel area as follows (Chow, 2010):

$$\eta_e = \frac{A_{cell} \eta_{cell}}{A_{Sol}} = \beta_{pack} \eta_{cell} \tag{5.13}$$

where β_{pack} is the packing factor, A_{cell} and A_{panel} are the cell and panel areas and ηe and η_{cell} are the cell and panel efficiencies. Parameters can once again be obtained from Anderson et al. (2009) in the absence of data.

5.3.1.4 Demand-Side: Occupancy Modelling

While the previous sections illustrated techniques for capturing the heat flows, the demand for heat is dependent on the occupants of the building. Appropriate methods for modelling different user types are needed here and are of particular relevance to a domestic setting. The building control literature is dominated by non-domestic buildings (such as office buildings), in which usage schedules are regular and predictable (e.g. office hours). This is not the case in a domestic setting whereby hours of absence and activity can vary depending on the types of occupant and their lifestyles. Despite this, uniform occupancy schedules (e.g. based on national averages) are prevalent in the literature. From the perspective of a useful building modelling framework, the use of different occupancy profiles should then be considered.

Given the wide range of potential occupant types and activities, defining realistic occupancy profiles can be a challenge. The occupancy-integrated archetype approach of Buttitta et al. (2019) seeks to overcome the complexity present by using a data-driven approach to categorise the user behaviour recorded in an extensive national Time Use Survey (TUS). With this method, activities were mapped to one of three states: absent from the dwelling, present but inactive or present and active. The authors then used a clustering approach to create five representative occupant activity types. In the modelling framework described here, a modelled dwelling is assigned a user-type corresponding to one of these archetypes. From these, the heating schedule of the building is determined.

5.3.1.5 Demand-Side: Domestic Hot Water Modelling

Aside from space-heating, the domestic hot water (DHW) demand can make up a significant proportion of the overall heat requirement of any building. In a well-insulated building, this hot water demand can exceed the space-heating demand. When considering predictive strategies that rely on forecasts of the heating requirement, this requires consideration (though in many cases, a separate heat source may be used to satisfy this demand). From a modelling perspective, DHW demand is usually characterised by demand spikes representing usage of taps, showers and baths. Such events could be modelled as stochastic processes whereby the likelihood of a DHW event varies through the day according to some average usage profile, such as that of Fig. 5.3, which is taken from a report by the Energy

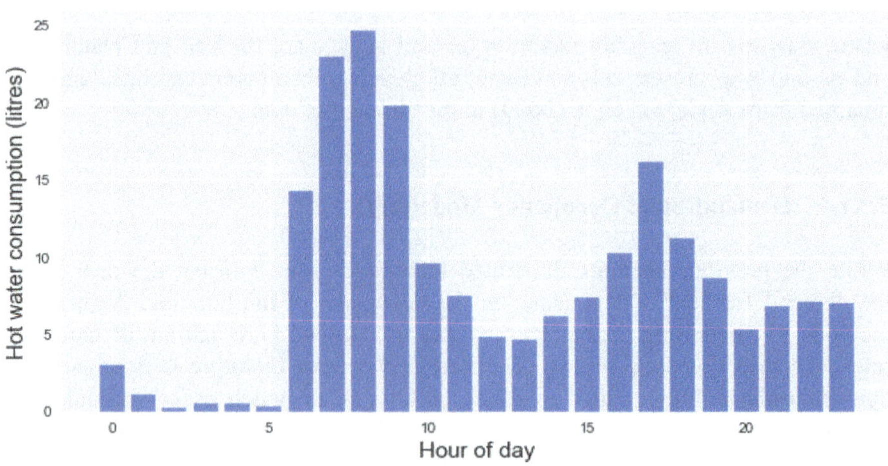

Fig. 5.3 Average daily DHW consumption

Savings Trust (2008) for average usage in the UK. The profile can be modified to ensure that events only occur during occupied periods and scaled depending on the number of dwelling occupants following the guidelines indicated in Henderson and Hart (2015). This relationship between occupant numbers (denoted N_{occ}) and water usage can be summarised as

$$DHW_{litres} = 38 + 25N_{occ}. \tag{5.14}$$

The energy consumption associated with each DHW event is related to the volume of water consumed and the temperature rise required to increase the cold inlet water temperature (given here as T_c) to a suitable outlet temperature (given here as T_h). Poisson distributions that describe typical volume requirements for different event types can be found as part of the Centre for Renewable Energy Systems Technology (CREST) model (McKenna and Thomson, 2016). Each DHW event can be assigned a required volume of water and the energy needed can then be calculated as:

$$DHW_{kWh} = 1000 DHW_{litres} \rho_w C_p \left(T_h - T_c\right), \tag{5.15}$$

where ρ_w and C_p are the density and specific heat capacity of water, respectively. Suitable values for T_c and T_h can be found in Energy Savings Trust (2008).

5.3.1.6 Demand-Side: Auxiliary Electrical Loads

Auxiliary electrical demands relating to lighting and other appliances can also be modelled. One simple method approach is to use two uniform daily profiles representing workdays and holidays/weekends. These can be taken from a report Zimmermann et al. (2012) in which measurements from 251 households were obtained and characterised in terms of dwelling and occupant type, then scaled based on the floor area.

Using an averaged approach such as this neglects the peaks and troughs that naturally occur during normal daily operation. Other approaches consider various appliance models and usage patterns to generate more realistic demand profiles (an excellent example can be found in the CREST model McKenna and Thomson, 2016). Here, however, these auxiliary loads are considered non-controllable from the perspective of the thermally-focussed controllers discussed here and as such, a more detailed approach is not elaborated further.

5.3.2 Implementation of Methods

Using the methodology developed in the previous sections, this section describes the creation and simulation of a model of a three-storey dwelling supplied by an air-source heat pump. The tools used are first described, followed by the outputs of the model.

5.3.2.1 Software Tools

To put the above methods into practice, different software tools are available with different purposes. A state-of-the-art review of commonly used software and tools for simulation/controller design in energy-efficient buildings and comparing their capabilities is given in Atam (2017). One such tool, particularly useful in control design, is the BRCM toolbox for Matlab developed at ETH Zurich (Sturzenegger et al., 2014). The toolbox allows for the generation of RC type thermal building models. Since it is Matlab-based, additional system models and statistical demand models, such as those described in Sects. 5.3.1.2–5.3.1.5, can be incorporated into a full system model. Furthermore, the functionality is provided in the BRCM models to import models created using Energyplus, a widely used tool for building energy performance evaluation (DOE, 2017). Further tools have been developed to assist a user in the definition of an Energyplus model, notably Openstudio, which is used here. The full toolchain is then as follows:

1. The building shape is drawn and building materials are defined using Openstudio.
2. The Openstudio model is then exported as an Energyplus model.
3. The BRCM toolbox is used to extract the model dynamics from the Energyplus model.
4. The model matrices are used in Matlab along with a statistical DHW model and a heat pump model to simulate the heat demand of the building for a given set of occupants.

5.3.2.2 Case Study: Three-Story Dwelling

To illustrate the implementation of the modelling methods described in the previous section, a simulation model of a three-storey residential terraced house with 14 rooms was created and used to simulate thermal behaviour. The building model is based on a modern dwelling in the Trent Basin development of Nottingham UK with building fabric materials chosen to ensure the structure corresponds to the Energy Performance Certificate (EPC) of the building. It should be noted that the choice of building and location used here is arbitrary. The workflow is not restricted to a specific set of building properties or location characteristics. The resulting thermal characteristics of the building are summarised in Table 5.1.

The linear state-space model representing the building consists of 223 states, 14 controlled inputs (heat supply to each zone), 14 outputs (zone air temperatures) and 20 disturbance channels. The disturbances include internal gains for each zone, the ground temperature, the air temperature and solar gains corresponding to each orientation. An air-source heat pump is used to supply heat via a buffer tank which

Table 5.1 Three-storey dwelling construction details

	Thermal transmittance W/m²K	Heat capacitance (per m² surface area) kJ/m²K	Solar factor (G-value)	Area m²
Roof	0.120	70.69	–	–
Floor	0.155	305.1	–	–
Ext. walls	0.128	187.5	–	–
Internal walls	2.581	33.1	–	–
Glazing	1	–	0.5	20.1

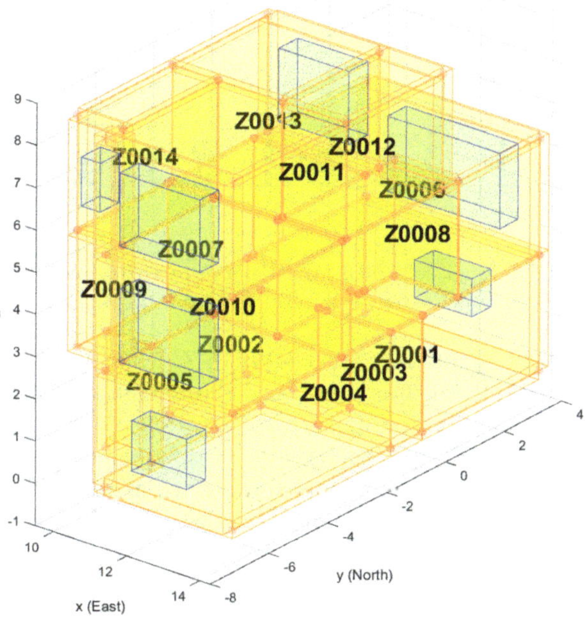

Fig. 5.4 Trent Basin terraced dwelling (floor area = 129.1 m²)

controlled to maintain a temperature of 60 °C at all times. A diagram of the building structure is shown in Fig. 5.4.

Prior to the development of the MPC strategy (which is detailed in Sect. 5.5), an on/off control strategy is applied whereby the controller of each radiator follows a hysteresis loop. The maximum heat supply is applied when the temperature falls below a specified low threshold and remains in this state until exceeding a specified high threshold at which point the heat supply is switched off. An occupancy schedule is chosen such that heating is required for a short period in the morning followed by a longer period in the evening each weekday. At weekends, a single longer occupied period is assumed. At night and during unoccupied periods, the

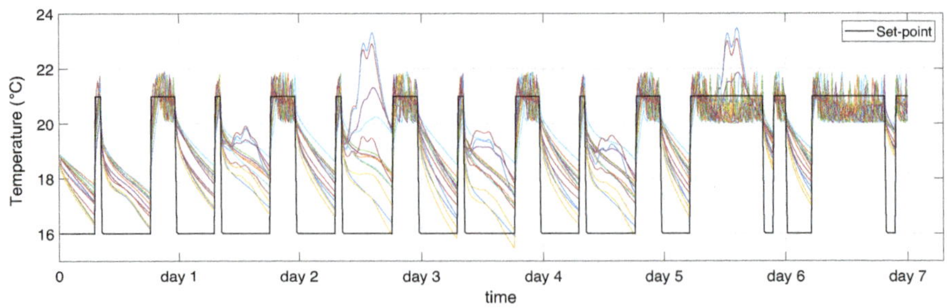

Fig. 5.5 Simulated zone temperatures for 7-day simulation with on/off radiator control strategy

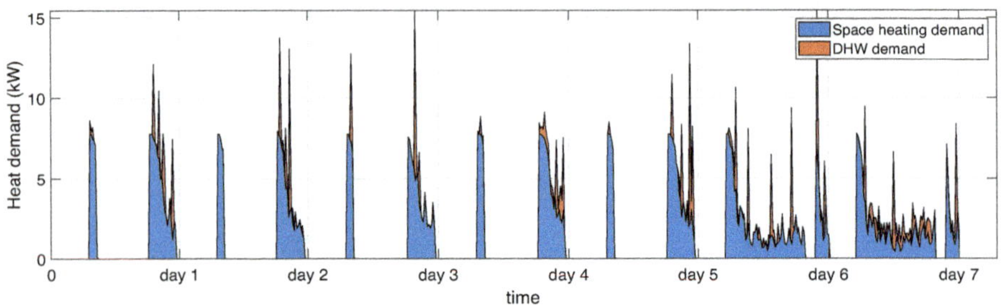

Fig. 5.6 Simulated heat demand (both space heating and hot water) for 1 week

desired temperature set-point is 16 °C while during occupied periods it is raised to 21 °C with a 1 °C comfort band allowed for the on/off hysteresis loop.

This set-up is simulated for a period of 7 days with weather data taken from January 2019 in London. The 14 zone temperatures are plotted in Fig. 5.5 with the room set-point temperature plotted as the thick black line. Since the strategy is purely reactive, it can be seen that when the lower set-point increases, a certain amount of time is needed before the zone temperatures begin to approach the desired comfort level. The influence of solar gains can be seen in the temperature spikes that occur in the third and sixth days.

The heat demand from the building for this 7-day scenario is shown in Fig. 5.6. This includes both the space heating demand and the DHW heat demand.

5.4 Low-Order Modelling for MPC Applications

With a simulation environment developed as outlined in the previous sections, the development of an MPC formulation is made more straightforward. In this section, modelling methods that can be used for MPC development are outlined. An optimisation problem lies at the heart of MPC-based strategies, whereby the behaviour of the building must be predicted with control choices made to optimise some pre-defined criteria. A model of the system behaviour is needed to carry out such an optimisation. Suitable models for MPC implementation differ from those used for simulation, however, in that the high degrees of complexity may lead to optimisation formulations that are not tractable in real-time. With this in mind, strategies for developing low-complexity models in the context of building energy systems are discussed here. The methods are implemented using the simulation environment developed in Sect. 5.3. By analysing the performance of the models in a simulated setting, the methods can be assessed prior to implementation in a real building.

5.4.1 A Model-Based Pathway

Using the methods outlined in Sect. 5.3, the production of a detailed simulation model can be achieved without significant modelling effort. The resulting model can contain a large number of equations and parameters, due to the need to characterise each individual structural component of the building fabric. Such a model may be unmanageable from an optimisation perspective. Nonetheless, model-reduction techniques can be applied to the detailed model to derive lower-complexity representations. Different model reduction approaches exist (see for example Astolfi, 2010; Deng et al., 2014), with the approach chosen here being balanced truncation.

5.4.1.1 Balanced Truncation

Using a balanced truncation approach, a linear transformation is carried out on the full state-space model and the system's Hankel singular values (HSVs) are identified. The HSVs of a system can be defined as the square-root of the eigenvalues of the product of the system's observability and controllability gramians. A transformation chosen to lead to equivalent, diagonal gramians results in what is referred to as a *balanced* model and the HSVs can be found as the main diagonal of either gramian. Relatively smaller HSVs imply lower energy states, the removal of which will have less effect on the more dominant system dynamics. As such, an appropriate number of states can be retained to ensure sufficient predictive performance with greatly reduced model-order (Lyons, 2020).

The general theory is briefly described here. A transformation $\tilde{x} = Tx$ can be applied to the model of (5.3) and (5.4):

$$\dot{\tilde{\boldsymbol{x}}}(t) = \tilde{A}\tilde{\boldsymbol{x}}(t) + \tilde{B}\boldsymbol{u}(t) \tag{5.16}$$

$$\tilde{\boldsymbol{y}}(t) = \tilde{C}\tilde{\boldsymbol{x}}(t), \tag{5.17}$$

whereby

$$\left(\tilde{A}, \tilde{B}, \tilde{C}\right) = \left(TAT^{-1}, TB, CT^{-1}\right). \tag{5.18}$$

The controllability gramian W_c and observability gramian W_o of this new system can be defined as:

$$\tilde{W}_c = TW_cT^T \tag{5.19}$$

$$\tilde{W}_o = \left(T^{-1}\right)^T W_0T^{-1}. \tag{5.20}$$

This system can be easily transformed into a balanced system with equal, diagonal gramians given by:

$$\tilde{W}_c = \tilde{W}_c = diag(\sigma_{hsv_1}, \sigma_{hsv_2}...\sigma_{hsv_{172}}), \tag{5.21}$$

arranged such that $\sigma_{hsv_1} \geq \sigma_{hsv_2} \geq \cdots \geq \sigma_{hsv_{172}} \geq 0$. σ_{hsv_p} denotes the pth HSV. In this transformed system, the states that correspond to smaller HSVs have less impact on system behaviour and can be removed.

5.4.1.2 Implementation in the Simulation Environment

Applying balanced truncation to the simulation model developed in Sect. 5.3 leads to a model with reduced order. If each zone is viewed as a separate entity, the states that best describe the interaction between the input and output of that zone can be found by looking at the HSVs. For the three-storey building model, the HSVs (found using Matlab's *hsvd* function) for zone 1 are shown in Fig. 5.7 in descending order of state energy. States that don't contribute greatly to the system behaviour have a low energy and can be discarded without significantly impacting the performance of the model.

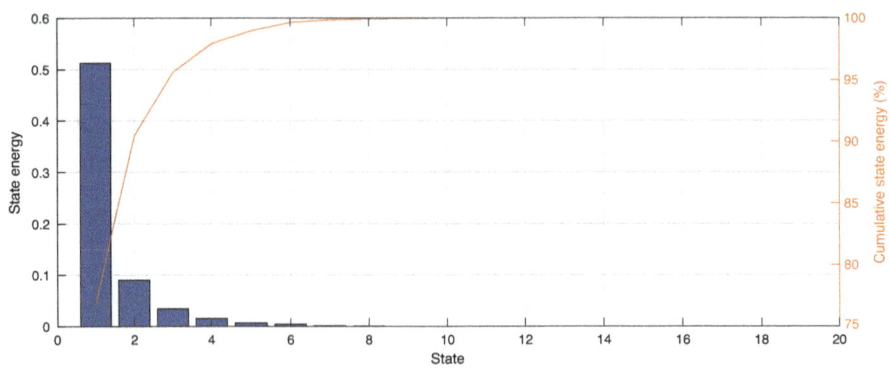

Fig. 5.7 Hankel singular values associated with the first zone of the simulation model

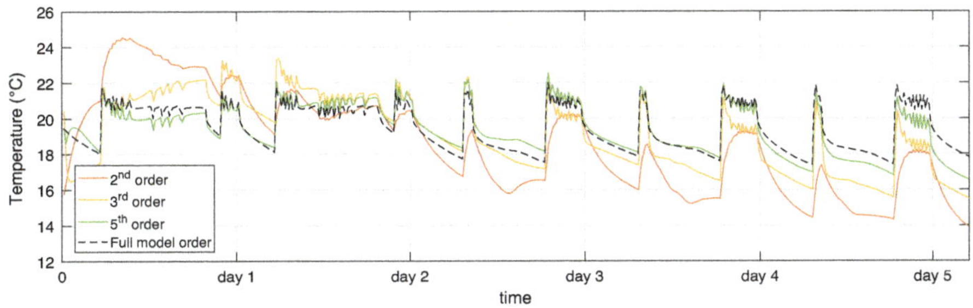

Fig. 5.8 Comparison of outputs for different model orders derived using truncation

It can be seen from this figure that the system behaviour can be mostly captured by a low number of states. As such, the full modelled system of Sect. 5.3, which was previously made up of 233 state equations, can now be replaced by zone models with as few as 2–5 states. A comparison of different model orders is shown in Fig. 5.8 where the same inputs and disturbances are applied to each. The difference between the full-order representation (the black dashed line) and the reduced-order representation increases with lower model-orders, though the 3rd and 5th order models appear to capture the behaviour quite well.

While methods exist for quantification of the difference between models, for example the gap metric approach (Georgiou, 1988), a true assessment of the usefulness of the model can be achieved by observing the performance of the model within an MPC formulation which will be carried out in Sect. 5.5.

5.4.2 A Data-Driven Pathway

Two disadvantages of the model-based pathway are that a detailed model must be available, and this detailed model must be a good representation of the real building. While having a detailed model is useful from an analysis perspective, from a scale-up perspective it may be unrealistic to assume that such a modelling effort would be applied to every building to which MPC is deployed. By directly deriving model parameters using data measurements taken from the building however, the need for a detailed simulation model is removed. The main prerequisite in this case is for sensors to be deployed appropriately in the building along with data handling and storage capabilities, which would most likely be required for MPC in any case. Typically measurements of the air temperature in the temperature-controlled rooms of the building would be used along with the heat flux into the rooms, or a proxy thereof. Examples would be the flow rate and temperature of the water flowing through a radiator-based heating system, or the temperature and airflow through an air-handling unit (AHU).

5.4.2.1 Parameter Identification

The general concept is to derive the relationships between the controlled inputs, measured or estimated disturbances and measured outputs of the building, but specific requirements depend on the control approach used. As to be expected, there are a large number of methods investigated in the literature for data-driven building model identification, covering a variety of machine-learning and system identification techniques (Maddalena et al., 2020; Li and Wen, 2014). For the purposes of MPC, the model structure is paramount as it must be suited to a receding horizon optimisation-based framework. While this narrows the range of possible choices, there remains no single accepted technique and the field is likely to evolve in the future. In the approach focussed on here, the model structure is chosen to be the same as for the model-based approach (i.e. the thermal dynamics of the building fabric are represented in a linear state-space form). As such, we refer to it as a grey-box strategy insofar as the model structure is assumed (unlike a black-box approach) but parameters are empirically derived (unlike a white-box approach).

5.4.2.2 Implementation in the Simulation Environment

Applying this technique using data obtained from the detailed simulation model, models for each zone can again be derived for different model orders. Inputs are the heat flows into each room/zone, outputs are the room/zone temperatures and external temperature, while solar radiation and internal gains can be included as disturbances. Inclusion of disturbance channels in a real strategy would depend on the availability of certain measurements. The model-order is user-specified, and the model parameters are identified using the N4SID function of Matlab's System Identification toolbox (Ljung, 1988).

Appropriate training data is crucial in the modelling process and the collection of such data is a challenging feature of the building energy application. Building heating schedules tend to be quite repetitive, leading to a low degree of excitation in the training data. Carrying out functional testing (e.g. pseudo-random-binary-sequence (PRBS) events) can be difficult if the building is occupied without violating comfort criteria. Furthermore, external uncontrollable disturbances (particularly the weather) play a large role in the thermal behaviour of the building. Long timescales may be needed to capture a wide range of weather events. In the case presented here, it is assumed that no functional testing is possible and training data is generated using the on/off strategy of Fig. 5.5.

In Fig. 5.9 the outputs generated by different model orders are plotted against the full-order model output where the same inputs have been applied to all. The data-set used for training the models was different to the data-set shown here. It can be seen that in all cases the behaviour has been captured quite well. Note, however, no noise is present in the full-order system or in the measurements taken. In reality, noise and other data issues can degrade the performance. The development of data-driven

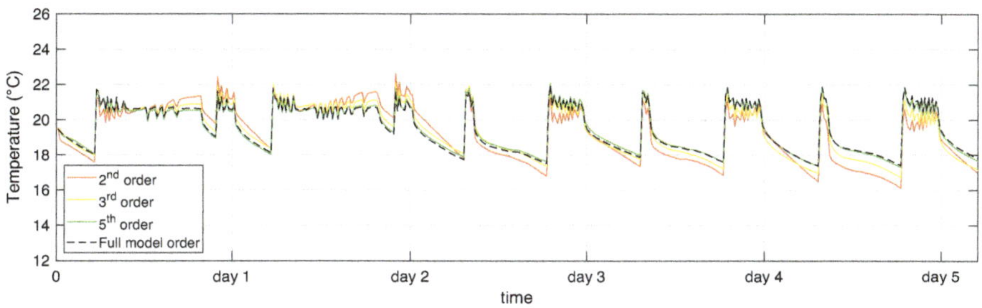

Fig. 5.9 Comparison of outputs for different model orders derived using system identification

methods that can provide performance guarantees in the face of uncertainty, without requiring intrusive training-data collection periods, remains an open challenge.

To illustrate the use of the modelling methods outlined, their implementation in an MPC strategy is required. Since methods are independent of the control architecture, two different control architectures, centralised and decentralised, are developed in the next section. The strategies will incorporate the low-complexity models of Sect. 5.4 with simulation and evaluation carried out in the simulation environment of Sect. 5.3.

5.5 Building Models within an MPC Framework

With a suitable modelling framework established along with suitable low-order models, the MPC strategy can now be developed. In this section, a suitable formulation is outlined and demonstrated using the simulation model of Sect. 5.3. The aim of the MPC formulation is to maintain the temperatures of air in each zone within a satisfactory comfort range while minimising energy consumption, subject to constraints in the system. At each sample, an optimisation problem is solved to determine such a set of inputs that satisfy the specified criteria over a given prediction horizon. The first elements of the optimal input trajectory are applied, and the optimisation is repeated at the next step in a receding horizon manner. The models developed in Sect. 5.4 form the basis of the optimisation problem.

Two architectures are considered here to highlight different possible modelling structures: centralised and decentralised. In both cases, the MPC seeks to find optimal heat inputs to each zone, while the heating system itself (e.g. the heat pumps and/or thermal stores) is not considered to keep the focus solely on the building models. Centralised (CMPC) and decentralised (DMPC) approaches have different advantages and disadvantages with suitability depending on context. CMPC can require a larger computation time, potentially exceeding the sample-time, whereas DMPC formulations, though more computationally lightweight, may neglect interaction between zones, potentially degrading performance (Lyons,

2020). Selection of a strategy requires careful consideration. This is a pressing challenge in the residential sector, where formulations that span multiple zones and multiple buildings are required to allow for aggregation of flexibility provision. A detailed comparison between centralised and decentralised approaches can be found in Pedersen et al. (2017), whereby the influence of the partition insulation level is considered. Examples of distributed strategies that seek to find a compromise between the two through the use of minimal information flows between otherwise decentralised controllers can be found in Morosan et al. (2010) and Worthmann et al. (2015).

5.5.1 *Centralised MPC: Zone-Level Control*

In the centralised formulation, all zones are considered within the same optimisation problem. The optimisation formulation takes measurements of the zone temperatures at each time step along with weather forecasts to determine optimal radiator heat supply settings for all zones using the low complexity models for prediction. The schematic of this is shown in Fig. 5.10. A key point to note is that in the centralised strategy, all zones are considered in the same model as well as the interactions between them. This necessitates greater complexity but may result in improved performance.

The low-complexity model matrices are given for this centralised strategy as A, B, C and E, with $\boldsymbol{y}(k)$ representing the vector of zone temperatures at time-step k, $\boldsymbol{x}(k)$ representing the vector of internal model states, $\boldsymbol{u}(k)$ representing the inputs to each zone and $\boldsymbol{v}(k)$ representing the disturbances. Note that this is a discrete-time model, as opposed to the continuous-time representation of (5.3) with k representing the discrete time-steps. The deviation outside the comfort set-point band in each zone is captured by the vector $\boldsymbol{\varepsilon}(k)$. Instead of applying a hard constraint on comfort (by using a constraint that states that the zone temperatures must stay within the specified comfort bounds), a soft constraint is used whereby comfort deviations are allowed, but are heavily penalised with a quadratic cost. This ensures that a

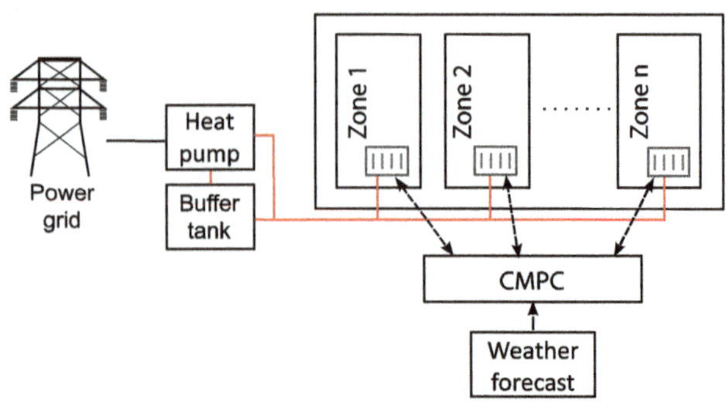

Fig. 5.10 Centralised building control strategy

feasible solution can be achieved, and larger deviations are penalised more heavily than smaller deviations. The comfort band is bounded by an upper set-point given as $T_{sp_{hi}}$ and a lower set-point $T_{sp_{lo}}$. The vector of maximum radiator heat supplies (one for each zone) is given as $\boldsymbol{q}^{max}$. Where R and S are weighting matrices balancing the energy minimisation and set-point tracking objectives, the objective and constraints of the centralised MPC problem are then given as:

$$\boldsymbol{u}^* = \arg\min \sum_{k=0}^{N-1} \left(\boldsymbol{\varepsilon}(k)^T R \boldsymbol{\varepsilon}(k) + S \left[\boldsymbol{u}(k)^T, \boldsymbol{\varepsilon}(k)^T \right]^T \right) \tag{5.22}$$

$$s.t. \quad \boldsymbol{x}(k+1) = A\boldsymbol{x}(k) + B\boldsymbol{u}(k) + E\boldsymbol{v}(k) \tag{5.23}$$

$$\boldsymbol{y}(k) = C\boldsymbol{x}(k) \tag{5.24}$$

$$0 \leq \boldsymbol{u}(k) \leq \boldsymbol{q}^{max} \tag{5.25}$$

$$T_{sp_{hi}}(k) - \boldsymbol{\varepsilon}(k) \leq \boldsymbol{y}(k) \leq T_{sp_{hi}}(k) + \boldsymbol{\varepsilon}(k) \tag{5.26}$$

$$0 \leq \boldsymbol{\varepsilon}(k) \tag{5.27}$$

The model-based or the data-driven pathway can be used to derive the A, B, C and E matrices. Using such a formulation, it is possible to influence the behaviour of the heating system by updating the objective and constraints accordingly. Perhaps the most common way in which this is achieved is by incorporating a time-varying energy cost in the weighting matrix S. Energy use at times of high cost is penalised more heavily than at times of low cost, resulting in solutions that shift the timing of the energy use while maintaining constraint satisfaction. Alternatively, input constraints can be tightened as required, to force curtailment of the energy supply. The quadratic form of the objective and the linear constraints ensure solutions can be obtained with standard quadratic programming solvers.

In either centralised or decentralised form, an estimation of the model state vector $\boldsymbol{x}$ at time k is needed to initialise the optimisation model at time k. These states are not representative of physical quantities (unlike in the full-order physics-based representation of the building, in which all states represent wall, air, floor or ceiling temperatures) unless a first-order model is used, in which case, the state is equivalent to the measured air temperature times a constant C. In the modelling framework, state estimation is carried whenever an MPC strategy is deployed by use of Kalman filtering.

5.5.2 Decentralised MPC: System-Level Control

In the decentralised formulation, a separate MPC problem is solved for each individual zone. Each zone has its own model that relates the zone's radiator heat

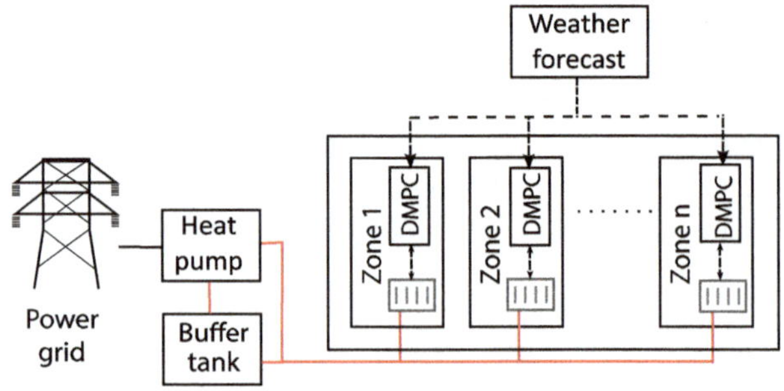

Fig. 5.11 Decentralised building control strategy

input and any relevant disturbances to the zone air temperature. Thermal interactions between different zones are not modelled. This is shown in the schematic of Fig. 5.11.

For a building with J zones, model matrices are given for zone $j \in J$ as A^z_j, B^z_j, C^z_j and E^z_j, with $y_j(k)$ representing the zone temperature at time k, $\boldsymbol{x}_j(k)$ representing the vector of internal model states, $u_j(k)$ representing the heat input to the zone and $\boldsymbol{v}_j(k)$ representing the zone-relevant disturbances. Once again, the deviation outside the comfort set-point band (bounded by $T_{sp_{lo}}$ and $T_{sp_{hi}}$) is penalised, with deviations denoted by $\varepsilon_j(k)$. The maximum radiator heat supply is given as q^{max}_j. The objective weighting matrices are given as R_j and S_j for the zone, leading to the following formulation:

$$\boldsymbol{u}^*_j = \arg\min \sum_{k=0}^{N-1} \left(\varepsilon^2_j(k)R + S\left[u_j(k), \varepsilon_j(k)\right]^T\right) \tag{5.28}$$

$$s.t. \quad \boldsymbol{x}_j(k+1) = A^z_j \boldsymbol{x}_j(k) + B^z_j u_j(k) + E^z_j v_j(k) \tag{5.29}$$

$$y_j(k) = C^z_j \boldsymbol{x}_j(k) \tag{5.30}$$

$$0 \le u_j(k) \le q^{max}_j \tag{5.31}$$

$$T_{sp_{lo},j}(k) - \varepsilon_j(k) \le y_j(k) \le T_{sp_{hi},j}(k) + \varepsilon_j(k) \tag{5.32}$$

$$0 \le \varepsilon_j(k) \tag{5.33}$$

The purpose of including different control structures is to show the adaptability of the proposed strategies to fit in multi-input-multi-output (MIMO) and single-input-single-output (SISO) contexts, rather than presupposing the existence of a one-size-fits-all solution.

5.5.3 Implementation in the Simulation Model

The behaviour of the control approach is demonstrated here by implementation of the decentralised formulation with 5th-order zone models derived using the model truncation method. A 15-min sample-time and a 20-step prediction horizon (5 h) were applied. The weighting matrices R_j and S_j were chosen to ensure that comfort violation is heavily penalised relative to financial cost. In this way, the controller attempts to keep the zone temperatures within or as close as possible to the comfort bounds. Of course, by increasing the relative weight on the financial penalty, larger deviations from the comfort bounds would be allowed for the sake of financial savings. Finding an appropriate balance for a specific user is part of the design process. Once again, by having a detailed simulation model available, this potentially abstract balance can be analysed in terms of specific financial savings and tangible comfort deviations.

The resulting simulated zone temperatures for the building over a 2-weekday simulation are shown in Fig. 5.12. The dark black lines indicate the designated comfortable temperature range for the building. It can be seen that the strategy successfully maintains the temperatures within these bounds without unnecessary overheating. The ability of the decentralised approach to maintain internal comfort indicates that the cross-zone interactions do not significantly impact the performance, which is unsurprising, since the temperature difference between zones is low so heat flow between zones can be assumed to be small.

A comparison of the MPC performance with that of the On/Off strategy from Sect. 5.3 is shown in Table 5.2. To quantify the level of comfort satisfaction, discomfort is measured here in K.hr, whereby a deviation outside the comfort bounds of 1K for 1 h corresponds to 1 K.hr. The reactive nature of the On/Off strategy leads to 3.8 K.hr deviation from the comfort bounds, compared to 0.7 K.hr for the MPC strategy. Interestingly, the energy requirement is greater for the MPC strategy, which reflects the conflicting nature of the comfort and energy goals. More energy was required to ensure the comfort goals were achieved. This can be seen clearly in Fig. 5.13 in which the average zone temperature of the buildings is plotted

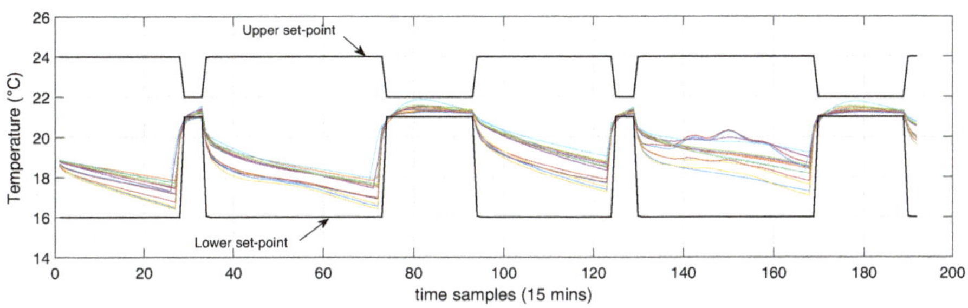

Fig. 5.12 Zone temperatures simulated in the building for a two-weekday simulation using a decentralised MPC control strategy

Table 5.2 Space heating performance of predictive and non-predictive strategies

	Energy demand kWh	Comfort deviation K.hr
On/off strategy	70.1	3.8
MPC strategy	72.4	0.7

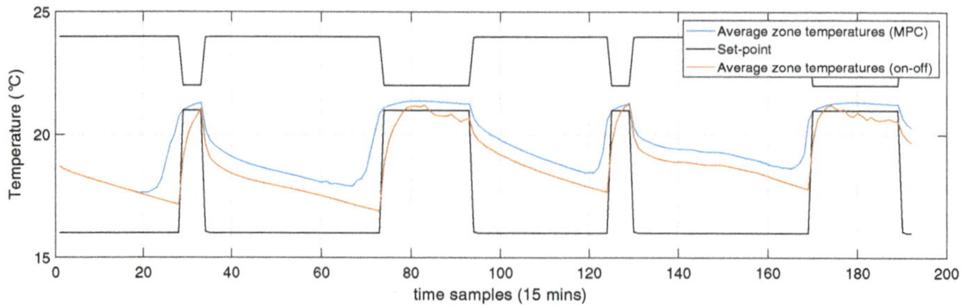

Fig. 5.13 Average zone temperatures for MPC (decentralised) and On/Off strategies for a two-weekday simulation

for the MPC and On/Off scenarios. To achieve comfort, the zones are maintained at a higher temperature when MPC is used.

The handling of multiple objectives (in this case, energy and comfort), is not necessarily trivial. The need to handle the objectives of different users across financial, environmental and resilience vectors, without the need for excessive design input, remains an open challenge, particularly given the diversity of stakeholder requirements in the residential sector.

5.5.3.1 Including Price-Awareness

Aside from a better thermal comfort performance, one of the key advantages of MPC when compared to traditional strategies is the ability to incorporate external factors in the control decisions, thus enabling the shifting of demands in a manner that benefits the power grid. A typical mechanism for achieving this is to introduce a variable electricity price, such as the *Agile* tariff used by Octopus Energy (2019). At times of high demand, the price can be increased to disincentivize energy use, encouraging a shift towards times of lower price. To show this using the modelling framework, the objective of the MPC strategy was adjusted to minimise electricity consumption cost instead of the energy consumption using a variable electricity tariff. The behaviour of this *price-aware* strategy is shown in Fig. 5.14 in which the average zone temperatures are plotted, as well as the electricity price. Two large price spikes can be seen in the two-weekday simulation. Prior to these spikes, an increase in the zone temperatures is visible. By shifting the energy use, less energy is consumed during the high-price period.

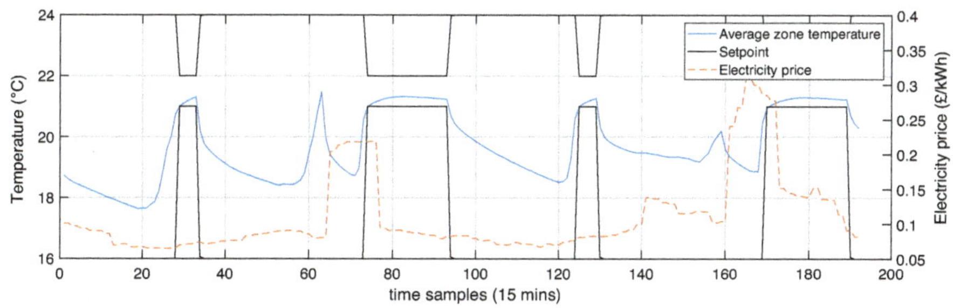

Fig. 5.14 Average zone temperature simulated in the building with price-aware MPC strategy for two-weekday simulation with decentralised MPC control strategy. The red, dashed line indicates the electricity price

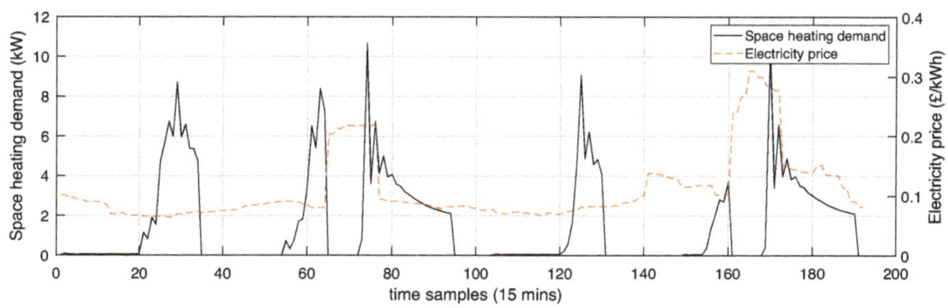

Fig. 5.15 Space heating demand and electricity price for 2-day simulation period

Table 5.3 Space heating performance of MPC with a time-varying price tariff strategy

	Electricity cost £	Energy demand kWh	Comfort deviation K.hr
Price-aware MPC	8.77	73.9	0.7
Non-price-aware MPC	9.82	72.4	0.8

This behaviour can be seen more clearly by observing the heat supplied to the building, as shown in Fig. 5.15 in which the electricity price is also plotted. During the two price spikes, the heat supply drops to zero, with a certain amount of pre-heating carried out before the price-spikes occur.

This behaviour results in a lower electricity cost (the total amount paid for electricity for the simulation length) when compared to a non-price aware strategy with little change in the comfort satisfaction, achieved by shifting energy use away from high-tariff times (Table 5.3). By supplying heat earlier the total heat requirement increases slightly. The amount of additional heat lost due to this shift in demand will depend on the level of insulation of the building. While it's generally understood that well-insulated buildings are more suited to this type of flexibility provision, the modelling framework can explicitly capture the trade-off. Simple methods for quantifying this trade-off without significant analysis remain a challenge.

5.5.3.2 Parameter Selection Considerations

The control design process can be streamlined using the methods and the workflow presented here, however, the design process is not completely eliminated. Any given choice of parameters and structures may not lead to a controller that behaves in a similar manner across all buildings and contexts. To show this, the impact of some of these parameters on the computation time is examined here.

Choices such as model-order and control structure will affect the resulting performance. For example, the impact of model order, prediction horizon and structure selection on the time taken to carry out a single optimisation using a 3.2 GHz 4-core processor is shown in the box-plots of Figs. 5.16 and 5.17. The use of box plots here is to capture the variation found in the optimisation times. In Fig. 5.16 it can be seen that although there is an improvement in using a 30th-order model in comparison to the full-order RC model, the significant time improvement is achieved when the low-order decentralised models are used.

In Fig. 5.16, a prediction horizon of 20 steps is used. The increased relevance of model order when longer horizons are used is shown in Fig. 5.17, with box plots once again used to compare the time taken to complete an optimisation, this time comparing a 20-step (5 h) and a 96-step (24 h) horizon. It can be seen that the time taken to carry out an optimisation using the full-order model becomes prohibitively large when the 1-day horizon is used (in the context of a 15-min sample time). For the low-order models, even though the shorter time-horizon again produces a much faster solution, the average time taken for the longer horizon is approximately 0.1 s (depending on the computational power available).

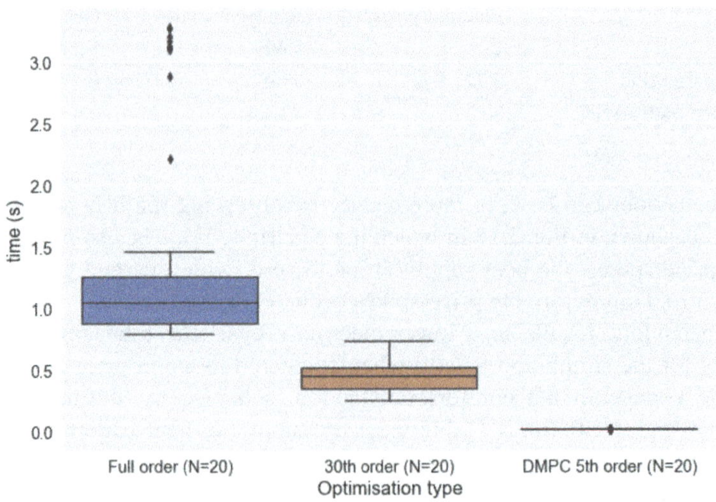

Fig. 5.16 Time taken to solve MPC optimisation formulations at each time-step for different model orders

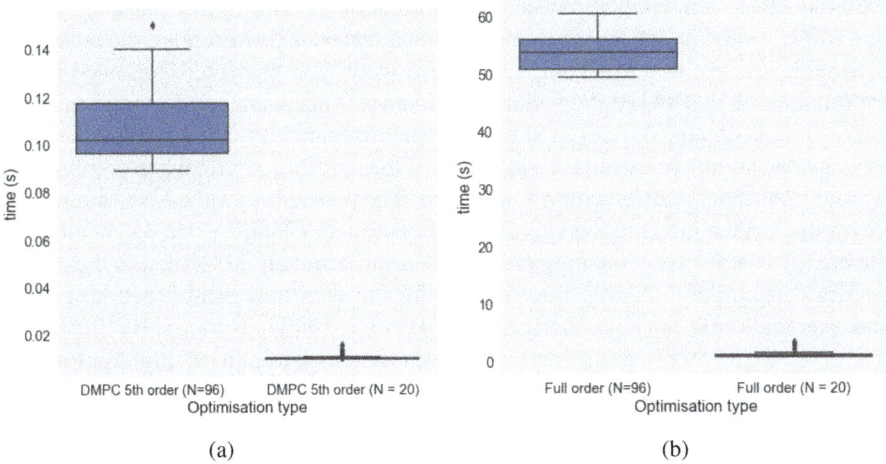

Fig. 5.17 Time taken to solve MPC optimisation formulations once for different model types and prediction horizons; (**a**) shows decentralised MPC optimisation times; (**b**) shows centralised MPC optimisation times

The purpose of this is to illustrate that the choices of prediction horizon, model order and control structure are all connected from the perspective of problem complexity and the correct choice of parameters may not be obvious but will depend on the context. High-order models may provide additional accuracy that is needed is some contexts, but not in others. Similarly, longer prediction horizons may or may not be required. The trade-off that stems from these decisions is all part of the design challenge as it stands. While the availability of a simulation environment can aid selection of such parameters, methods for automating the design process further could offer significant benefits.

5.6 Conclusions

This chapter outlines a workflow for designing and evaluating MPC strategies in the residential building sector to act as a guide for practitioners in the area. With electrification of the heating and transport sectors well underway, the need for intelligent control approaches that can be applied to the domestic building sector to enable effective demand-side management is urgent. The development of appropriate models for this purpose acts as a significant barrier to the scale-up of such methods. The workflow introduced here incorporates the establishment of a detailed physics-based simulation environment, the development of low-complexity control-oriented models, the formulation of a suitable optimisation strategy and the evaluation of the strategy using the simulation environment. Suitable methods and tools for simulating the thermal behaviour of a building were first detailed, with the production of a detailed physics-based model of a three-storey building carried out. Model-based and data-driven methods that can be used to capture the building's thermal behaviour in a low-complexity form were then presented, followed by the formulation of an MPC strategy that seeks to ensure occupant comfort with minimal energy use. Using the developed simulation environment, it was shown that the MPC strategy can significantly reduce violations of the comfort criteria when compared to a more traditional reactive control approach. The possibility of the MPC strategy to incorporate price information was also demonstrated. Using a price-aware strategy, the energy cost for space-heating reduced by approximately 11%, though the overall energy consumption slightly increased. With the workflow established here, such strategic trade-offs can be properly analysed prior to implementation. By illustrating the process in detail in the manner presented, in addition to highlighting the modelling principles required in the sector, some of the challenges that remain are emphasised, including parameter selection, the impact of which was shown in terms of computational performance.

Acknowledgments This work has received funding from the EPSRC (Engineering and Physical Sciences) under the Active Building Centre project (reference number: EP/V012053/1).

References

Al Essa, M. J. M. (2019). Home energy management of thermostatically controlled loads and photovoltaic-battery systems. *Energy, 176*, 742–752.

Anderson, T. N., Duke, M., Morrison, G. L., & Carson, J. K. (2009). Performance of a building integrated photovoltaic/thermal (BIPVT) solar collector. *Solar Energy, 83*, 445–455.

Astolfi, A. (2010). Model reduction by moment matching for linear and nonlinear systems. *IEEE Transactions Automatic Control, 55*, 2321–2336.

Atam, E. (2017). Current software barriers to advanced model-based control design for energy-efficient buildings. *Renewable and Sustainable Energy Reviews, 73*, 1031–1040.

Atam, E., & Helsen, L. (2016). Control-oriented thermal modeling of multizone buildings: Methods and issues: Intelligent control of a building system. *IEEE Control Systems Magazine, 36*(3), 86–111.

Barone, G., Buonomano, A., Calise, F., Forzano, C., & Palombo, A. (2019). Building to vehicle to building concept toward a novel zero energy paradigm: Modelling and case studies. *Renewable and Sustainable Energy Reviews, 101*, 625–648.

Bünning, F., Huber, B., Heer, P., Aboudonia, A., & Lygeros, J. (2020). Experimental demonstration of data predictive control for energy optimization and thermal comfort in buildings. *Energy and Buildings, 211*, 109792.

Buttitta, G., Turner, W. J. N., Neu, O., & Finn, D. P. (2019). Development of occupancy-integrated archetypes: Use of data mining clustering techniques to embed occupant behaviour profiles in archetypes. *Energy and Buildings, 198*, 84–99.

CDBB. (2018). *The gemini priciples* (vol. 15). Cambridge: University of Cambridge.

Coakley, D., Raftery, P., & Keane, M. (2014). A review of methods to match building energy simulation models to measured data. *Renewable and Sustainable Energy Reviews, 37*, 123–141.

Coulson, J., Lygeros, J., & Döfler, F. (2020). Distributionally robust chance constrained data-enabled predictive control. *arXiv: 2006.01702 [math.OC]* (pp. 1–14).

Chow, T. T. (2010). A review on photovoltaic/thermal hybrid solar technology. *Appl Energy, 87*, 365–379.

Deng, K., Goyal, S., Barooah, P., & Mehta, P. G. (2014). Structure-preserving model reduction of nonlinear building thermal models. *Automatica, 50*, 1188–1195.

DOE. (2017). EnergyPlus | EnergyPlus. U.S. Department of Energy's. Retrieved May 10, 2018. https://www.energyplus.net/

Drgoňa, J., Arroyo, J., Cupeiro Figueroa, I., Blum, D. Arendt, K., Kim, D., Perarnau Oll, E., Oravec, J., Michael, W., Vrabie, D. L., & Helsen, L. (2020). All you need to know about model predictive control for buildings. *Annual Reviews in Control, 50*, 190–232.

Duffie, J. A., & Beckman, W. A. (2013). *Solar engineering of thermal*. Hoboken: Wiley.

EASME (2017). TABULA WebTool, Retrieved Dec. 06, 2021, from https://webtool.building-typology.eu/#bm

Energy Savings Trust. (2008). Measurement of domestic hot water consumption in dwellings. Retrieved Dec. 06, 2021, from https://assets.publishing.service.gov.uk/government/uploads/system/uploads/attachment_data/file/48188/3147-measure-domestic-hot-water-consump.pdf

Florschuetz, L. W. (1976). Extension of the Hottel-Whillier-Bliss model to the analysis of combined photovoltaic/thermal flat plate collectors. *Solar Energy, 22*, 361–366.

Georgiou, T. T. (1988). On the computation of the gap metric. In: *Proceedings of the IEEE Conference on Decision and Control*.

Gonzato, S., Chimento, J., O'Dwyer, E., Bustos-Turu, G., Acha, S., & Shah, N. (2019). Hierarchical price coordination of heat pumps in a building network controlled using model predictive control. *Energy and Buildings, 202*, 109421.

Greater London Authority. (2018). Zero carbon London: A 1.5oC compatible plan.

Harish, V. S. K. V., & Kumar, A. (2016). A review on modeling and simulation of building energy systems. *Renewable and Sustainable Energy Reviews, 56*, 1272–1292.

Henderson, J., & Hart, J. (2015). BREDEM 2012: A technical description of the BRE Domestic Energy Model - Version 1.1. Retrieved Dec. 06, 2021, from https://www.bre.co.uk/filelibrary/bredem/BREDEM-2012-specification.pdf

Kathirgamanathan, A., De Rosa, M., Mangina, E., & Finn, D. P. (2021). Data-driven predictive control for unlocking building energy flexibility: A review. *Renew Sustain Energy RevRenewable and Sustainable Energy Reviews, 135*, 110120.

Killian, M., & Kozek, M. (2016). Ten questions concerning model predictive control for energy efficient buildings. *Building and Environment, 105*, 403–412.

Li, X., & Wen, J. (2014). Review of building energy modeling for control and operation. *Renewable and Sustainable Energy Reviews, 37*, 517–537.

Lin, Y., Middelkoop, T., & Barooah, P. (2012). Issues in identification of control-oriented thermal models of zones in multi-zone buildings. In: *Proceedings of the IEEE Conference on Decision and Control* (pp. 6932–6937). Piscataway: IEEE.

Ljung, L. (1988). System identification toolbox. Matlab user's Guide.

Lyons, B., O'Dwyer, E., & Shah, N. (2020). Model reduction for model predictive control of district and communal heating systems within cooperative energy systems. *Energy, 197*, 117178.

Ma, Y., Kelman, A., Daly, A., & Borelli, F. (2012). Predictive control for energy efficient buildings with thermal storage. *IEEE Control Systems Magazine, 32*, 44–64.

Maddalena, E. T., Lian, Y., & Jones, C. N. (2020). Data-driven methods for building control – A review and promising future directions. *Control Engineering Practice, 95*, 104211.

McKenna, E., & Thomson, M. (2016). High-resolution stochastic integrated thermal-electrical domestic demand model. *Applied Energy, 165*, 445–461.

Morosan, P.-D., Bourdais, R., Dumur, D., & Buisson, J. (2010). Distributed model predictive control for building temperature regulation. In: *American Control Conference (ACC), 2010* (pp. 3174–3179)

Octopus Energy agile pricing explained (2019). Retrieved Oct. 26, 2020, from https://octopus.energy/blog/agile-pricing-explained/

O'Dwyer, E., Pan, I., Acha, S., Gibbons, S., Shah, N., DeJong, M. J., Schooling, J. M., & Viggiani, G. M. B. (2019). Modelling and evaluation of multi-vector energy networks in smart cities. In: *International Conference on Smart Infrastructure and Construction 2019 (ICSIC)* (pp. 161–168). London: ICE Publishing.

O'Dwyer, E., Pan, I., Charlesworth, R., Butler, S., & Shah, N. (2020). Integration of an energy management tool and digital twin for coordination and control of multi-vector smart energy systems. *Sustainable Cities and Society, 62*, 102412.

Oliveira Panão, M. J. N., Mateus, N. M., & Carrilho da Graça, G. (2019). Measured and modeled performance of internal mass as a thermal energy battery for energy flexible residential buildings. *Applied Energy, 239*, 252–267.

Pedersen, T. H., Hedegaard, R. E., Knudsen, M. D., & Petersen, S. (2017). Comparison of centralized and decentralized model predictive control in a building retrofit scenario. In: *Energy Procedia, 122* (pp. 979–984).

Pepe, D., Bianchini, G., & Vicino, A. (2018). Estimating PV forecasting models from power data. In: *2018 IEEE International Energy Conference, ENERGYCON 2018* (pp. 1–6). Piscataway: IEEE.

Rätz, M., Javadi, A. P., Baranski, M., Finkbeiner, K., & Müller, D. (2019). Automated data-driven modeling of building energy systems via machine learning algorithms. *Energy and Buildings, 202*, 109384.

Shaikh, P. H., Nor, N. B. M., Nallagownden, P., Elam, I., & Ibrahim, T. (2014). A review on optimized control systems for building energy and comfort management of smart sustainable buildings. *Renewable and Sustainable Energy Reviews, 34*, 409–429.

Simulationresearch. (2016). FrontPage - bcvtb. Retrieved May 13, 2018 from https://simulationresearch.lbl.gov/bcvtb

Staffell, I., Brett, D., Brandon, N., & Hawkes, A. (2012). A review of domestic heat pumps. *Energy & Environmental Science, 5*, 9291.

Sturzenegger, D., Gyalistras, D., Semeraro, V., Morari, M., & Smith, R. S. (2014). BRCM matlab toolbox: Model generation for model predictive building control. In: *American Control Conference* (pp. 1063–1069).

Sturzenegger, D., Gyalistras, D., Morari, M., & Smith, R. S. (2016). Model predictive climate control of a Swiss office building: Implementation, results, and cost–benefit analysis. *IEEE Transactions on Control Systems Technology, 24*, 1–12. Retrieved from https://doi.org/10.1109/TCST.2015.2415411

Worthmann, K., Kellett, C. M., Braun, P., Grüne, L., & Weller, S. R. (2015). Distributed and decentralized control of residential energy systems incorporating battery storage. *IEEE Transactions on Smart Grid, 6*, 1914–1923.

Zimmermann J-P, Evans M, Griggs, J., King, N., Harding, L., Roberts, P., & Evans, C. (2012). Household electricity survey: A study of domestic electrical product usage (pp. 213–214.). Intertek Testing & Certification Ltd.

Zwickel, P., Engelmann, A., Groll, L., Veit Hagenmeyer, Sauer, D., & Faulwasser, T. (2019). A comparison of economic MPC formulations for thermal building control. In *2019 IEEE PES Innovative Smart Grid Technologies Europe (ISGT-Europe)* (pp. 1–5). Piscataway: IEEE.

Chapter 6
MPC and Optimal Design of Residential Buildings with Seasonal Storage: A Case Study

P. Falugi, E. O'Dwyer, M. A. Zagorowska, E. Atam, E. C. Kerrigan, G. Strbac, and N. Shah

6.1 Introduction

Achieving net-zero carbon emissions by 2050 will entail significant changes to the way electrical energy is generated, transmitted and used. Future electricity systems will be characterised by substantial volumes of intermittent renewable generation that will displace conventional plants, thus increasing the need for backup reserves and ancillary services. In addition, decarbonising the heating and transport sectors through electrification will lead to increases in peaks that are disproportionately higher than the corresponding increases in annual electricity demand. In this new reality, leveraging operational flexibility will be a critical challenge for developing a cost-efficient net-zero carbon system.

Optimal technology sizing is required to ensure that the system design suits the dwelling under consideration in terms of efficiency, cost and flexibility. Simultaneous optimisation of design and operation of a system is called *co-design*. The effect of a control strategy on the design process is frequently overlooked. Incorporating

This research was supported by the EPSRC under the grant EP/V012053/1.

P. Falugi (✉) · M. A. Zagorowska · E. Atam · G. Strbac
Department of Electrical and Electronic Engineering, Imperial College London, London, UK
e-mail: p.falugi@imperial.ac.uk; m.zagorowska@imperial.ac.uk; g.strbac@imperial.ac.uk

E. C. Kerrigan
Department of Electrical and Electronic Engineering, Imperial College London, London, UK

Department of Aeronautics, Imperial College London, London, UK
e-mail: e.kerrigan@imperial.ac.uk

E. O'Dwyer · N. Shah
Department of Chemical Engineering, Imperial College London, London, UK
e-mail: e.odwyer@imperial.ac.uk; n.shah@imperial.ac.uk

V. Vahidinasab, B. Mohammadi-Ivatloo (eds.), *Active Building Energy Systems*, Green Energy and Technology, https://doi.org/10.1007/978-3-030-79742-3_6

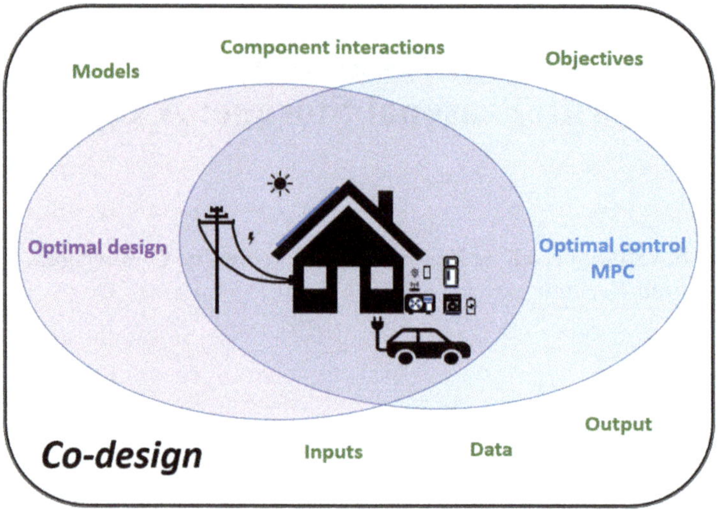

Fig. 6.1 Co-design for residential buildings

the behaviour of a controller in the loop leads to a potentially computationally demanding problem. Ignoring the control behaviour, on the other hand, may lead to an ill-suited set of technologies for a given situation and unnecessary capital expenditure.

In this work, we present an approach to sizing the optimal technology mixes while incorporating the effect on the operation of the optimised system. The results are presented in a case study for residential buildings (see Fig. 6.1). The new approach uses residential building models and technology libraries developed along with MPC strategies and system-level optimization frameworks, leveraging on the capabilities of the ICLOCS2 toolbox (Nie et al., 2018, 2020), combining short- and long-term dynamics by solving a single finite-horizon optimal control problem.

Currently, residential buildings account for about 22% of the global final energy use as documented in International Energy Agency and United Nations (2019), and with the electrification of the heating and transport sector, their consumption is bound to increase drastically. Model predictive control (MPC) has excellent potential for improving energy usage in buildings and increasing their flexibility and adaptation capability to support the whole energy system since it enables straightforward integration of state and input constraints. MPC has been successfully applied to the control of heating, ventilation and air conditioning (HVAC) systems, as shown in the review (Hameed Shaikh et al., 2014).

The chapter includes a detailed case study on optimal sizing of battery storage and the roof area used by photovoltaic panels (PV) and thermal solar to maximize self-sufficiency, while considering optimal management of the building assets. Energy storage technologies are an essential element for enhancing the utilization of existing assets and improving operational flexibility. In the present study, we include storage devices operating at different time scales to enable the possibility of shifting energy between seasons.

The rest of the chapter is structured as follows. Section 6.2 introduces preliminaries on co-design and optimal control, focusing in particular on co-design and MPC. Section 6.3 presents a literature review on co-design and model predictive control applied to residential buildings. The model of the residential building used in the chapter is introduced in Sect. 6.4. Section 6.5 shows the results of the case study and discusses the mutual influences between design and model predictive control. The chapter ends with conclusions in Sect. 6.6.

6.2 Status Quo, Challenges and Outlook

Co-design consists in simultaneous optimisation of design and operation of systems (Garcia-Sanz, 2019). The idea of co-design stems from a multidisciplinary analysis of interactions between subsystems. The first contributions to co-design were focused on optimising a fixed structure using multi-objective constrained formulations (Rao, 1988; Sensburg et al., 1989). Recent works moved forward including in the co-optimisation of the design of the structure (Fathy et al., 2003; Maraniello & Palacios, 2016; Herber & Allison, 2019). This section introduces the background on co-design and puts it in the context of optimal control problems, in particular model predictive control.

6.2.1 Co-design and Optimisation

Control co-design has been recently recognised to be "an engineering game-changer" (Garcia-Sanz, 2019). The author of Garcia-Sanz (2019) discusses the concept of control co-design in a broad sense and envisages its crucial role in embedding multidisciplinary subsystem interactions under a unified framework to deliver radically new products and systems. The concept is identified as a game-changer in recognition of its ability to design the entire system to achieve optimal solutions otherwise unobtainable. In the optimisation context, such ideas appeared at the end of the 1980s to simultaneously identify the plant and control parameters Rao (1988). Recently, the authors of Diangelakis et al. (2017) provided an overview of control approaches used in co-design application in process systems engineering.

Given the challenges coming with the solution to complex optimisation problems, the application of co-design has been limited until recently. In the last few decades, applications of optimal system design under optimal operation have flourished in the design of multi-energy systems, where the time resolution spans from hours to years. In Pecci et al. (2017), the authors described an application of co-design to optimal operation and placement of control valves in a water distribution network. The co-design framework enabled finding a trade-off among competing objectives, such as the cost of the components, the operation of the system and external demand.

In particular, Yan et al. (2017) indicated the advantages of simultaneous optimisation of the design and operation of energy systems connected to electrical grid. The authors showed that adjusting the size of photovoltaic panels improves the operation of the system and extends the lifetime of the equipment. Further examples of the use of co-design in optimisation of energy systems can be found in Xu et al. (2020).

6.2.2 Co-design and Closed-Loop Control

The applications of co-design from Sect. 6.2.1 are primarily focused on the operation of open-loop systems. As indicated in Malcolm et al. (2007), steady-state and open-loop approaches are insufficient for ensuring flexibility. In particular, they showed that the integration of design and closed-loop control improves the performance of systems operating close to their limits.

The importance of co-design approaches aiming at technology sizing under closed-loop operation has also been demonstrated in Martelli et al. (2018) for commercial building and in Atam et al. (2013) for determination of ground-coupled heat pump systems. In Martelli et al. (2018), an optimization framework for optimally designing and operating multi-energy systems with seasonal storage is presented. A set of representative design days are selected by solving a nonlinear mixed-integer optimisation problem. The use of typical days simplifies the solution of the problem. The inclusion of seasonal storage requires introducing coupling between design days or discriminating between variables that are related to discrete decision variables. Afterward, to assess the quality of the achieved design, the operation of the system is simulated with hourly resolution using real data.

A co-design approach where simultaneous optimal sizing and control input determination for ground-coupled heat pump systems (which are widely used for energy-efficient residential buildings) was considered in Atam et al. (2013). The control framework used in the paper consisted in optimal control with periodic boundary conditions. A parameterized state-space model for borehole thermal dynamics in terms of borehole length was developed and in the optimal control formulation operational constraints such as bounds on the circulating fluid temperature, and heat/cold demand satisfaction were taken into account. The cost function in the optimal control problem consisted of a weighted sum of investment and operational costs with hourly time resolution.

6.2.3 Co-design and Predictive Control

A further extension of the co-design can be achieved by explicitly including predictive control in the optimisation problem. Model predictive control is one of the most common control schemes that lends itself to the idea of co-design as a way of simultaneous optimisation of technology and operation of a system. Figure 6.1 shows how co-design encompasses both optimal design (left) and optimal control, such as MPC (depicted on the right). The intersection of optimal design and control leads to a comprehensive framework that enables a holistic approach to improved operation of a system. The co-design approach draws from both areas to satisfy often conflicting objectives.

6.2.3.1 Model Predictive Control

Model predictive control (MPC) is an advanced control framework, which stems from the field of optimal control. It enables a straightforward integration of constraints on state and input variables, which makes it a useful tool in real-life applications, as indicated in Maciejowski (2002). A comprehensive description of MPC and its applications can be found in Maciejowski (2002) and Rawlings et al. (2017), for example.

Figure 6.2 presents an example of the MPC framework applied for reference tracking (red solid line). MPC uses a model of a dynamic system at time k to predict the behaviour of a system (dashed black) over the prediction horizon of length N, if the output at time k is known (solid black). The control action that minimizes a selected objective function over the prediction horizon is found. Then the first part of the control input (long-dashed black) is applied to the system over the control horizon of length M. In the next time instant, $k+1$, the prediction horizon is shifted, and the procedure is repeated.

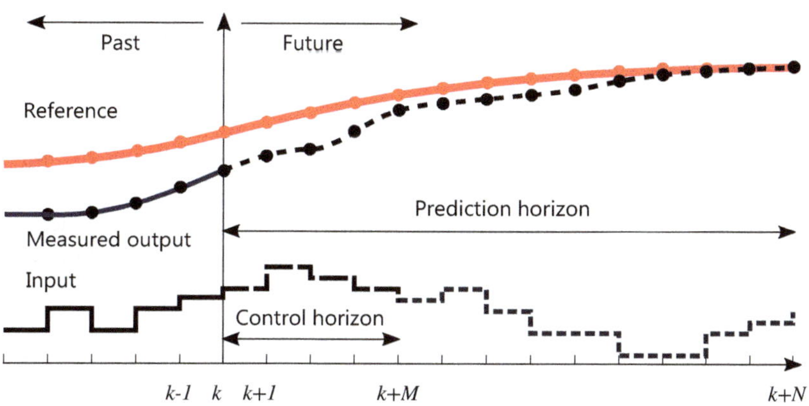

Fig. 6.2 MPC scheme where the measured output (solid black with dots) tracks the reference signal (red line with dots). An optimisation problem is solved at the k-th time instance over a prediction horizon of length N (dashed lines) and provides a prediction of the trajectory of the system (dashed black line with dots). The resulting control is then applied on the control horizon (large dashes)

6.2.3.2 Co-design and MPC

In the context of MPC, co-design applications must consider a high complexity of the system description requiring a time resolution of minutes or even seconds.

Co-design MPC frameworks were proposed in Suardi et al. (2013) and Khusainov et al. (2019) to co-optimise closed-loop software implementation and hardware performance. The co-design process in MPC boils down to be a multi-objective formulation where the nature of the optimised components belongs to different areas of competence and operate at different time scales, as pointed out in Khusainov et al. (2016), while the main focus is maintained on the closed-loop performance of the system.

The authors of Moura et al. (2010) proposed a stochastic model predictive control to optimise battery sizing and power management in electric vehicles. In particular, they indicated that it is necessary to find a trade-off between the size of the battery and the estimated usage of the car. A battery that is too small or too large for the need of a driver would decrease the performance of the car and increase power consumption.

In Diangelakis et al. (2017), the developed predictive control framework for simultaneous optimisation of the size of the equipment and its operation was applied to a range of applications in process system engineering. The authors indicated the importance of long-term horizon for optimal operation of a system.

6.2.4 *Software Tools for Modelling and Analysis*

The co-design framework presented in this chapter has its foundation on building models and technology libraries developed along with MPC strategies and system-level optimization frameworks leveraging on the capabilities of the ICLOCS2 toolbox (Nie et al., 2020).

ICLOCS2 is a comprehensive software suite for solving nonlinear optimal control problems in Matlab. ICLOCS2 comprises a wide selection of numerical methods and automated tools assisting the design and implementation of a large variety of optimization problems. The toolbox allows formulating and solving a variety of optimisation problems, including dynamic equations with equality and inequality constraints, through the direct definition of general differentiable functions.

The toolbox allows defining problems of the following form

$$\min_{\substack{x(\cdot),u(\cdot)\\ p,t_0,t_f}} \int_{t_0}^{t_f} \ell(x(t), u(t), p, t)dt + V(x(t_0), x(t_f), p, t_0, t_f) \tag{6.1a}$$

subject to a continuous state trajectory $x : \mathbb{R} \to \mathbb{R}^{n_x}$ and the constraints

$$f(\dot{x}(t), x(t), u(t), p, t) = 0, \forall t \in \mathcal{T} \text{ a.e.}, \tag{6.1b}$$

$$g(\dot{x}(t), x(t), \dot{u}(t), u(t), p, t) \leq 0, \forall t \in \mathcal{T} \text{ a.e.}, \tag{6.1c}$$

$$c(\dot{x}(t), x(t), \dot{u}(t), u(t), p, t) = 0, \forall t \in \mathbb{T}, \tag{6.1d}$$

$$\psi_E(x(t_0), x(t_f), p, t_0, t_f) = 0, \tag{6.1e}$$

$$\psi_I(x(t_0), x(t_f), p, t_0, t_f) \leq 0, \tag{6.1f}$$

where 'a.e.' stands for 'almost everywhere' in the Lebesgue sense on the interval $\mathcal{T} := [t_0, t_f] \subset \mathbb{R}$, where t_0 and t_f denote, respectively, the initial and final time. The set $\mathbb{T} \subset \mathcal{T}$ is a finite subset of $\mathcal{T}$. The input trajectory $u : \mathbb{R} \to \mathbb{R}^{n_u}$ is allowed to be piece-wise differentiable. The vector $u(t)$ is used to model the whole set of time varying decision variables, while the decision vector $p \in \mathbb{R}^{n_s}$ describes all the decision variables that are constant in $\mathcal{T}$.

For instance, in the present framework, the parameter p is used to model the size of technology to be determined. In particular, p includes the size of the battery and the roof area covered by PV and thermal solar. The differential-algebraic equations describing the dynamic of the system are defined by (6.1b), where $f : \mathbb{R}^{n_x} \times \mathbb{R}^{n_x} \times \mathbb{R}^{n_u} \times \mathbb{R}^{n_s} \times \mathbb{R} \to \mathbb{R}^{n_f}$. The cost function allows one to model a variety of different objectives since it comprises the so-called Mayer cost term $V : \mathbb{R}^{n_x} \times \mathbb{R}^{n_x} \times \mathbb{R}^{n_s} \times \mathbb{R} \times \mathbb{R} \to \mathbb{R}$ and the integral of the running cost $\ell : \mathbb{R}^{n_x} \times \mathbb{R}^{n_u} \times \mathbb{R}^{n_s} \times \mathbb{R} \to \mathbb{R}$. The inequality constraints (6.1c), where $g : \mathbb{R}^{n_x} \times \mathbb{R}^{n_x} \times \mathbb{R}^{n_u} \times \mathbb{R}^{n_u} \times \mathbb{R}^{n_s} \times \mathbb{R} \to \mathbb{R}^{n_g}$, describes upper and lower bounds on variables and functions of them. General boundary equality and inequality constraints can be imposed through (6.1e)–(6.1c), where $\psi_E : \mathbb{R}^{n_x} \times \mathbb{R}^{n_x} \times \mathbb{R}^{n_s} \times \mathbb{R} \times \mathbb{R} \to \mathbb{R}^{n_E}$ and $\psi_I : \mathbb{R}^{n_x} \times \mathbb{R}^{n_x} \times \mathbb{R}^{n_s} \times \mathbb{R} \times \mathbb{R} \to \mathbb{R}^{n_I}$. The resulting optimisation problem is infinite-dimensional, because the decision variables are defined on an infinite dimensional space subject to an uncountable set of constraints.

The solution of the problem requires the application of numerical methods to compute an approximate solution. ICLOCS2 offers a set of different methods and procedures called direct transcription methods to approximate the infinite-dimensional problem. Once the continuous-time optimisation problem has been transcribed, the resulting finite-dimensional optimisation problem can be efficiently solved with a state-of-art solvers as IPOPT (Wächter & Biegler, 2006) and WORHP (Büskens & Wassel, 2013) since the resulting problem is highly structured. For a comprehensive tutorial, see Kerrigan et al. (2020).

6.3 Co-design and Predictive Control in Residential Buildings

According to the British Standard Institution (BSI, 2020), the term *residential housing* encompasses both *apartment blocks* with multiple *housing units*, as well as *detached houses* (or *single-family houses*). Control and optimisation of residential housing is therefore a challenging task, which requires a multidisciplinary approach. The flexibility of a co-design approach from Sect. 6.2 makes it an attractive framework for optimisation and control of residential housing. As shown in Fig. 6.1, co-design encompasses multiple aspects of optimisation of residential buildings. These include daily operation of household appliances, transportation, as well as energy generation with photovoltaics, and energy storage enabling further interactions with the electrical grid.

This section introduces the existing approaches to optimisation and control of residential housing using co-design and model predictive control. The residential housing used in this work will be described in Sect. 6.4.

6.3.1 Co-design for Residential Buildings

Recently, the need for increased flexibility at all levels of the system has stimulated research on optimal system operation and configuration at the residential building level. Most of the works on optimal sizing and operation in residential buildings focus on the sizing of batteries coupled with PV and heat pumps. Optimal design approaches for battery sizing with PV and energy management of smart home are proposed in Wu et al. (2017) and Beck et al. (2016). The authors of Koskela et al. (2019) further emphasized the impact of the size of PVs and battery on the profitability from the economic perspective. In particular, Koskela et al. (2019) analysed the interactions between the size of the PVs and battery storage.

An optimization framework for sizing of PV and electric heat pump in dwellings oriented to self-consumption has been proposed in Beck et al. (2017). The authors analysed the sizing and configuration of a building system under various operating scenarios for the load profile and exogenous data. The paper stresses that simplified sizing approaches not taking into account the operation of the system lead to oversized systems. Similarly, Linssen et al. (2017) indicates that optimal sizing of PVs is crucial to correct calculation of the operational cost in a building.

Several authors have considered the co-design problem of residential buildings in relation to the interactions of the building with electrical grid. In particular, the influence of financial incentives for the operator of the building on the operation of the system was analysed. A co-design approach considering the relation among financial incentives, system sizing, and operation of residential battery, PV and heat pump was proposed in von Appen (2018). Recently, the authors of Koskela et al. (2019) presented the influence of economic incentives on the sizing of PVs and battery storage.

The authors of Fischer et al. (2016) analysed the sizing of a thermal storage and heat pump in a residential building to quantify the impact of on-site PV generation and variable electricity prices. Even though the proposed frameworks relied on a coarse static representation of the system operation, the authors find that the use of variable prices and the presence of PV generation may generate a different decision with respect to the current sizing procedures.

A co-design framework to size thermal and electric storage systems in residential buildings has been proposed in Baniasadi et al. (2020). The authors show that a co-design approach has the potential of reducing the annual electricity bill by 80%. The proposed framework uses daily representative blocks for each season with a time resolution of an hour.

6.3.2 *MPC for Residential Buildings*

Most of the literature focuses on commercial and industrial buildings, while not many studies have been devoted to the application of MPC for residential buildings (Thieblemont et al., 2017). Recently the interest in applying MPC to residential buildings has enormously increased due to stringent decarbonization targets, because MPC can improve the energy efficiency, so that it reaches the desired limits. MPC can also handle conflicting objectives, multiple assets, and thermal and physical constraints. MPC also naturally includes the possibility of imposing occupant preferences, even allowing human-in-the-loop control architectures (Schütte et al., 2019).

A comprehensive review of applications of MPC to building control and optimisation taking into account both the objective of the control, as well as the modelling aspect, has been provided in Drgoňa et al. (2020). Recently, Kathirgamanathan et al. (2020) provided an in-depth analysis of how predictive control can be used to ensure flexibility of the operation of a building, with particular focus on modelling of the building itself. The current chapter builds on these reviews and provides insights from the perspective of including storage facilities in the model used for control.

6.3.2.1 MPC with Battery Storage

The authors of Sun et al. (2016) considered the problem of energy management in residential buildings with integration of PV and battery energy storage. The used PV and battery models were nonlinear and hence nonlinear MPC was chosen as a control method, which was solved at each step using dynamic programming. Building load demand prediction was done using artificial neural networks. In addition, the authors of Sun et al. (2016) studied the trade-off between battery aging and cost minimization and explored the sensitivity of the controller performance to control horizon length and load forecast accuracy. Authors found through simulation studies that the proposed control framework achieves 96–98% of the best possible performance, given perfect predictions over a long time period.

Similar works have been performed in Cui et al. (2016, 2017). In Cui et al. (2016), an adaptive optimal co-scheduling approach was developed for HVAC control and battery management in residential buildings. The approach was tested based on detailed simulations. In the problem formulation, the battery degradation was explicitly taken into account. It was concluded that, compared to baseline approaches, the developed approach resulted in up to a 15% reduction in the total electricity cost.

In Cui et al. (2017), the same authors extended the work in Cui et al. (2016) by considering a hybrid electrical energy storage system instead of a single battery system. The hybrid system consisted of a battery bank and a supercapacitor bank. The rationale behind considering a hybrid energy storage system was the desire of exploiting the strengths of the two different storage technologies to come up with a better solution. This new solution provided a reduction up to 10% in the total cost compared to the results in Cui et al. (2016).

In Maasoumy et al. (2014), three interacting components, namely, HVAC control, EV charging and battery usage in residential buildings, were co-scheduled using an integrated and holistic MPC framework to reduce building energy bill while providing the required thermal comfort for the occupants. In the MPC cost function, the power grid energy consumption cost, battery usage cost (battery charge cost based on the off peak price plus the battery depreciation cost) and power grid peak demand cost were included to form the total cost to be minimized. Detailed simulations showed significant reductions in total cost compared to no co-scheduling cases.

6.3.2.2 MPC with Seasonal Storage

The work in Darivianakis et al. (2017) indicated that the main challenge related to predictive control of buildings with seasonal storage lies in the mismatch in the timescales. Seasonal storage systems operate in days or months, whereas day-to-day control of a building is usually considered in minutes or hours. The authors of Darivianakis et al. (2017) showed that using historical data enabled them to provide bounds on the uncertainty in the operation of the building and make a better use of the seasonal storage. The cost of operating the building was not considered.

In a similar way, Kumar et al. (2019) proposed a hierarchical MPC scheme to take into consideration the difference in timescales. They explored the periodic behaviour of seasonal storage to formulate an upper-level MPC controller operating on long timescales and a lower-level MPC controller which takes care of faster dynamics. The size of the available seasonal storage was not optimised.

The authors of Lago et al. (2020) concluded that MPC was well suited for applications with day-ahead pricing. They developed a control algorithm that takes into account both the long-term seasonal storage and varying prices and interactions of the building with electricity market.

Neither of the authors considered the sizing of the seasonal storage in their MPC framework. The sizing problem, due to the seasonal storage, requires imposing cyclic boundary conditions and modelling the operation of the system throughout the full year to evaluate a representative annual system behaviour.

6.4 Modelling of Residential Buildings

The basic active building is equipped with a generation element and storage devices that enhance flexibility.

6.4.1 Temperature Modelling

To represent the thermal behaviour of the building, the commonly used single-zone lumped-capacitance approach is taken (Hazyuk et al., 2012). This form allows for the thermal mass of the building to be considered. The parameters are chosen to reflect the physical properties of a well-insulated building with a floor area of approximately 90 m^2. In this form, the building is represented by the first-order temperature model

$$\dot{T}_t = (UA + \rho_{air} V C^p_{air} n_{ac})/C_{build}(T^A_t - T_t) + 1/C_{build}(Q^{HP}_t + Q^{SS}_t - Q^{CP}_t), \tag{6.2}$$

where Q^{SS}_t is the heat supplied by seasonal thermal storage, Q^{HP}_t and Q^{CP}_t are, respectively, the heat and cooling provided by electric-driven heat pumps. The model parameters for a three bedroom dwelling with a high level of insulation are reported in Table 6.1.

The temperature needs to satisfy user-dependent comfort constraints

$$\underline{T}_t \leq T_t \leq \overline{T}_t, \tag{6.3}$$

Where $\underline{T}_t$ and $\overline{T}_t$ are the desired thermal comfort limits. Even if the thermal comfort is a user-dependent choice, the design process needs to satisfy predefined standards as documented in BSI (1999) and CIBSE (2015).

Table 6.1 Dwelling parameters

Description	Parameter	Value	Unit
Average U-value	U	0.93195	W/(m^2 K)
Wall surface area	A	82.06959707	m^2
Air density	ρ_{air}	1.225	kg/m^3
Building volume	V	224.05	m^3
Air heat capacity	C^p_{air}	1.005	kJ/(kg K)
Air changes per hour	n_{ac}	1	h^{-1}
Building thermal mass	C_{build}	15,286.6114	kJ/K
Floor surface area	S_F	89.62	m^2
HP capacity	$\overline{u}^{eH}$	6	kW
CP capacity	$\overline{u}^{ceH}$	6	kW

6.4.2 Energy Supply Technologies

The heat pump draws heat from ambient air, and its key modelling aspect is its coefficient of performance (COP), which is the ratio between supplied heat Q_t^{HP} and the consumed electric power u_t^{eH}

$$Q_t^{HP} = COP(T_t^A)u_t^{eH}. \tag{6.4}$$

The COP is often assumed to be a constant for simplicity. Since the output of a heat pump reduces with external temperature, in the present study we have modelled the COP as a linear function of the external temperature. It is practice to test heat pumps at an external temperature of 7 °C and indoor of 20 °C. Under such operating conditions, the COP is about 3. The coefficient is modelled as follows:

$$COP(T_t^A) = m_{COP}(T_t^A - 7) + 3. \tag{6.5}$$

The parameters of all assets are reported in Table 6.2. The cooling pump coefficient COP_{cool} has been assumed constant.

Table 6.2 Asset parameters

Description	Parameter	Value	Unit
HP coefficient of performance slope	m_{COP}	0.067	°C
CP coefficient performances	COP_{cool}	0.7	–
Thermal solar efficiency	η	0.7	–
Seasonal storage efficiency	η^S	0.4	–
Seasonal Max discharging rate	$\overline{P}^{eS}$	0.4175	kW
Battery cyclic efficiency	η^b	0.88	–
Battery Max state of charge	$\overline{SoC}$	60	kWh
Power/irradiance gain	θ_1	0.12	kW/m^2
Power/irradiance correction	θ_2	−0.0001345	–
Power/(temperature irradiance) correction	θ_2	−0.00325	–

6.4.3 Energy Storage Technologies

The seasonal storage can be modelled as a storage element in which energy can be stored through reversible reactions or sorption processes. In this study, the thermochemical storage is based on fully reversible hydration and dehydration reactions of chemical salts such as calcium chloride ($CaCl_2$). To store energy, hot air is passed over the salt in matrix (SIM), removing moisture and storing the energy via an endothermic reaction. To recover energy, humid air is passed over the salts, causing an exothermic reaction, releasing heat that can be directed to the zones. The heat generated by thermal solar panels can be stored for use when it is cold. Shorter-term diurnal operation is also possible. The state of charge SoC_t^{SS} of seasonal storage is modelled as

$$\frac{dSoC_t^{SS}}{dt} = -Q_t^{SS} + \eta^S Q_t^{In}, \tag{6.6}$$

where Q_t^{SS} and Q_t^{In} denote, respectively, the discharging and charging heat flows and η^S is the efficiency of the seasonal storage. Since for every unit of electricity, 5.5 units of heat are generated, the extracted heat Q_t^{SS} is proportionally related to the used power u_t^{eS} as

$$Q_t^{SS} = 5.5u_t^{eS}. \tag{6.7}$$

The extracted heat is required to satisfy the physical limits, and it induces bounds on the consumed power

$$0 \leq u_t^{eS} \leq \overline{P}^{eS} \tag{6.8}$$

where $\overline{P}^{eS}$ takes into account the ability of SIM technologies to boost air temperature up to 20 °C.

The building also includes a rechargeable battery to add flexibility in the system. (Li-ion) batteries exhibit high energy and power density along with no memory effect and low self-discharge relative to other cell chemistries. Currently, lithium batteries, compared to other technologies, have high power and energy density and have a low self-discharge. They are also highly configurable to create a wide range of power ratings and capacity sizes. It is also predicted that lithium phosphate LiFePO4 batteries will have the highest share (Pillot, 2015) in 2025. A simple model of the battery state of charge SoC_t is

$$\frac{dSoC_t}{dt} = -u_t^{dch} + \eta^b u_t^{ch}, \tag{6.9}$$

where u_t^{dch} and u_t^{ch} denote the discharging and charging rates, respectively. The parameter η^b models the discharging and charging efficiencies. The capacity and charge/discharge limits are

$$
\begin{aligned}
&0 \leq SoC_t \leq S^B \leq \overline{SoC}, \\
&0 \leq u_t^{dch}, u_t^{ch} \leq \frac{S^B}{T_{ds}}
\end{aligned}
\tag{6.10}
$$

where $\overline{SoC}$ is the possible maximum capacity size of the battery and T_{ds} is the time it requires to fully discharge at the maximum discharge rate.

6.4.4 Energy Generation Technologies

The present case study includes the option of installing thermal solar and photovoltaic panels. The useful energy Q_t^{TS} produced by a thermal solar panel is modelled as

$$
Q_t^{TS} = \eta I_t S^{TS} \tag{6.11}
$$

where I_t denotes solar irradiance, S^{TS} the roof area covered by thermal solar panels and η the conversion efficiency. The maximum power produced by PV panels can be accurately modelled as a nonlinear function (Dows & Gough, 1995; Pepe et al., 2018) of the solar irradiance and the external temperature as follows:

$$
P_t^{PV} = \theta_1(1 + \theta_2 I_t + \theta_2 T_t^A) I_t S^{PV}, \tag{6.12}
$$

where S^{PV} denotes the roof area covered by the PV panel. The operating limits refer to the design specs for the multi-crystalline JAP6 4BB module range manufactured by Solar (2020). The parameter values for a location in the south of the UK are reported in Table 6.2.

The system dynamics can be compactly represented in a state space representation as

$$
\begin{bmatrix} \dot{T}_t \\ \dot{SoC}_t^{SS} \\ \dot{SoC}_t \end{bmatrix} = \begin{bmatrix} -(UA + \rho_{air} V C_{air}^p n_{ac})/C_{build} & 0\,0 \\ 0 & 0\,0 \\ 0 & 0\,0 \end{bmatrix} \begin{bmatrix} T_t \\ SoC_t^{SS} \\ SoC_t \end{bmatrix} +
$$

$$
\begin{bmatrix} COP(T_t^A)/C_{build} & 5.5/C_{build} & -COP_{cool}/C_{build} & 0 & 0 \\ 0 & -5.5 & 0 & 0 & 0 \\ 0 & 0 & 0 & -1 & \eta^b \end{bmatrix} \begin{bmatrix} u_t^{eH} \\ u_t^{eS} \\ u_t^{CeH} \\ u_t^{dch} \\ u_t^{ch} \end{bmatrix} + f_t, \tag{6.13}
$$

where the vector of exogenous signal is

$$f_t = \begin{bmatrix} (UA + \rho_{air} V C^p_{air} n_{ac})/C_{build} T^A_t \\ \eta^S I_t S^{TS} \\ 0 \end{bmatrix} \tag{6.14}$$

and the vectors $u_t := [u^{eH}_t \; u^{eS}_t \; u^{CeH}_t \; u^{dch}_t \; u^{ch}_t]'$ and $x_t := [T_t \; SoC^{SS}_t \; SoC_t]'$ denote the input and the state, respectively, where the symbol $'$ stands for the transpose of a vector.

6.5 Case Study: Optimal Technology Sizing for Residential Buildings with Seasonal Storage

The sizing problem requires imposing cyclic boundary conditions and modelling the operation of the system throughout the full year to correctly evaluate the behavior of the seasonal storage. The proposed co-design problem minimises an economic cost comprising the electricity bill, carbon emissions and annualised capital cost of batteries, PVs and thermal solar panels. Let $p := [S^{PV}, \; S^{TS}, \; S^B]$ and $C^p := [C^{PV}, \; C^{TS}, \; C^B]$ denote, respectively, the vector of the parameters and the associated vector cost. The problem is required to satisfy operational and physical constraints and it is defined as

$$\min_{u_t, x_t, p} \int_{t_0}^{t_f} (c^B_t u^B_t + c^m_t u^B_t - c^S_t u^S_t) \mathrm{d}t + C^p p' \tag{6.15}$$

$$\text{subject to (6.3), (6.4), (6.5), (6.7), (6.8), (6.10), (6.11), (6.12), (6.13)} \tag{6.16}$$

$$u^B_t - u^S_t + u^{dch}_t - u^{ch}_t + P^{PV}_t = u^{eH}_t + u^{eS}_t + u^{CeH}_t \tag{6.17}$$

$$0 \le u^B_t \le 30, \; 0 \le u^S_t \le 30 \tag{6.18}$$

$$0 \le u^{eH}_t \le \overline{u}^{eH}, \; 0 \le u^{CeH}_t \le \overline{u}^{CeH} \tag{6.19}$$

$$0 \le p(1) + p(2) \le S_F, \quad p(1), p(2) \ge 0 \tag{6.20}$$

$$0 \le p(3) \le \overline{SoC} \tag{6.21}$$

$$x(t_0) = x(t_f) \tag{6.22}$$

for all $t \in [t_0, t_f]$, where u^B_t and u^S_t are the bought and sold power, respectively, with a bound of 30 kW determined by the connection contract. The boundary condition (6.22) enforces the system to operate respecting the cyclic behaviour of the seasons. This condition corresponds to include the periodic trajectories in the set of possible steady-state optimal solutions. The physical limitation that the total area occupied by the generation technologies must not exceed the available roof area is expressed by (6.20). The capacity S^B of the battery is limited by the constraint

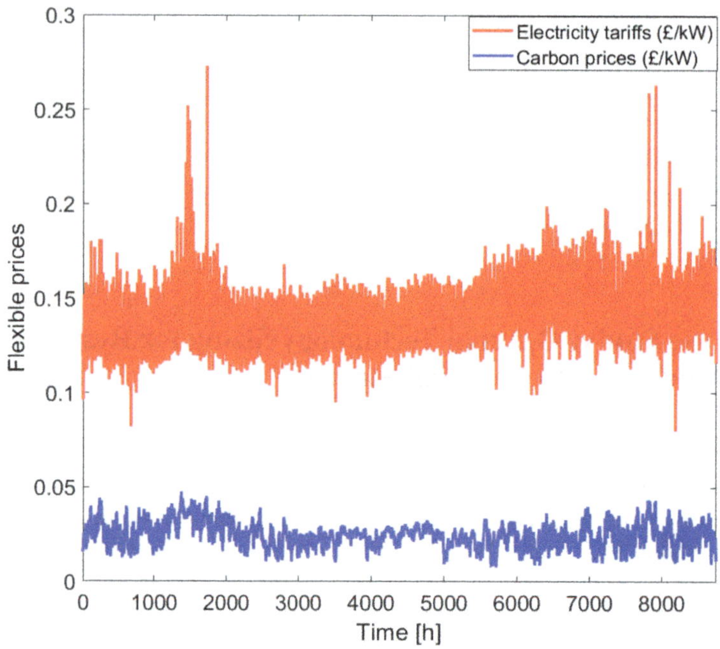

Fig. 6.3 The costs for carbon emissions and electricity prices throughout the whole year considering a carbon price cost of 100£/(ton CO2e)

(6.21). The co-design problem (6.15) subject to the constraints (6.16)–(6.22) is a Linear Programming (LP) problem.

6.5.1 Case Study Formulation

The presented case study considers time-varying electricity prices (Fig. 6.3) with 15 min resolution based on Nordpool data with a pricing mechanism used as described by Octopus Energy (2020). It has been assumed that at each time instant the price of the sold power c_t^S satisfies $c_t^S = 0.9c_t^B$. Meteorological data profiles for different UK locations have been obtained from a database maintained by the European Commission Joint Research Centre (JRC, 2012).

In order to analyse the benefits of designing the system considering how it will be operated and the interplay between the different technologies in enhancing overall system performance, we have also used a modified capital cost denoted as CAPEX_1 in Table 6.3.

The annualised capital costs have been computed considering the technology lifespan in Table 6.3 and an interest rate $r = 2\%$. In particular, the equivalent annual cost of each technology is obtained dividing the capital expenses (CAPEX) by the

Table 6.3 Asset investment parameters

	C^{PV}	C^{TS}	C^{BS}
$CAPEX_1$	180£/m^2	30£/m^2	50£/kWh
$CAPEX_2$	300£/m^2	500£/m^2	500£/kWh
Technology lifespan (years)	30	25	15

'present value of annuity factor'

$$a_{y,r} = \frac{1 - \frac{1}{(1+r)^y}}{r}$$

with y denoting the year.

6.5.2 Results and Discussion

The co-design problem (6.15) has been implemented using ICLOCS2 described in Sect. 6.2.4 and solved with IPOPT (Wächter & Biegler, 2006). The adopted transcription method is the explicit Euler with 15 min time resolution. The Euler method enables analysis of systems with discontinuities in the inputs. The maximum time resolution is dictated by the piecewise constant electricity prices which update every 15 min. Higher time resolutions up to 2.5 min did not significantly affect the obtained optimal solution for the size of the battery under consideration.

We have solved the co-design problem with the two different capital costs in Table 6.3. Furthermore, a third case has been created, sizing all the technologies separately from how the building is operated. In particular, the thermal solar has been made capable of providing to the seasonal storage almost the whole heat required for thermal comfort in winter. The third case has been investigated for two different terminal conditions on the state.

Table 6.4 reports the obtained optimal technology size and cost for the co-design problems in columns 1 and 2. The third and fourth columns in Table 6.4 report the annual operational cost achieved by a predictive controller with predetermined technologies. The model predictive controller is based on problem (6.15), where the technologies are determined a priori.

Terminal state conditions in MPC are used to enforce optimality of the long-term dynamics (Darivianakis et al., 2017). In case 3.1, the terminal condition has been left free, while in case 3.2 the condition is equated to the optimal value of $x(t_f)$ returned by the co-design problem $CAPEX_1$. Note that the cases $CAPEX_1$ and $CAPEX_2$ report doubled bill savings and emission reductions up to 86% compared to case 3.1.

The total system cost for $CAPEX_2$ is higher than for $CAPEX_1$ since the technologies are expensive. Instead, the operation cost and the net carbon emissions are of the same order of magnitude. The net carbon emissions represent the emission

Table 6.4 Comparison results

Case	Problem (6.15) CAPEX$_1$	Problem (6.15) CAPEX$_2$	Operation 3.1	Operation 3.2
Battery capacity (kWh)	4.82	0	30	30
Area PV (m^2)	84.07	89.62	44.81	44.81
Area TS (m^2)	5.5	0	44.81	44.81
Total optimal cost (£/year)	−90.25	406.56	–	–
Operational cost (£/year)	−793.26	−793.89	−387.51	193.225
Investment cost (£/year)	703.01	1200.46	–	–
Net carbon emissions (kg/year)	−1548.2	−1603.9	−831.26	83.54

variation contributed by the active building across 1 year. The emission index accounts for a positive contribution due to purchased electricity and a negative one accrued by selling electricity.

Note that the solution obtained for CAPEX$_2$ excludes the storage technologies and the performance obtained for CAPEX$_1$ and CAPEX$_2$ are comparable under the operating conditions used in the sizing process. However, the performance of the design that excludes storage assets may deteriorate considerably under different operating conditions. This situation happens because, in the cost, the value of flexibility is not precisely quantified. Also, the occurrence of different future scenarios is ignored.

Figures 6.4a and 6.6a show how the predictive controller satisfies the thermal comfort requirements over the year for both the co-design problems. However, the dwell temperature of the design performed using CAPEX$_2$ is subject to more fluctuations as highlighted by also comparing Figs. 6.5a and 6.6b. The more frequent and larger fluctuations are due to the fact that in the co-design with CAPEX$_2$, the only source of flexibility is given by the stored heat in the building. The room temperature profiles in summer (Fig. 6.5b) do not show substantial difference in these two cases.

Figure 6.7a, b illustrates the achieved thermal comfort in case 3 under the two different terminal conditions. The home temperature in Fig. 6.7a presents fewer fluctuations than all the other cases, due to multiple sources of flexibility available. However, the predictive controller with a too stringent terminal constraint at the end of the period is forced to use more power and the dwelling thermal mass to satisfy the boundary constraint. This situation causes severe performance deterioration, as documented in Table 6.4.

The study shows a high sensitivity of the problem to the boundary conditions. Figure 6.8a, b show the optimal trajectories of the predictive controller for the two different terminal conditions. These results are consistent with the result reported in Darivianakis et al. (2017).

Figures 6.9 and 6.10 illustrate the management of the battery in case 3.1 and CAPEX$_1$, respectively. For case 3.1, the battery is underutilised in both seasons, and it is oversized for the current operating conditions. Conversely, the battery obtained with the co-design process are fully and efficiently used in both seasons.

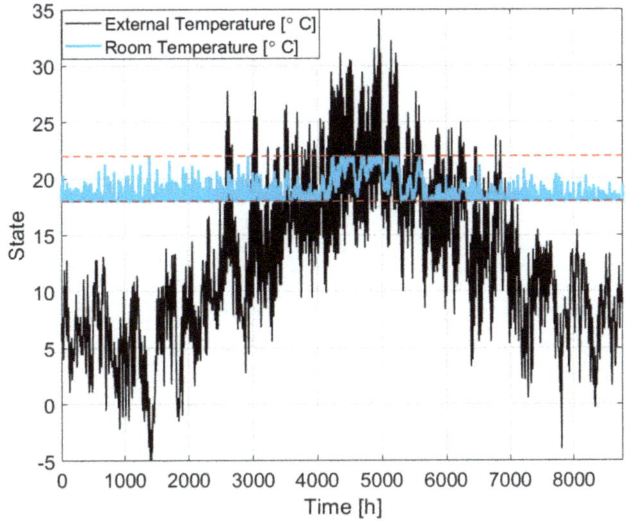

(a)

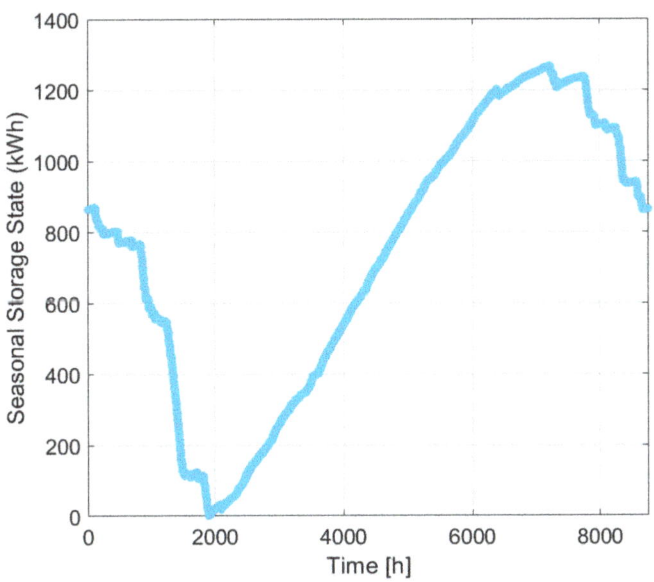

(b)

Fig. 6.4 Annual trajectories for the co-design for CAPEX_1. (**a**) Temperatures. (**b**) State of charge of seasonal storage

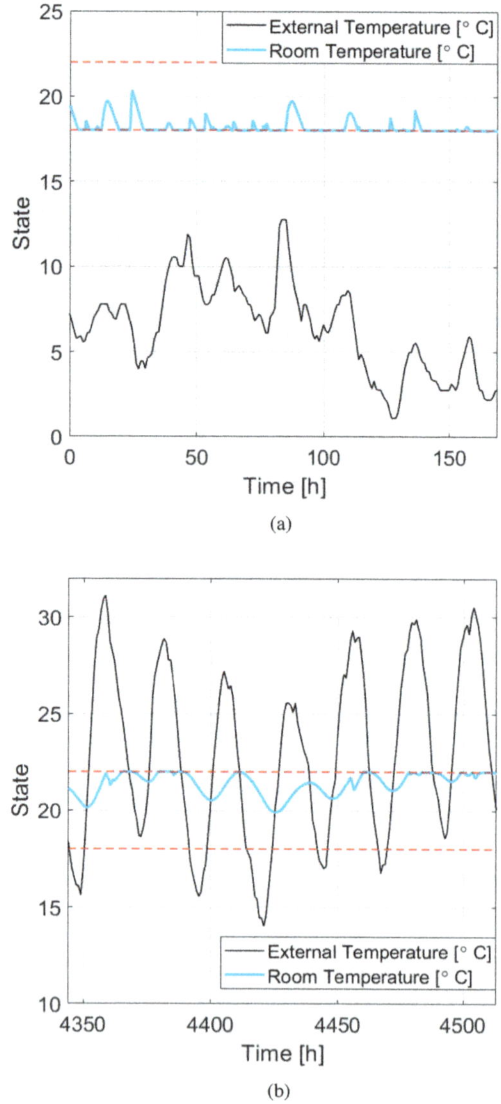

Fig. 6.5 Weekly trajectories for the co-design for $CAPEX_1$. (**a**) Temperatures in winter. (**b**) Temperatures in summer

To appreciate the discharge and charge patterns in reducing the bill, the electricity and carbon tariffs during the weeks under analysis are reported in Fig. 6.11a, b.

The electricity consumption of the heat and cooling pumps is shown in Fig. 6.12a, b, and in the selected weeks, there is not a substantial difference between $CAPEX_1$ and case 3.1. Conversely, the power profiles in Figs. 6.13 and 6.14 are substantially different. The oversized battery in case 3.1 trades larger quantities of energy with the grid causing higher peaks of bought and sold power.

Fig. 6.6 Trajectories for the co-design for $CAPEX_2$. (**a**) Annual temperatures. (**b**) Temperature in a winter week

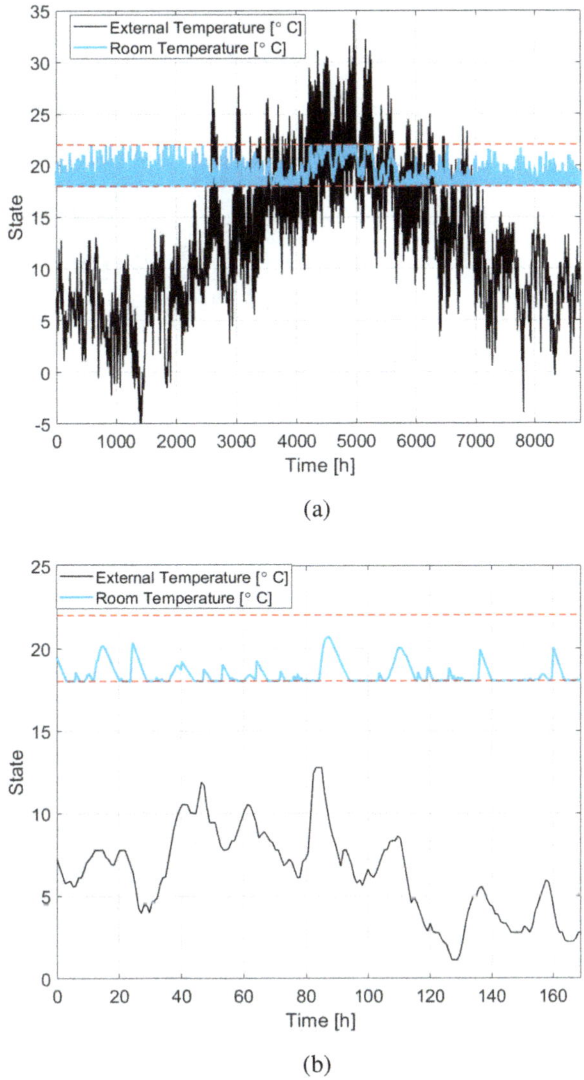

6.6 Conclusions

Achieving net-zero carbon emissions by 2050 will change the way electrical energy is generated, transmitted and used. One of the approaches to tackle the challenges related to decarbonization is to improve the design and operation of buildings. Most of the literature focuses on office and industrial buildings, whereas residential housing is often neglected.

However, since the energy challenge of this century requires balancing multiple and opposing goals, it is essential to translate them in a precise and effective

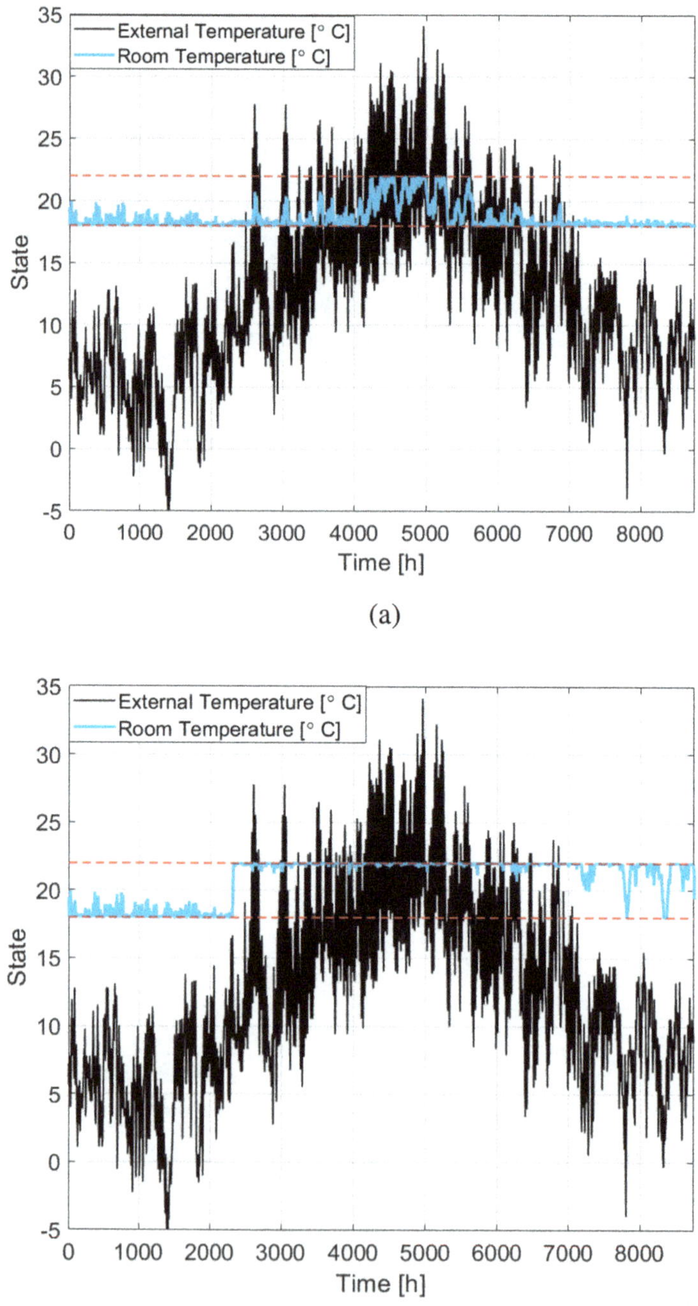

(a)

(b)

Fig. 6.7 Annual trajectories for case 3. (**a**) Temperatures in case 3.1. (**b**) Temperatures in case 3.2

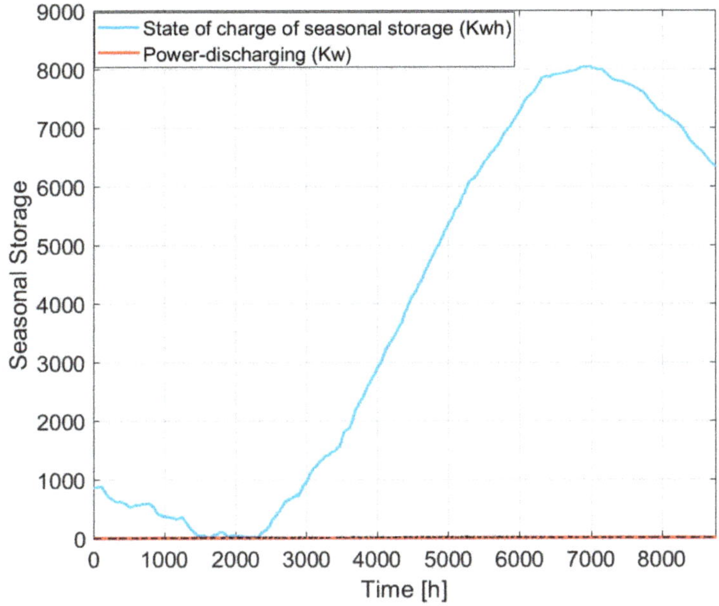

(a)

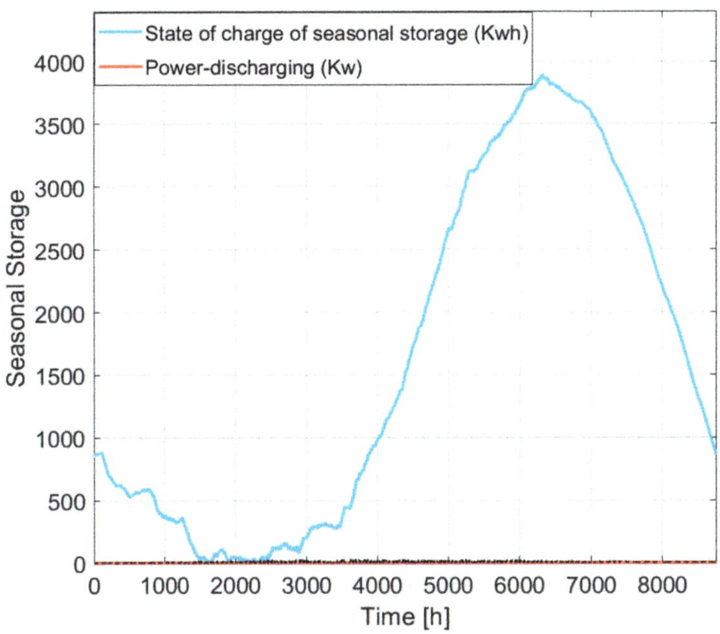

(b)

Fig. 6.8 Annual state of charge of the seasonal storage for case 3. (**a**) State in case 3.1. (**b**) State in case 3.2

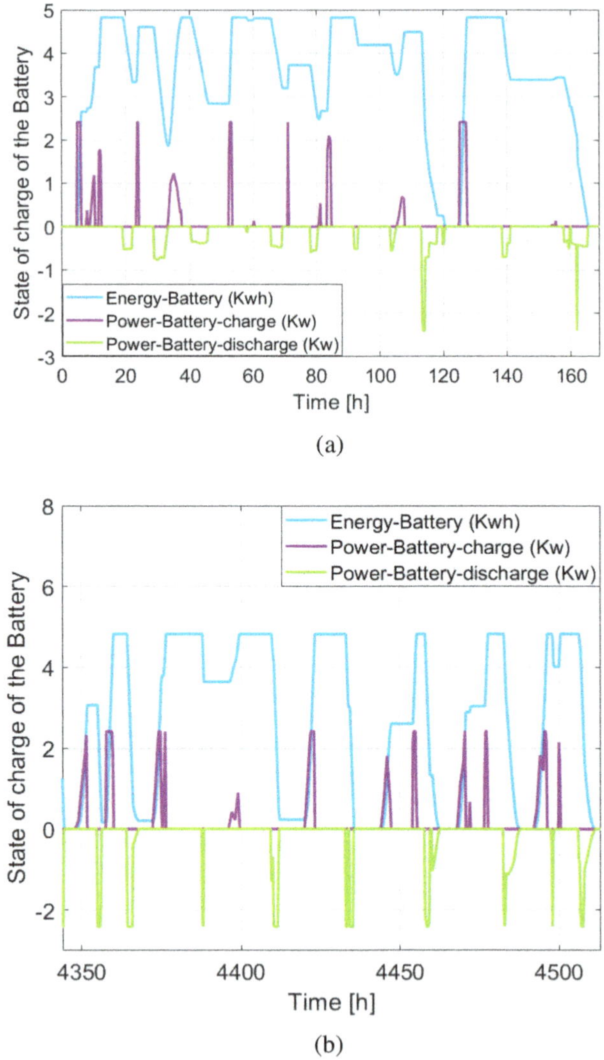

Fig. 6.9 Weekly trajectories describing the battery for $CAPEX_1$. (**a**) Battery behaviour in a winter week. (**b**) Battery behaviour in a summer week

way into quantitative objectives suitable for an optimisation problem. This issue is complicated because the interplay between different goals is not always well understood. For instance, the value of flexibility against potential cost savings, carbon emission reductions and robustness of the system is not precisely quantified.

The challenge partly arises in correctly identifying all relevant system parts, their interactions, inputs and outputs to describe the sought trade-off accurately. At the same time, once the modelling aspects have been taken care of, one needs to set an optimisation criterion that affords robust decisions in the face of uncertainty and is amenable to resolution, despite the complexity of a problem whose temporal resolution spans from years to minutes.

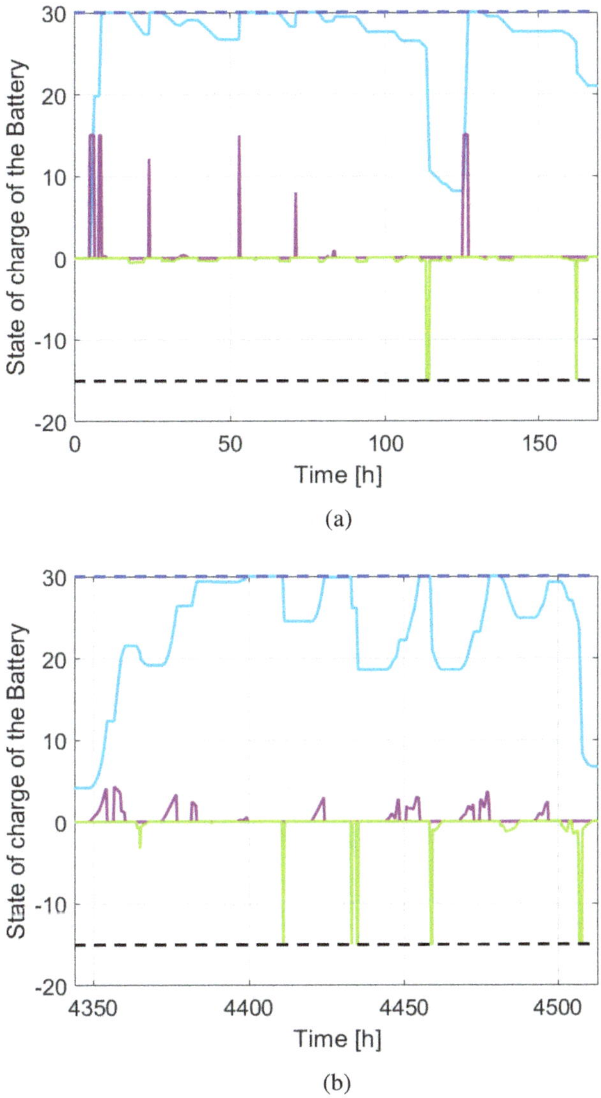

Fig. 6.10 Weekly trajectories describing the battery in case 3.1. (**a**) Battery behaviour in a winter week. (**b**) Battery behaviour in a summer week

This chapter presented an approach to simultaneous optimisation of design and operation of a residential building taking into account external weather conditions, as well as varying energy prices. In particular, the chapter presents an efficient way of applying Model Predictive Control to satisfy the requirements of the occupants of a building equipped with seasonal storage.

The chapter discussed the connections between MPC and co-design in the context of residential buildings. A case study was presented to showcase the potential of the co-design framework in enhancing flexibility and self-sufficiency and achieving cost-effectiveness of the system. The results from the chapter indicate that

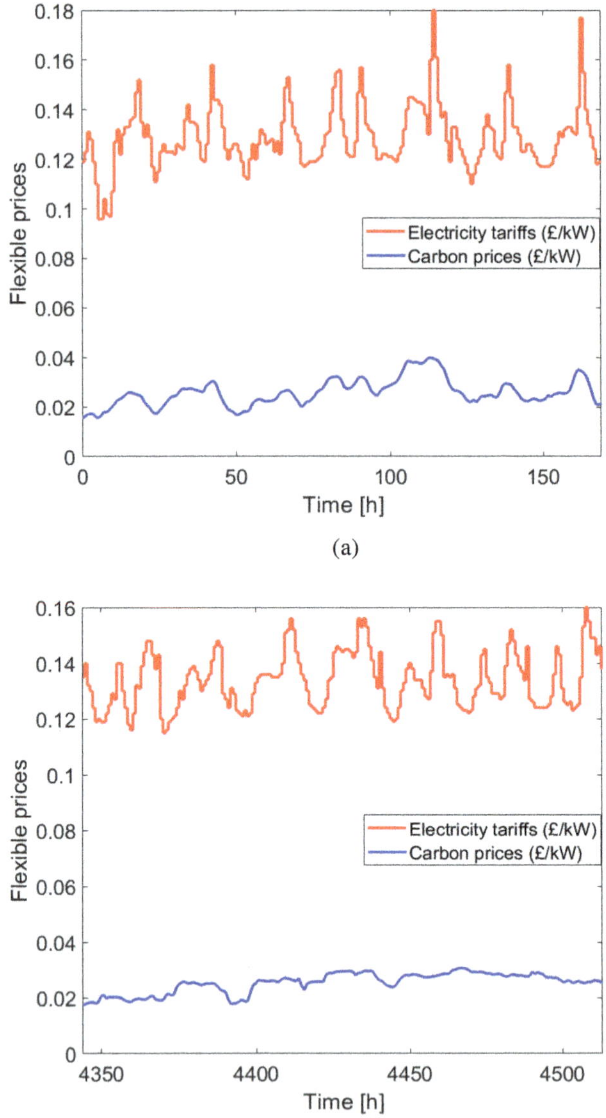

(a)

(b)

Fig. 6.11 Weekly electricity and carbon prices. (**a**) Tariffs in a winter week. (**b**) Tariffs in a summer week

Model Predictive Control enables tackling the challenges related to simultaneous optimisation of design and operation of a residential building. The presented co-design framework enables a straightforward integration of the constraints related to thermal comfort of the occupants and inclusion of cyclic constraints to take into account the seasonal storage. Batteries, PV panels and seasonal thermochemical storage, coupled with thermal solar, show benefits in increasing seasonal, intermittent renewable hosting capability. In particular, the case study based on a

Fig. 6.12 Weekly electricity consumption for thermal comfort for $CAPEX_1$. (**a**) Heating in a winter week. (**b**) Cooling in a summer week

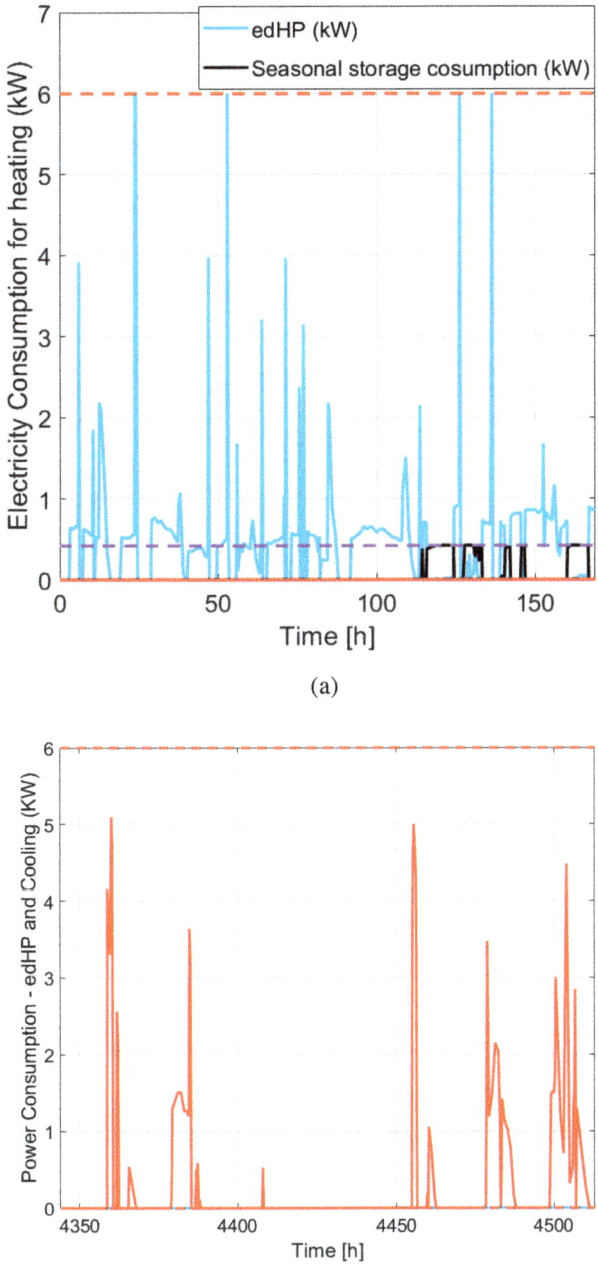

(a)

(b)

low-fidelity model reported significant bill savings and emission reductions. The achieved improvements show that the co-design framework integrated with MPC

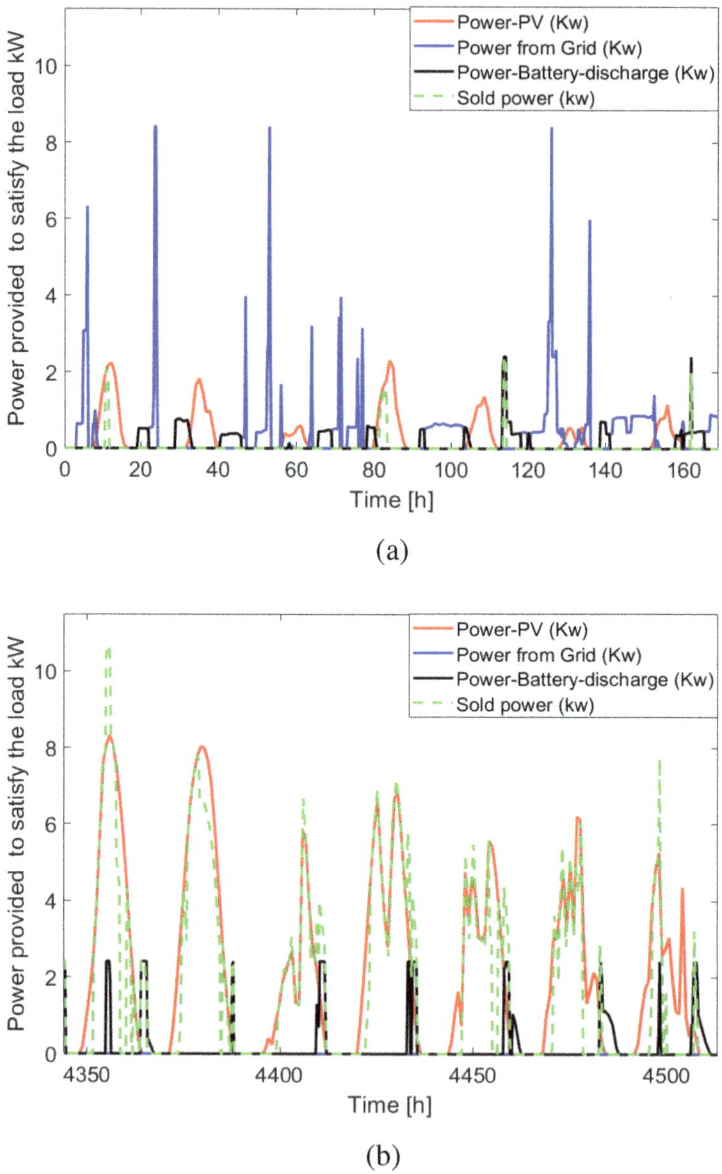

Fig. 6.13 Weekly trajectories for $CAPEX_1$. (**a**) Power profiles in a winter week. (**b**) Power profiles in a summer week

is a promising candidate for tackling the challenges related to decarbonization of residential building. However, the limitations of the case study will be further explored taking into account improved accuracy of the model, as well as the effect of the uncertainty.

Fig. 6.14 Weekly trajectories for Case 3.1. (**a**) Power in a winter week. (**b**) Power profiles in a summer week

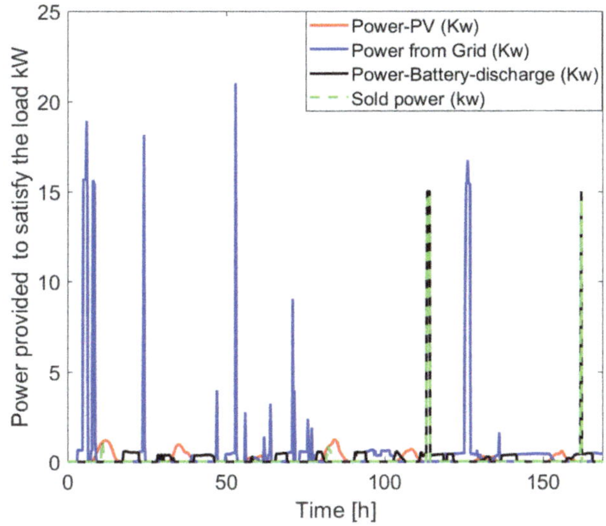

(a)

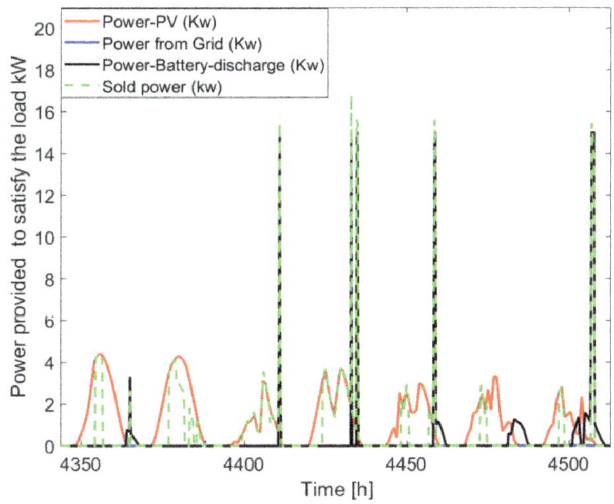

(b)

Active buildings can address the challenge of balancing demand and supply of a system where a massive penetration of renewables will induce large variability in generation. In the presented case study, we used historical price-data; however, it is estimated that future prices' volatility will increase three times and, consequently, the potential benefits of this approach will increase tremendously. Active buildings can address the challenge of balancing demand and supply of a system where a massive penetration of renewables will induce large variability in generation.

The proposed methodology is extendible to include the co-optimisation of the energy prices of multiple ancillary services. In particular, residential buildings can participate in the system balancing services and frequency regulation for large outages in real time, in the security of supply and network congestion management. Their potential sits in the provision of flexibility for reserves spanning time scales of minutes and hours down to frequency services deployed within seconds. Moreover, active buildings have a strong potential to provide security against extreme events through the smart management of electric heating and vehicles.

Overall, the flexibility of MPC and optimal design of residential buildings indicate that the framework presented in this chapter is a good candidate for future work, such as data-driven control and robust optimisation to handle uncertainty inherent in real-life applications.

References

Atam, E., Verhelst, C., Helsen, L. (2013). Borehole dynamics parameterization for thermal performance optimization under operational constraints. In *European Geothermal Congress*.

Baniasadi, A., Habibi, D., Al-Saedi, W., Masoum, M. A., Dasa, C. K., & Mousavia, N. (2020). Optimal sizing design and operation of electrical and thermal energy storage systems in smart buildings. *Journal of Energy Storage, 28*, 101186.

Beck, T., Kondziella, H., Huard, G., & Bruckner, T. (2016). Assessing the influence of the temporal resolution of electrical load and PV generation profiles on self-consumption and sizing of PV-battery systems. *Applied Energy, 173*, 331–342.

Beck, T., Kondziella, H., Huard, G., & Bruckner, T. (2017). Optimal operation, configuration and sizing of generation and storage technologies for residential heat pump systems in the spotlight of self-consumption of photovoltaic electricity. *Applied Energy, 188*, 604–619.

BSI. (1999). PD CR 1752:1999. Ventilation for buildings. Design criteria for the indoor environment. Technical report, British Standard Institution.

BSI. (2020). BS ISO 18523-2:2018. Energy performance of buildings. Schedule and condition of building, zone and space usage for energy calculation. Residential buildings. Technical report, British Standard Institution.

Büskens, C., & Wassel, D. (2013). The ESA NLP solver WORHP. In G. Fasano & J. Pintér (Eds.), *Modeling and optimization in space engineering*. New York: Springer.

CIBSE. (2015). CIBSE Top Tips Information Sheet 8: Temperature in Indoor Workplaces (Thermal Comfort). Technical report.

Cui, T., Chen, S., Wang, Y., Zhu, Q., Nazarian, S., & Pedram, M. (2016). Optimal co-scheduling of hvac control and battery management for energy-efficient buildings considering state-of-health degradation. In *Proc. Asia South Pac. Design Autom. Conf.* (pp. 775–780).

Cui, T., Chen, S., Wang, Y., Zhu, Q., Nazarian, S., & Pedram, M. (2017). An optimal energy co-scheduling framework for smart buildings. *Integration: The VLSI Journal, 58*, 528–537.

Darivianakis, G., Eichler, A., Smith, R. S., & Lygeros, J. (2017). A data-driven stochastic optimization approach to the seasonal storage energy management. *IEEE Control Systems Letters, 1*(2), 394–399.

Diangelakis, N. A., Burnak, B., Katz, J., & Pistikopoulos, E. N. (2017). Process design and control optimization: A simultaneous approach by multi-parametric programming. *AIChE Journal, 63*(11), 4827–4846.

Dows, R., & Gough, E. (1995). PVUSA procurement, acceptance, and rating practices for photovoltaic power plants. Tech, 1995. Pacific Gas and Electric Company, San Ramon, CA.

Drgoňa, J., Arroyo, J., Cupeiro Figueroa, I., Blum, D., Arendt, K., Kim, D., Ollé, E. P., Oravec, J., Wetter, M., Vrabie, D. L., & Helsen, L. (2020). All you need to know about model predictive control for buildings. *Annual Reviews in Control*. https://www.sciencedirect.com/science/article/pii/S1367578820300584

Fathy, H. K., Papalambros, P. Y., Ulsoy, A. G., & Hrovat, D. (2003). Nested plant/controller optimization with application to combined passive/active automotive suspensions. In *American Control Conference*.

Fischer, D., Lindberg, K. B., Madani, H., & Wittwer, C. (2016). Impact of PV and variable prices on optimal system sizing for heat pumps and thermal storage. *Energy and Buildings, 128*, 723–733.

Garcia-Sanz, M. (2019). Control co-design: An engineering game changer. *Advanced Control for Applications: Engineering and Industrial Systems, 1*, 1–10.

Hameed Shaikh, P., Bin Mohd Nor, N., Nallagownden, P., Elamvazuthi, I., & Ibrahim, T. (2014). A review on optimized control systems for building energy and comfort management of smart sustainable buildings. *Renewable and Sustainable Energy Reviews, 34*, 409–429.

Hazyuk, I., Ghiaus, C., & Penhouet, D. (2012). Optimal temperature control of intermittently heated buildings using Model Predictive Control: Part I - Building modeling. *Building and Environment, 51*, 379–387.

Herber, D. R., & Allison, J. T. (2019). Nested and simultaneous solution strategies for general combined plant and control design problems. *Journal of Mechanical Design, 141*. https://doi.org/10.1115/1.4040705

International Energy Agency and United Nations. (2019). Global status report for buildings and construction, towards a zero-emissions, efficient and resilient buildings and construction sector. https://www.unenvironment.org/resources/publication/2019-global-status-report-buildings-and-construction-sector, Accessed October 26, 2020.

JRC. (2012). Photovoltaic geographical information system. https://ec.europa.eu/jrc/en/pvgis, Accessed October 26, 2020.

Kathirgamanathan, A., De Rosa, M., Mangina, E., & Finn, D. P. (2020). Data-driven predictive control for unlocking building energy flexibility: A review. *Renewable and Sustainable Energy Reviews, 135*, 110120.

Kerrigan, E., Nie, Y., Faqir, O., Kennedy, C. H., Niederer, S. A., Solis-Lemus, J. A., Vincent, P., Williams, S. E. (2020). *Direct Transcription for Dynamic Optimization: A Tutorial with a Case Study on Dual-Patient Ventilation During the COVID-19 Pandemic*. IEEE.

Khusainov, B., Kerrigan, E. C., & Constantinide, G. A. (2016). Multi-objective co-design for model predictive control with an FPGA. In *European Control Conference* (pp. 110–115).

Khusainov, B., Kerrigan, E. C., & Constantinide, G. A. (2019). Automatic software and computing hardware codesign for predictive control. *IEEE Transactions on Control Systems Technology, 27*, 2295–2304.

Koskela, J., Rautiainen, A., & Järventausta, P. (2019). Using electrical energy storage in residential buildings–sizing of battery and photovoltaic panels based on electricity cost optimization. *Applied Energy, 239*, 1175–1189.

Kumar, R., Wenzel, M. J., Ellis, M. J., ElBsat, M. N., Drees, K. H., & Zavala, V. M. (2019). Hierarchical MPC schemes for periodic systems using stochastic programming. *Automatica, 107*, 306–316.

Lago, J., Suryanarayana, G., Sogancioglu, E., & De Schutter, B. (2020). Optimal control strategies for seasonal thermal energy storage systems with market interaction. *IEEE Transactions on Control Systems Technology*, 1–16. https://doi.org/10.1109/TCST.2020.3016077

Linssen, J., Stenzel, P., & Fleer, J. (2017). Techno-economic analysis of photovoltaic battery systems and the influence of different consumer load profiles. *Applied Energy, 185*, 2019–2025.

Maasoumy, M., Wei, T., & Zhu, Q. (2014). Co-scheduling of HVAC control, EV charging and battery usage for building energy efficiency. In *Proc. of the 2014 IEEE/ACM International Conference on Computer-Aided Design* (pp. 191–196).

Maciejowski, J. M. (2002). *Predictive control: With constraints*. Pearson Education.

Malcolm, A., Polan, J., Zhang, L., Ogunnaike, B. A., & Linninger, A. A. (2007). Integrating systems design and control using dynamic flexibility analysis. *AIChE Journal, 53*(8), 2048–2061.

Maraniello, S., & Palacios, R. (2016). Optimal vibration control and co-design of very flexible actuated structures. *Journal of Sound and Vibration, 377*, 1–21.

Martelli E., Gabrielli, P., Gazzani, M., & Mazzotti, M. (2018). Optimal design of multi-energy systems with seasonal storage. *Applied Energy, 129*, 408–424.

Moura, S. J., Callaway, D. S., Fathy, H. K., & Stein, J. L. (2010). Tradeoffs between battery energy capacity and stochastic optimal power management in plug-in hybrid electric vehicles. *Journal of Power Sources, 195*(9), 2979–2988.

Nie, Y., Faqir, O., & Kerrigan, E. C. (2018). ICLOCS2: Solve your optimal control problems with less pain. In *Proc. 6th IFAC Conference on Nonlinear Model Predictive Control.*

Nie, Y., Faqir, O., & Kerrigan, E. C. (2020). ICLOCS2 (version 2.5). http://www.ee.ic.ac.uk/ICLOCS/, Accessed October 26, 2020.

Octopus Energy. (2020). Frequently asked questions about Octopus Tracker - Octopus Energy. https://octopus.energy/tracker-faqs/, Accessed October 26, 2020.

Pecci, F., Abraham, E., & Stoianov, I. (2017). Outer approximation methods for the solution of co-design optimisation problems in water distribution networks. *IFAC-PapersOnLine, 50*(1), 5373–5379. 20th IFAC World Congress.

Pepe, D., Bianchini, G., & Vicino, A. (2018). Estimating PV forecasting models from power data. In *IEEE International Energy Conference, ENERGYCON* (pp. 1–6).

Pillot, C. (2015). Battery market development for consumer electronics, automotive, and industrial: Materials requirements and trends, June 2015. Qinghai EV Rally.

Rao, S. S. (1988). Combined structural and control optimization of flexible structures. *Engineering Optimization, 13*, 1–16

Rawlings, J. B., Mayne, D. Q., & Diehl, M. (2017). *Model predictive control: Theory, computation, and design.* Madison, WI: Nob Hill Publishing.

Schütte, M., Hota, A. R., Eichler, A., & Lygeros, J. (2019). Dynamic mechanism design for human-in-the-loop control of building energy consumption. In *Proceedings of the American Control Conference* (pp. 247–252).

Sensburg, O., Schmidinger, G., & Fuellhas, K. (1989). Integrated design of structures. *Journal of Aircraft, 26*, 260–270.

Solar, J. A. (2020). Product datasheets. https://www.jasolar.com/, Accessed October 26, 2020

Suardi, A., Longo, S., Kerrigan, E. C., & Constantinide, G. A. (2013). Energy-aware MPC co-design for DC-DC converters. In *European Control Conference* (pp. 3608–3613).

Sun, C., Sun, F., & Moura, S. J. (2016). Nonlinear predictive energy management of residential buildings with photovoltaics & batteries. *Journal of Power Sources, 325*, 723–731.

Thieblemont, H., Haghighat, F., Ooka, R., & Moreau, A. (2017). Predictive control strategies based on weather forecast in buildings with energy storage systems: A review of the state-of-the art. *Energy and Buildings, 153*, 485–500.

von Appen, J. (2018). Incentive design, sizing and grid integration of residential pv systems with heat pumps and battery storage systems. In *IEEE 15th International Conference on the European Energy Market.*

Wächter, A., & Biegler, L. (2006). On the implementation of an interior-point filter line-search algorithm for large-scale nonlinear programming. *Mathematical Programming, 106*, 25–57.

Wu, X., Hu, X., Yin, X., Zhang, C., & Qian, S. (2017). Optimal battery sizing of smart home via convex programming. *Energy, 1*, 444–453.

Xu, Y., Yan, C., Liu, H., Wang, J., Yang, Z., & Jiang, Y. (2020). Smart energy systems: A critical review on design and operation optimization. *Sustainable Cities and Society*, 102369. https://doi.org/10.1016/j.scs.2020.102369

Yan, B., Luh, P. B., Warner, G., & Zhang, P. (2017). Operation and design optimization of microgrids with renewables. *IEEE Transactions on Automation Science and Engineering, 14*(2), 573–585.

Chapter 7
Control and Management of Active Buildings

Ben Wooding, Vahid Vahidinasab, Milad Kazemi, and Sadegh Soudjani

7.1 Introduction to Control of Active Building Energy Systems

7.1.1 Control Approaches for Enabling Energy Services Provision from Buildings

The power network is becoming increasingly intermittent as the contribution from renewable energy generation rises. To maintain stability and functionality of the power network, storage of renewable energies and demand-side control techniques are required. *Smart grids* provide the communication infrastructure to accomplish this goal. Smart grid control originated from the idea that the demand-side of the power grid can shift or shed load to reduce the strain on the network, while also maintaining consumer satisfaction and other specialist requirements (Roche et al., 2013). The main benefit of a *smart city*, is to help its citizens by making city-related decisions. A smart grid differs from a smart city since its communication network revolves around optimising the power network, while a smart city also considers other city-level features as well as energy provision. Both smart grid and smart city benefit from timely and relevant information transfer.

B. Wooding · M. Kazemi · S. Soudjani (✉)
School of Computing, Newcastle University, Newcastle Upon Tyne, UK
e-mail: b.wooding1@newcastle.ac.uk; milad.kazemi2@newcastle.ac.uk; sadegh.soudjani@newcastle.ac.uk

V. Vahidinasab
Department of Engineering, School of Science and Technology, Nottingham Trent University, Nottingham, UK
e-mail: vahid.vahidinasab@ntu.ac.uk

V. Vahidinasab, B. Mohammadi-Ivatloo (eds.), *Active Building Energy Systems*, Green Energy and Technology, https://doi.org/10.1007/978-3-030-79742-3_7

To give a supporting example, consider the energy management system of a country. The power usage of each house in a city is measured regularly, and an aggregate model can be created to show the energy demands of different cities. The flow of power to those cities is also calculated, and expected needs are considered. From this data, an energy model can be created, and the power system can be controlled so the needs of each city are met. The population of the city is usually proportional to the city's energy usage. Services such as hospitals, schools, factories and universities, are also more frequent for higher population numbers. City centres and industrial zones will use more resources than residential areas, so a control scheme needs to consider all of these aspects when being created. One control could be that higher use prices should be charged to areas that use more energy, with the intent of bringing their usage down (Rathore et al., 2016). We will look at these energy management concepts in greater depth in the following sections.

7.1.2 Pros and Cons of Control Approaches

Smart cities and smart grids are valuable as they allow society to make intelligent and efficient decisions, with the support of regularly updating information. Control is the central principle for a smart grid and is essential for safe and efficient operation. It is a broad term that covers various processes and system objectives.

At a district or city level, energy management requires incorporating the growth of the network, more energy generating resources and storage options, and increased decentralisation caused by devices becoming increasingly interconnected, e.g. over the internet. On top of this, the grid demand is flexible as devices can connect and disconnect at any moment, and there is also flexibility in the energy prices. All of these lead to additional control complexities (Reynolds et al., 2019).

One difficulty for control is the non-linearity of the system being controlled. Non-linear system optimisations require complex calculations and are time consuming. The required computations most often rely on models of the control system, but building these models with significant accuracy is another difficulty (Miao et al., 2015). Available techniques to tackle these challenges include system identification (Ljung, 2010), linearisation around a fixed point (Ławryńczuk, 2017), and model-free data-driven control (Zhang et al., 2017). But this is still a significant challenge to overcome for most control systems.

Another potential challenge associated with smart grid control is security. Smart grid control revolves around two-way power and information flow of the communication networks. Should an attacker gain the correct privileges, they may be able to deliberately jeopardise grid stability with induced load surges (Arnaboldi et al., 2020), or other methods. There are several reasons this attack is unlikely, simply the more distributed and diverse the network is the more security protocols and devices that would need to be exploited. But the negative use of control schemes for destabilisation should be considered due to the potential large-scale damages if such incidents happen.

Smart infrastructures, such as smart buildings in a smart city, are increasing in prevalence. Smart infrastructure is self-monitoring, self-governing and able to communicate with other aspects of the smart grid. From this, three research topics have emerged surrounding their control. Firstly, consumers have been empowered to interact with the new resources available to them and systems with multiple decision-makers have emerged. Secondly, incentives are created to enable flexible consumption, such as cheaper prices when consuming electricity at night. This is known as *transactive control* (Hu et al., 2017). Finally, resilient control is a term given to protecting the system from large system failures, and does so by leveraging the communication between the smart devices. These areas each support the higher level smart grid control, but also the end-users who receive incentives and negotiating powers in the energy market (Annaswamy et al., 2016).

Moreover, control of the smart grid enables the increased integration of renewable energies. Renewable energy generation is uncertain and intermittent, but if properly controlled can be a big step forward to the global aim of net-zero carbon. Ultimately, all the positives and negatives presented in this section hinge on the quality of the control schemes. High quality control, from accurate system models and security defences, has no negatives of note. However, poor quality control from inaccurate models and flawed security, would be more harmful than helpful. Deciding which category a system falls into is a different challenge entirely and requires appropriate research to quantitatively characterise *quality* of a given control scheme applied to a system.

7.2 Coordination Structures for Management and Control of the Energy Systems

In this section, we introduce the coordinating structures available in smart grids. There are two key characteristics of such coordination frameworks, these are the *system structure* and the *energy resource type*. System structure considers the configuration of the power units included within the system and their interconnections. The energy resource type may be renewable or non-renewable energy, and this has influence over which coordination framework would optimise the system performance (Vezzoli et al., 2018).

7.2.1 Coordinated Structures from the System Perspective

The possibility of large-scale energy transfer became a reality with the rise of the industrial revolution. The process involved extracting natural resources, transporting them and then converting the resource to energy via large turbine-generators. Even today, all three steps of the process are expensive and involve operations to a scale that only large corporations or bodies of government can handle. Delivering the

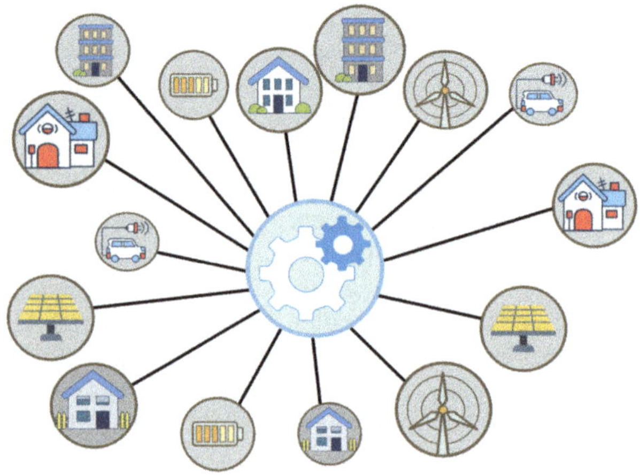

Fig. 7.1 Centralised energy systems scheme

energy to users was done through a *centralised framework*: a central large-scale generation unit sending the energy to global users far from the original generator, see Fig. 7.1.

As technology advances, generation units have become more readily available and at a reasonable price. This has created an opportunity for smaller companies and individuals to share in the market. A *decentralised framework* is shown in Fig. 7.2, where small generation units provide energy to local users and communication channels are formed between generators to share excess energy or store it.

In the decentralised and centralised approaches, the local users only consume energy from the power network. However, it is becoming more common for those users to also take part in power generation. For example, a user may decide to install a *photovoltaic (PV) panel* at their house to generate some electricity and reduce energy bill costs. This contributed to the development of the *distributed framework*. In this framework, local users can become *prosumers*, producing and consuming energy, simultaneously. A prosumer could become energy self-sufficient, but by forming a distributed network with other prosumers, any excess energy can be shared. Distributed networks can then connect to share that energy wider, see Fig. 7.3. As there is added complexity in the distributed framework, it is also more complex when designing control schemes compared with centralised controls (Zonetti et al., 2019).

The three schemes discussed above and shown in Figs. 7.1, 7.2, and 7.3 could reflect possible energy flow between component of the network. They could also reflect possible communications between these components. Interpreted differently, a lack of link between two components means lack of direct energy transfer or communication between these two components. As such, all these constraints should be taken into account when designing a control scheme for the whole network.

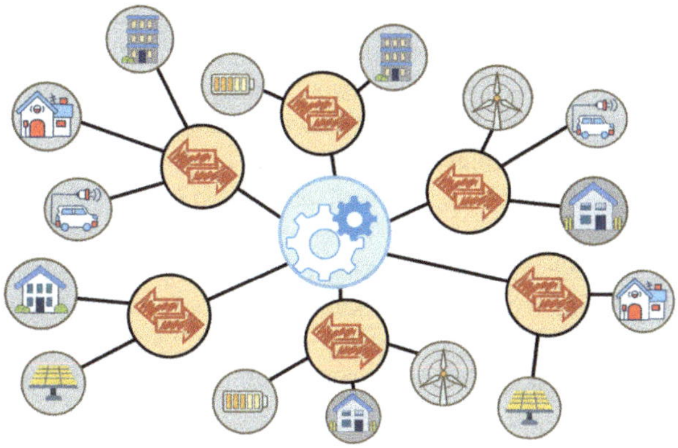

Fig. 7.2 Decentralised energy systems scheme

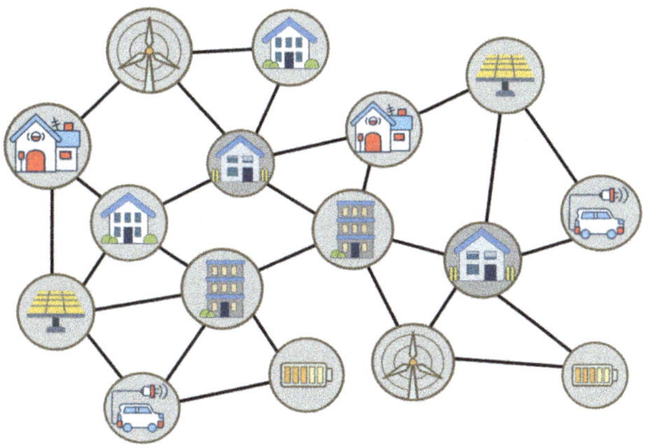

Fig. 7.3 Distributed energy systems scheme

7.2.2 Coordinated Structures from the Perspective of Energy Resources

Energy resources can be divided into two categories of renewable and non-renewable, and influence the choice of framework to be implemented. Extraction of non-renewable fossil fuels such as oil as well as the required infrastructure for their transportation are expensive. Therefore, a large-scale centralised approach suits non-renewable generation. In contrast, a small wind turbine or photovoltaic panel does not provide much energy on its own, but the total energy can scale to large quantities of generation when these are connected within a distributed framework in large numbers. Decentralised networks, being a middle ground between centralised and distributed networks, are therefore suited to smaller generators or larger renewable generation (e.g. a solar farm).

Managing resources used for energy generation presents its own challenges. Sustainability of a particular energy resource will depend on location, legal policies and economics of a region. Wood, for example, is used in biomass, which may become sustainable with well-planned schemes to replant uprooted trees. Uprooting trees without replanting them will lead to the complete depletion of that resource.

The use of fossil fuel energy in centralised systems is unsustainable in the long term as eventually the planet will be depleted of those resources. In contrast, renewable energy relies on sustainable energy sources (sunbeams, winds, geothermal stones, tides, etc). As well as sustainability, countries are beginning to manage their resources with respect to carbon emissions which are significantly lower in renewable energy generation. Overall, a shift towards distributed frameworks is expected, with non-renewable resources eventually going to be depleted and net-zero carbon energy goals forming. In the UK, legislation requires 100% carbon emission reduction relative to 1990 levels by 2050 (Stark et al., 2019).

7.2.3 Coordinated Structures from the Lens of Security

When considering security features of the frameworks, it is worth noting that should a power fault occur in a centralised or decentralised network that all or some of the network may be negatively impacted by the generation loss. A real example is from 9th August 2019. Initially triggered by a thunder strike, generation unit losses affected one million customers, health care buildings, transport and water facilities up to two days after the initial event (Energy Emergencies Executive Committee , 2019). The distributed framework is more forgiving when experiencing a fault. Each unit in the network is connected to multiple other units. A power unit can fail and rather than affecting the rest of the network it can be disconnected so the failure does not propagate through the network.

7.3 Proper Control Methods and Techniques for Active Buildings

In the previous sections, we have discussed the purpose of control and systems that might need controlling. In this section, the control methods and techniques that are applicable for active buildings are introduced.

7.3.1 Conventional Control and Monitoring Methods

7.3.1.1 Supervisory Control and Data Acquisition

Supervisory control and data acquisition (SCADA) is a control system architecture used in industry for monitoring and control of power systems, smart grids and also power generation and transmission. It is also used in building control and for

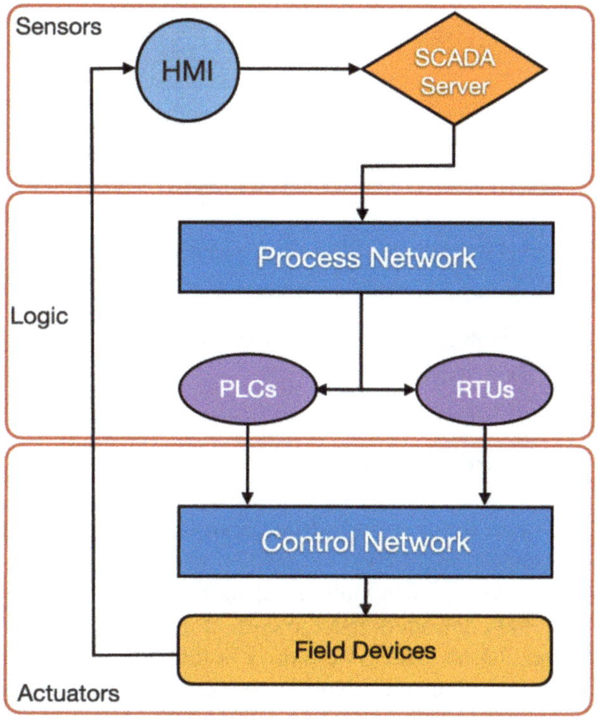

Fig. 7.4 Control system architecture of SCADA

other public infrastructures such as traffic lights and water management systems. SCADA uses wired and wireless communication across four network layers. At the highest level, an operator can communicate with the process layer via the *Human-machine Interface* (HMI). The process layer forwards this onto logic devices; the *Remote Terminal Units* (RTUs) and *programmable logic controllers* (PLCs) which can use the given information for aggregate control of field devices. At the lowest level *field devices* control and monitor the physical processes being observed by the system. These are sensors, pumps and other low-level pieces of equipment, that the observer may use to control the system as a whole. These devices provide feedback to SCADA via the HMI which helps check if the actual behaviour matches the desired behaviour (Figueiredo and Sá da Costa, 2012), see Fig. 7.4.

As automation within industry increases and costs of operation reduces, the use of SCADA systems is expected to keep rising. However, the rise of the *Internet of Things* (IoT) has also impacted SCADA systems, and a transition from onsite and standalone systems to internet connected and remotely accessible systems is occurring. SCADA systems were never designed for network connectivity or network security. The focus was on reliability and device protection by isolating devices. SCADA is a complex system and with many inter-dependencies, so as devices go online, SCADA may become vulnerable. Another security concern, is that due to high installation costs, they remain operational for 8–15 years (Nazir et

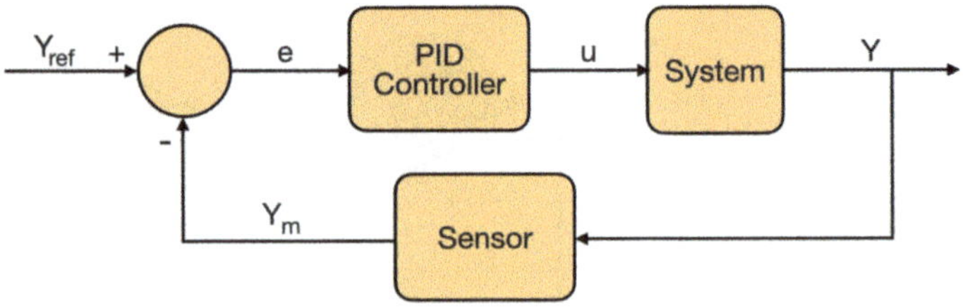

Fig. 7.5 PID controller—the system output is measured and checked against a reference value

al., 2017). Relying on possible outdated or legacy systems could leave entry points to cyber-attacks.

7.3.1.2 PID Control

The most common industrial controller choice is the *PID controller*, an acronym for proportional (P), integral (I), derivative (D) control (Franklin et al., 2002). This is due to the simplicity of its operation and tuning, as well as its widespread use. PID controllers use the current and previous error measurements of a system for regulation. The system error is the difference between a reference value of the system and the equivalent measured value. Tuning of the system can be completed by adjusting the constant values, P, I and D, of the controller, which affect each of proportional, integral and derivative areas of the control equation respectively, as in Eq. 7.1.

In Fig. 7.5, an example plant, is shown with a PID controller. The system error, e, is fed into the PID controller. The PID controller uses e, the time step, t, and constants P, I, and D, to choose a new system input u, as shown in Eq. 7.1. The new input is passed to the system which produces the new output Y. The values Y_{ref} and Y_m represent the expected output of the system and the measured output respectively

$$u(t+1) = Pe(t) + I\sum_{t=1}^{t} e(i) + D(e(t) - e(t-1)) \tag{7.1}$$

PID control is valuable for *Single-Input Single-Output (SISO)* systems and uncoupled two input two output systems. For *Multi-Input Multi-Output (MIMO)*, other techniques such as Model Predictive Control are more valuable (Khaled & Pattel, 2018; Gasparyan, 2008).

7.3.2 Model Predictive Control

Model Predictive Control (MPC), also called *Receding Horizon Control (RHC)* (Boyd et al., 2004), is a control technique with some freedom involved in its implementation and is one of the fastest growing control techniques. MPC has a broad range of applications such as clinical anaesthesia, the cement industry, and robotics (Camacho & Bordons, 2007). MPC algorithms use a model to represent a system, and attempt to minimise a cost function. Although, the implementation can be different, there are three important consistent ideas involved in MPC:

1. Use of a process model to create a prediction horizon;
2. Calculate a set of future control signals;
3. Use a receding strategy—the first value in the control sequence is applied to the process as it moves forward in time. At each time step a new prediction horizon and set of future control signals are calculated.

MPC uses a model to predict the future plant outputs. This is also known as the prediction horizon, $y(t + k \mid t)$ for $k = 1, \ldots, N$, where t is the current time step, and y is a set of future outputs for a time horizon, N. This is calculated using past control signals, u, past outputs, y, up until current time step, t, as well as predicted future control signals (Fig. 7.6).

The set of future control signals is represented by $u(t + k \mid t)$ for $k = 0, \ldots, N$. From this, a control sequence which minimises the objective function at each time step, $t + k$, is also known. The objective function consists of a cost function and system constraints. It attempts to keep the process near to the reference value, $w(t + k)$. The optimisation uses the error between the reference values and the measured values.

Calculating the optimal control signal considers the system constraints and the cost function. The overall cost function consists of the sequence of manipulated

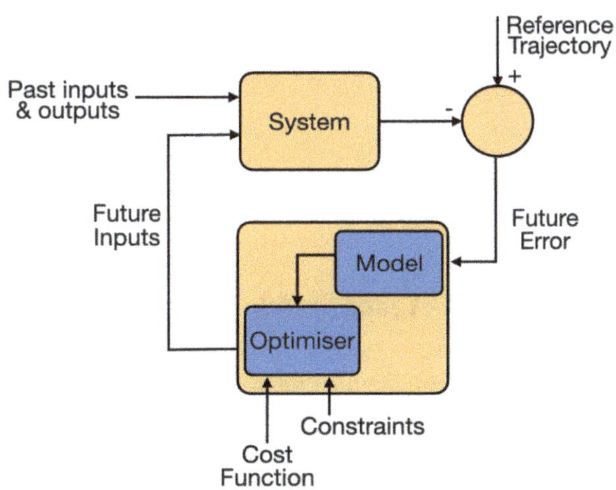

Fig. 7.6 Model predictive control—a model based control approach which predicts the next input that should be used. The model then fully updates for the next time step

values (Z_k), a cost function for tracking error (J_y), a cost function for manipulated variable tracking (J_u), a cost function for change in manipulated variables ($J_{\delta u}$) and a cost function for constraint violations (J_ϵ) (see Eq. 7.2).

$$J(Z_k) = J_y(Z_k) + J_u(Z_k) + J_{\delta u}(Z_k) + J_\epsilon(Z_k) \tag{7.2}$$

The real-time solver computes the future sequence of manipulated variables, but only one control signal is exported for the next time step. At time t the control signal $u(t \mid t)$ is applied to the process. The remaining values of the sequence are discarded. For the next time step $t + 1$, a new prediction horizon and sequence of control signals are calculated. Therefore, at time $t + 1$ the control signal $u(t + 1 \mid t + 1)$ is used. This is more optimal than using the value $u(t + 1 \mid t)$ computed in the first sequence since it considers the newly measured output value $y(t+1)$ which was unknown to the sequence at time step t. This is known as the receding strategy. Therefore as t increases the accuracy of MPC should improve as it has its predicted values corrected by real measurements.

MPC is advantageous as it is relatively simple to understand for those without control system experience and has applications in a wide variety of systems including unstable systems or those with complex dynamics. MPC deals easily with multi-variable processes, and can compensate for dead time. Using feed-forward control it is able to compensate for uncertainties in the system.

There are two major drawbacks. The first is that the required computation in the MPC algorithm can be very heavy. Implicit MPC is run online in the microprocessor, and there is no benefit in computing the best control signal after the time step it was needed has already passed. It is possible to compute the MPC algorithm offline, and store these values in lookup tables. For explicit MPC this is done and then the tables are imported to the microprocessor (Khaled & Pattel, 2018). The second drawback, is the reliance on an accurate process model. MPC is computed a priori, with prior knowledge of the process. If the model used to compute the MPC algorithm has the wrong system dynamics, then the difference in the model and the real system will create discrepancies between the predicted control and the real control of the process. Although MPC is designed to adapt to errors in the model, the more significant the errors are the more difficult it will be for the MPC to correct them.

For those wishing to investigate MPC in more depth in the context of buildings, the review article Drgoňa et al. (2020) is helpful.

7.3.3 Multi-Agent System Approaches

An energy system like the one in an active building could be alternatively described as a group of components which interact to produce a reliable service at the lowest cost to consumers. This description involves decomposing an energy system into smaller structures which interact with one another. Multi-agent system (MAS)

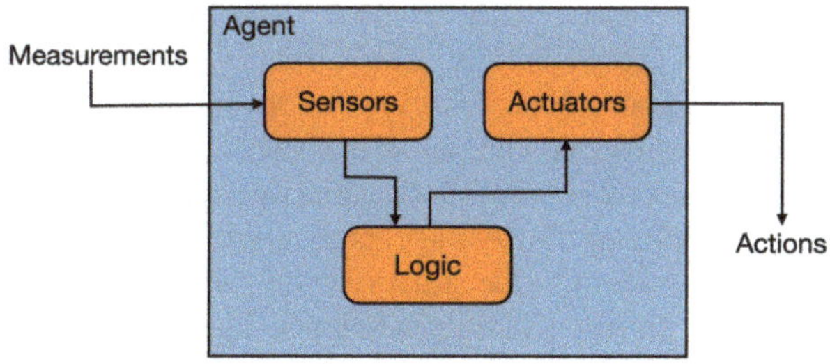

Fig. 7.7 A model of an agent which uses measurements of the environment to impact the agent's choice of actions

approaches support this framework for both modelling and control, and in Roche et al. (2013) a definition for MAS is given:

> A multi-agent system is a system composed of a collection of autonomous and interacting entities called agents, evolving in an environment where they can automatically perceive and act to satisfy their needs and objectives.

Agents receive information from the environment precepts through sensors. Such as a building with a PV panel on it's roof might have a sensor which detects the amount of solar irradiance currently being absorbed. This information is then passed through the agents logic which decides what action should be taken, in our example this might be whether to sell that energy to the grid if it is in excess of the buildings own current energy requirements. This choice is passed to the actuators which complete the task so the decided action occurs within the environment, potentially changing the global state of the environment. An example of this can be seen in Fig. 7.7. As the agents are separated from the environment they are in, MAS is a *distributed* approach to the energy management. MAS can also be described as *proactive*, the agents follow their own objectives or ones that cover the whole system, and *social*, the agents interact via cooperation or competition dependent on their objectives. For the MAS approach to be effective, an understanding of agent-agent and agent-environment interactions is paramount.

There are three main types of agents within the MAS framework. The first is the *reactive agent* which is the most basic and has minimal responses to environment precepts. They provide fast responses when needed and also have useful results when modelled as a large group of simple agents, rather than a single large agent. An example of a reactive agent is an EV which connects or disconnects to the grid without using any smart charging scheme. Modelling charging behaviour of a large group of EVs provides interesting results.

An *intelligent agent* has more functionality than a reactive agent since it is able to use its resources to achieve some objective. An example of an intelligent agent is a smart building with an energy management system (EMS), the agent uses its resources to satisfy the control objectives of its EMS.

The final agent is a *learning agent*, this agent gains information by analysing the outcome of their actions. The learning agent will have a good grasp of the environment it is in, which is used for decision making. The aforementioned smart building example, could become a learning agent by predicting the behaviour of those who use the building, and utilising this for its resource management.

MAS approach to control is bottom-up, due to the local knowledge of its agents and the flexible interactions which they have. The agents only know what they need to know, and this reduced data transfer across the network, making MAS scalable. It is unnecessary for two agents that will never interact to be aware of one another. The agents adapt to the situations they are in, making them flexible to faults where neighbouring agents fail or when new agents are added to the neighbourhood.

Other system approaches rely on predicting network changes and have costs associated with maintenance or redesign the network. MAS can avoid these costs. As these agents are cooperating or competing with one another autonomously the control approach required for the system is to distribute the control tasks between the agents. Decisions are made locally, with neighbouring agents grouping to form microgrids. The microgrids can then act as agents themselves to satisfy the global control objectives of the whole grid in a decentralised structure.

Difficulties of MAS revolve around the proactivity of the system. As each agent has a local objective and groups produce global objectives, it may occur that competing objectives arise. These challenges make distributed control more complex than the centralised control approach where an action would be demanded to suit the global specification. To help with this a clear communication framework between agents is needed, and definitions of the roles of the agents within the network to resolve objectives. This also benefits future hardware that will be incorporated into the networks.

7.3.4 Artificial Intelligence and Data Driven Control

Data driven control (DDC) approaches, including *machine learning* (ML) control are growing in prominence. Having been originally developed for static data, the methods and algorithms have been shown to be equally valuable when considering dynamic systems. The broad range of techniques allows ML control to optimise both linear and nonlinear systems. The most notable techniques that will be discussed in this section are: a) *reinforcement learning (RL)* and *artificial neural networks (ANN)* for designing control strategies; and b) *genetic algorithms (GA)* and *genetic programming (GP)* for finding a parameterised controller.

A major benefit of ML, compared to other control techniques already discussed in this chapter, is the possibility for model free control. Real-world control problems are especially difficult because they involve highly nonlinear dynamics, with an objective of maximising or minimising a certain property. System identification techniques may be impractical due to cost, complexity or other reasons (human based systems have ethical considerations). ML relies on sensor data only to optimise an objective function. It is a powerful tool when system models are unavailable. Some example real-world fields that ML techniques can help include epidemiology, robotics and fluid dynamics, but there are many more (Brunton & Kutz, 2017).

For the system that will be controlled, a cost function is minimised to satisfy the system objective. The ML controller will pass inputs to the physical system which reduce this cost value. To do this, a best strategy must be learned by the ML controller. This is completed offline, with the outputs of the system and the cost function passing to the ML controller. Using this data and differing ML algorithms a control strategy is formed. With more data the controller can gain more experience and provides better control functions.

7.3.4.1 Reinforcement Learning: Experience Based Control

RL is an algorithm that acquires experience over time to improve its control policy. *Markov decision processes (MDPs)* are the most commonly used framework for RL. MDPs incorporate uncertainty in their description of system dynamics and control laws. This promotes optimisation and exploration of the state space in equal measure.

RL is run by an agent who is in charge of choosing the control policy. Typically, the solution to an RL control strategy is binary, either the strategy is successful or it is not. To improve this, algorithms have formed a *value function*, these are known as Q-learning algorithms. This function denotes the value of success represented by the current state, and can be considered proportional to the likelihood of a winning strategy. For example, if a good control policy is in effect, the value function for the overall system state should be high. As RL learns from experience, the algorithm might initially choose low scoring strategies, but after many iterations will learn to choose higher scoring control strategies with more chance of ultimately being successful (Sutton & Barto, 1998).

7.3.4.2 Artificial Neural Networks: Data Driven Control

ANN were developed from the *perceptron*, a mathematical model of a synapse in the brain. Perceptrons are quite limited, only returning a binary value, but when layered on top of one another they can learn system behaviours. ANN are easy to create and implement. As more research into ANN is done, other techniques such as *recurrent neural networks* and *deep neural networks* have been developed. In essence, ANN is used to predict an output for any given input based on all previously known inputs and outputs.

Neural networks consist of three sections, see Fig. 7.8. The first section is the input layer, where the system inputs are passed into the algorithm. The inputs are then passed to the hidden layer, with N-layers of neurons. The neurons in the first layer receive all the normalised inputs from the input layer and compute an output value which is then passed to each neuron in the next layer. Those neurons complete the same process, with the inputs they receive, until all layers have computed

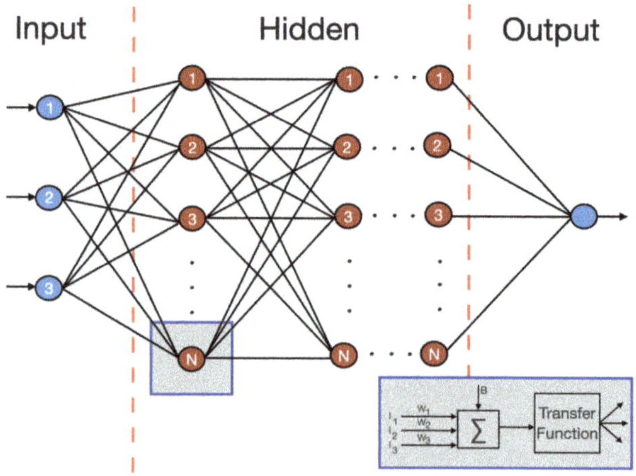

Fig. 7.8 Artificial neural network with multiple hidden layers. A model representation of Eq. 7.3 is also given with inputs, I, bias, B, and weights, w

some output. The final output values are then passed to the output layer for the overall output prediction. From previous prediction errors, the ANN can adjust weights associated with each layer to improve future performance, this is known as *backpropogation* (Rumelhart et al., 1986).

$$y = tansig(B + \sum_{n=1}^{n}(I_n \times w_n)) \tag{7.3}$$

At an individual neuron the output formula is given in Eq. 7.3. y is the neurons output, B is a bias value, I_n is the input and w_n is the weight associated to the respective input. The summation of inputs multiplied by their weights is calculated, then a bias value is added. This result is then passed through a transfer function such as $tansig()$, $logsig()$ or another, depending on the specific algorithm in use. In practical terms, there are many useful tools already available to help with the creation of an ANN. One example is the neural network toolbox for MATLAB (Demuth, 2004).

To run the ANN algorithm, a large data set is required of measured input and output values. The data is divided into a training data set (majority of the data) and a validation data set (remaining data). The strength of ANN lies in its training algorithm, which decides the weights and the bias to apply to each neuron. The algorithm optimises the values of w and b for the data, using a mean square error or sum of errors in the process. Once the training has finished forming the model, the algorithm can predict the expected output values of a given input.

Occasionally, the ANN model focuses on smaller details present in the training set over the general trends of the input to output mapping. This is known as *overfitting* and to avoid this, validation data is used. The model is run using the inputs from the validation data, but the output values remain hidden. The model predicts outputs from the validation data that can be checked against the actual outputs from that data which had been hidden. If the there is a significant error in these predictions then it is likely that overfitting has occurred within the model, and the training should be redone.

7.3.4.3 Parameter Synthesis Using the Genetic Algorithm

When the structure of a control law is given but the parameters are unknown, ML control becomes a parameter identification. Genetic algorithms (GAs) are meta-heuristic optimisation approaches based on the laws of natural selection from evolutionary biology, and also have applications outside of ML control. With each new generation, the most successful (or fittest) individuals climb to the top of the rankings. An individual is a member of a generation, k, with a random assignment of parameter values. This combination of parameters is called its DNA, and are written as a numeric sequence. These parameters are what the algorithm tries to optimise via a set of *genetic rules*.

The fittest individuals in a generation have minimised their cost function or error scores. Once a generation has been computed, all the individuals are ordered by cost function value. A probability is assigned to each individual, relative to this cost value. A lower cost will equate to a higher probability of being selected for the next generation, $k + 1$. Optionally, the top x number of individuals can be immediately moved to $k + 1$, with probability 1. This is known as *elitism* and for the remaining individuals, any who are selected for generation $k + 1$ will undergo one of three genetic rules:

1. *Replication*—The individual is moved to generation $k + 1$ immediately with no modifications.
2. *Crossover*—Two individuals swap a portion of their DNA, then both move to generation $k + 1$ in their new modified forms.
3. *Mutation*—An individual has its DNA randomly modified before moving to generation $k + 1$.

The remaining individuals are then discarded. As the GA iterates through the generations, the fittest individuals with lowest cost function scores appear. GA stops when these individuals have converged or a stopping condition is met.

GA is useful as it does not require the iterations of a brute force algorithm, trying every possible set of parameters, and it scales to high dimension spaces better than other algorithms, such as Monte Carlo sampling. However, GA does not guarantee converging on an optimal solution. Additionally, choosing the size of the population and the number of generations affect the algorithms performance.

7.3.4.4 Genetic Programming: Control Law Identification

GP is an extension of GA, but can be treated as its own control technique. It is used to optimise the parameters of the system and even the structure of the system. GP can also find appropriate control laws. It does this by completing a GA approach for different structures as well as different parameters for those structures.

GP uses a recursive tree structure to encode the complex functions of sensor signals. By using such a generalised framework it is possible to identify the structures of nonlinear control laws for highly nonlinear systems. GP works especially well on problems where the solutions can be checked quickly, this allows GP to test a large number of individual control laws for their suitability. A successful GP approach will find not just the optimal parameters for the model but also the optimal model to use.

7.3.5 Game Theoretic Approaches

Game theory is an optimisation method with mathematical foundations. Entities, or players, attempt to achieve individual objectives and take actions to complete them. Deciding on the best actions to take, and the outcome of those actions, is the essence of game theory. A player's utility is the value of their success in the game, relative to the other players. Depending on the strategy, a players utility can increase or decrease. Game theoretic approaches usual try and manage the utility of the different players to satisfy an overall global objective. There are two main areas of game theory; noncooperative and cooperative approaches.

Under these two main umbrellas, games can also be either dynamic or static. Dynamic games consider time when playing, and thus allow for consecutive player moves. Static games on the other hand do not consider time, and so player moves are either simultaneous or alternating.

In *noncooperative game theory*, players cannot communicate with one another and so must choose actions without coordinating their choices (Başar & Olsder, 1998). The global solution to the game often takes the form of a *Nash equilibrium*. A Nash equilibrium is a state where no player can improve their utility. Essentially, this is a draw between all players, and changing the strategy will only result in worsened performance for the player who changes strategy. Finding such an equilibrium is not always simple and certainly not guaranteed (Roughgarden, 2010).

Cooperative game theory allows communications between the players. Since the players can communicate, they have the option of whether they wish to cooperate with one another or not. This leads to two main cooperative strategies. Firstly, the *Nash bargaining strategy* where the players communicate with one another to determine a contract under which they agree to cooperate. This strategy allows for competition, but with some agreed trading. Secondly, the *coalition strategy* where the players group together as one coalition. This strategy is fully cooperative, and the players unite under the same objective. Once it is clear which type of game is being played by the players, the in-game strategies can be considered (Curiel, 2013).

There are three essential parts to a players turn, no matter which game they are under. The player must consider the global game state to understand the current utility and set of actions that player has. The player must estimate their prospective utility for their actions. The player updates their strategy based on those observations and chooses an action. There is some variety regarding the algorithms which are used to complete these three steps:

- *Best response dynamics*—simplest approach, chooses the action which maximises the players utility but does not guarantee convergence to an equilibrium point.
- *Fictitious play*—considers the actions of all players before choosing an action. For zero-sum games, it will always converge to a Nash equilibrium.
- *Regret matching*—a strategy which chooses the least detrimental action, as opposed to the most beneficial action.

There are also other algorithms such as reinforcement learning and stochastic learning that can be chosen (Silver et al., 2017).

An assumption from classical game theory is that the players are rational, in reality this is not necessarily the case. Small changes from the optimal strategy via non-optimal play could have disastrous knock-on consequences. To combat this there are analytical techniques to avoid such faults from occurring. Overall, the robustness of the algorithm design and model is essential to safe operations.

Game theory approaches have many power system applications. Some examples include; cooperative energy exchange, distributed control of microgrids and smart grid communication technologies. Classical approaches of game theory for demand-side management focus on the relationship between individual buildings and the operator and the optimal strategy for buildings is to use an aggregate behaviour when negotiating with the operator (Saad et al., 2012).

7.4 Coordinated Control of Active Buildings: Case Studies

With the development of increased renewable technologies, we are seeing a transition from centralised coordination and control strategies to distributed ones. Power networks have increased complexity and decreased predictability, proportional to the increased distribution. For control systems to remain smart, the system intelligence must come from the information sharing and cooperation of the components inside the network (Roche et al., 2013). This is appropriate when we consider active buildings as they are able to take part in both the generation and demand as prosumers.

In this section, case studies of active buildings within the three different coordination frameworks are provided.

7.4.1 Centralised

For this case study, we will analyse the frequency response from the *load-frequency control* (LFC) model as well as the response from active buildings included in the network that can be used for frequency response. As this is a centralised network, we consider a single controller which is able to cut the power to buildings when necessary in order to respond to load disturbances. Figure 7.9 shows a simple model of the active building behaviour; when the frequency leaves the deadband interval the controller cuts the power to buildings where necessary to reduce their load on the system and provide frequency response services.

In this scenario, the buildings are only considered to be loads on the network. Due to privacy concerns, the controller does not have permission to regulate the smart devices within an active building unless the building is part of a scheme such as aggregation. Aggregation is not a centralised control scheme and therefore the buildings can only contribute to frequency response by being removed from the network entirely and tasked with relying on their own energy generation.

Finding an optimal network controller can reduce the likelihood of frequency disturbances causing the load shedding of active buildings. In this case study, a PID controller will be used to provide an improved frequency response on the simple controller given in Bevrani (2009). The system values used for the simulation can be seen in Table 7.1. The step load disturbance for the simulations is 0.02 pu.

Power systems are highly nonlinear and uncertain, and regulation combines fast voltage and rotor angle responses with slower frequency responses. By only considering frequency response, a simpler linear model under a load disturbance can be used for frequency control synthesis. The relationship of the generator-load dynamic can be described as:

$$\Delta P_m(t) - \Delta P_L(t) = 2H\frac{\delta}{\delta t}\Delta f(t) + D\Delta f(t) \tag{7.4}$$

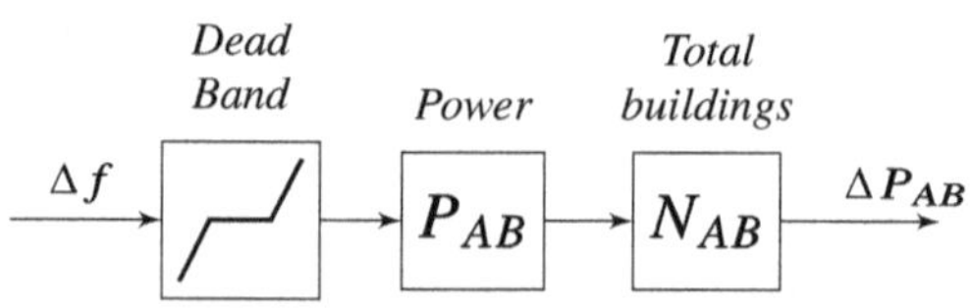

Fig. 7.9 A simple model for participation of active buildings in frequency response services

Table 7.1 Values used for simulation adapted from Bevrani (2009)

Symbol	D	$2H$	R	T_g	T_t	P_{AB}	N_{AB}	deadband
Value	0.015	0.1667	3.00	0.08	0.40	2000	25	0.0 ± 0.01
Units	pu/Hz	pu s	Hz/pu	pu s	pu s	kW	–	Hz

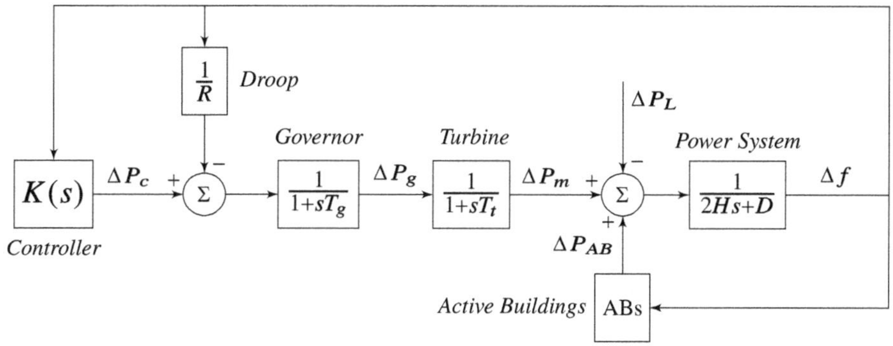

Fig. 7.10 A simple LFC model, with generator dynamics taken from Bevrani (2009) with added active buildings for centralised control

Δf is the frequency deviation, ΔP_m is the mechanical power changes, ΔP_L is the load changes, H is the inertia constant and D is the load damp coefficient, where 1% change in frequency is equivalent to D% change in load. By applying the Laplace transform on Eq. 7.4, a new equation is formed.

$$\Delta P_m(s) - \Delta P_L(s) = 2Hs\Delta f(s) + D\Delta f(s) \tag{7.5}$$

Equation 7.5 can be used in Simulink to produce a graphical block diagram of the system, Fig. 7.10. The block diagram, taken from Bevrani (2009), represents the LFC synthesis of a non-reheat steam generator unit and includes governor, generator, turbine, load and secondary control characteristics. The slow boiler dynamics, and fast generator dynamics are ignored. A speed droop characteristic, R, is included which shows the speed regulation due to governor actions. K is the system controller.

For the baseline case the controller $K(s)$ is $-0.3/s$ and for PID control the values are set to $P = -0.3$, $I = -0.3$ and $D = -0.1$. Another PID controller omitting the derivative, known as a *proportional-integral* (PI) *controller*, is also provided in Fig. 7.11. It is clear to see from the figure that the PID controller improves the frequency response of the system, having a lower minimum frequency value and also reducing the need to use the active building power as part of the control. This leads to less disconnections of buildings from the network which is beneficial to both building users and the network. At the minimum frequency measurement, the baseline controller requires the equivalent P_{AB} power of 25 ABs, while for the PID controller this is nearly halved. Even the simpler PI controller provides a large improvement on the baseline controller taken from Bevrani (2009). The PID controller is simple to implement and for SISO systems such as the basic frequency response model it is a valuable control technique.

Realistically, after a frequency disturbance of only -0.06 Hz to have a *low frequency demand disconnection* (LFDD) scheme trigger would mean the network was very unstable. From literature, in the UK it is at -0.8 Hz where the LFDD

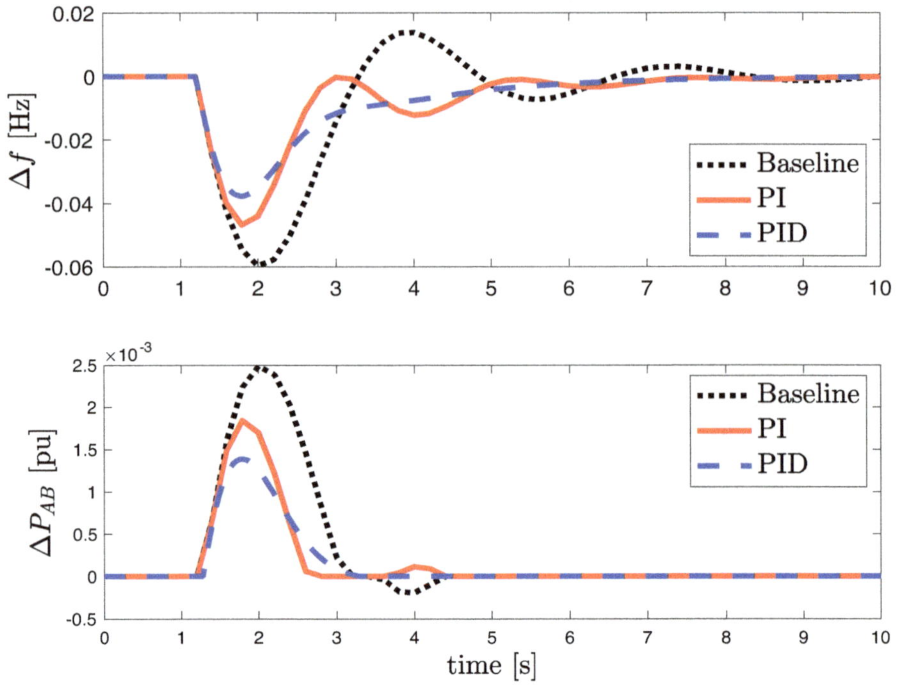

Fig. 7.11 Frequency response using PID, PI and baseline controller from Bevrani (2009) **(top)**. Active building load change **(bottom)**

schemes are imposed (National Grid, 2020). We will adjust the values of the simulation so that the deadband is 0 ± 0.8 Hz and the load disturbance is 0.3 pu. Now the active buildings will only be disconnected from the network when the frequency falls to -0.8 Hz or below. In Fig. 7.12, it can be clearly seen that for the baseline controller the LFDD scheme triggers at -0.8 Hz, while for both the PI and PID controllers the network protects the active buildings from needing to be disconnected. In this case study, it can be seen that control and management of ABs from the side of the power grid can reduce the need for LFDD schemes. Additionally, it can be implied that ABs which manage their consumption from the power grid will become less of a burden as they have reduced the load needing to be controlled.

7.4.2 Decentralised

In the second case study, we will consider a system with a decentralised structure and analyse the role of active buildings in frequency control of such a system. In the decentralised structure, the system is divided into smaller areas, which each have

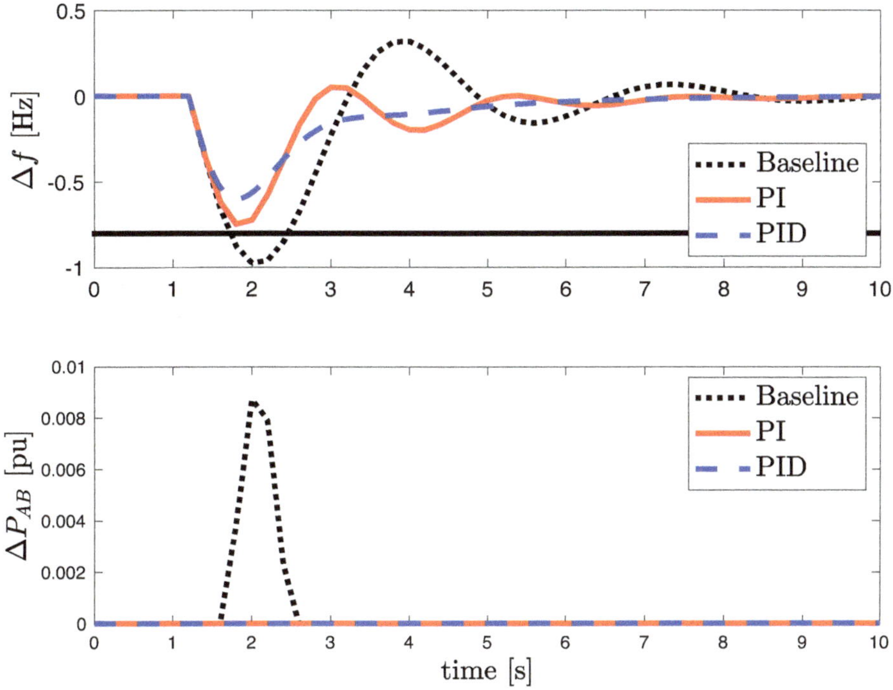

Fig. 7.12 Frequency response using PID, PI and baseline controller, the horizontal bar marks where the LFDD scheme triggers **(top)**. Active building load change **(bottom)**

their own control method of the frequency. In this way, a global control is completed by the interactions of the smaller control areas. For this case study, we will consider 3 areas which each use MPC as their control technique.

In the centralised structure, it was not possible for the active buildings to respond to the frequency events since the central controller did not have the permissions to adjust the consumption of the buildings. In decentralised control, buildings can sign up to the schemes run by aggregators. The aggregator is able to negotiate in the energy market on behalf of a significant number of buildings, thus reducing the energy costs of the buildings but also providing services to the power network. Having opted into the scheme, the buildings have given aggregators permissions to adjust their consumption values and can be controlled for demand-side response services. In this case study we consider that each area has 50 large buildings available for control with 1000 kW of controllable load.

A basic 3-area system for frequency response is provided in Bevrani (2009). Figure 7.13 shows one area indexed by i of the system. As well as maintaining the frequency of the area, a controller is needed to deal with the interchange of power between areas. This is done by adding the tie line flow deviations to the secondary control of the frequency area, ΔP_{tie_i}. The value of ΔP_{tie_i} is also included in the summation before the Power System block inside the diagram. For space, this value

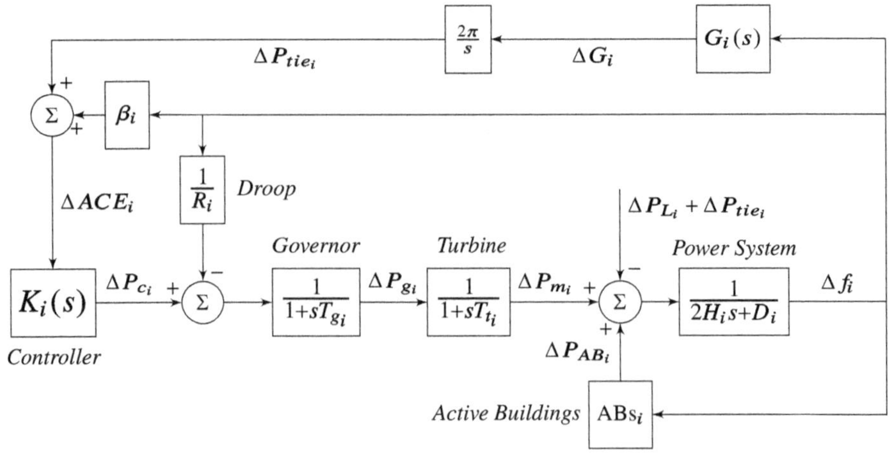

Fig. 7.13 A simple LFC model, with generator dynamics taken from Bevrani (2009) with added active buildings for centralised control. The diagram shows area i, one of the areas inside the 3-area system

Table 7.2 Values used for simulation adapted from Bevrani (2009)

Area	D_i	$2H_i$	R_i	T_{g_i}	T_{t_i}	β_i	ΔP_{L_i}
$i=1$	0.015	0.1667	3.00	0.08	0.40	0.3483	0.3
$i=2$	0.016	0.2017	2.73	0.06	0.44	0.3827	0.0
$i=3$	0.015	0.1247	2.82	0.07	0.3	0.3692	0.3

is combined with P_{L_i} in the diagram. $G(s)$ is used to calculate the amount of power given or received from the other areas, its formula is given in Eq. 7.6, where N is the number of areas, T_{ij} is the synchronising torque coefficient between areas i and j, and Δf is the frequency of area i or j respectively. The values used in the following simulations are shown in Table 7.2. In the simulations $T_{12} = T_{21} = 0.2$, $T_{13} = T_{31} = 0.25$, $T_{23} = T_{32} = 0.12$, $K_1(s) = -0.3/s$, $K_2(s) = -0.2/s$ and $K_3(s) = -0.4/s$.

$$\Delta G_i = \sum_{j=1,j\neq i}^{N} T_{ij}\Delta f_i - \sum_{j=1,j\neq i}^{N} T_{ij}\Delta f_j \tag{7.6}$$

Now that we defined the decentralised frequency response system that will be used, we can discuss the control technique. We will compare the performances when an aggregator uses a robust MPC controller and when it uses a simple controller based on a saturation of the frequency, see Fig. 7.14. The simple saturation controller, in future it will be referred to as the base controller, chooses an input u to apply to the system. u is the participation value of the buildings in the control, and in each area $P_{AB_i} = 1000$ kW and $N_{AB_i} = 50$. The saturation sets $u = -2 \times \Delta f$

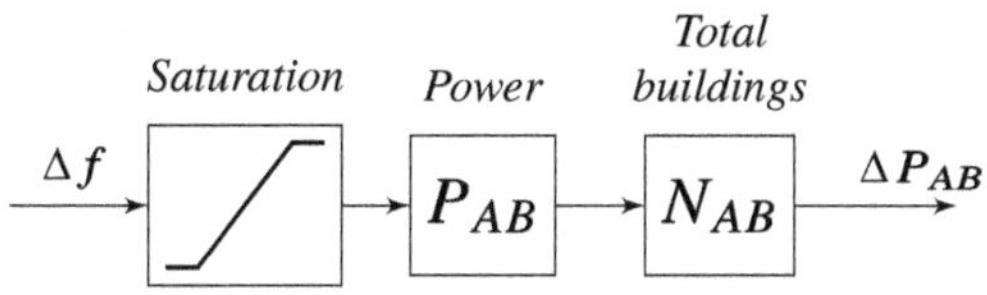

Fig. 7.14 The base controller for active building frequency response for the three area system.

Hz, but also if $\Delta f < -0.5$ Hz then $u = 1$ and if $\Delta f > -0.015$ Hz then $u = 0$. A buffer zone of 0.015 Hz is given to stop unnecessary fluctuations around the nominal value.

For this case study, we are considering a system experiencing a disturbance and desire to return the system back to the nominal value as quickly and smoothly as possible. An MPC control algorithm which suits this scenario is the robust optimisation framework of the MPC algorithm. Using YALMIP (Löfberg, 2012) and MATLAB, we will derive a controller for each area, that is robust enough for certain levels of disturbance.

The first step is to discretise the system. It is fairly simple to convert the transfer function model shown in Fig. 7.13 to a state space model. Consider each area has 5 states, f_i,P_{c_i},P_{g_i},P_{m_i}, and P_{tie_i}, each area has one input, the percentage participation of active buildings, and a disturbance, P_{L_i}. This gives a system with 15 states, 3 inputs and 3 disturbances that can be described in the form:

$$\dot{x}(t) = Ax(t) + Bu(t) + B_w w(t)$$

where $x(t)$ is the state, $u(t)$ is the input and $w(t)$ is the system disturbance.

To discretise the system, and transform the continuous differential equations to discrete difference equations, a clever transformation is computed.

$$e^{\begin{bmatrix} A & B \\ 0 & 0 \end{bmatrix} T} = \begin{bmatrix} A_d & B_d \\ 0 & I \end{bmatrix}$$

where T is the sample step, A_d and B_d are the discretised state space matrices of A and B respectively, and I is the identity matrix. By treating the disturbance as an input, B_w can also be discretised using the same equation.

As previously described, the MPC algorithm has three steps.

1. Create a prediction horizon.
2. Calculate a set of future control signals.
3. Use a receding strategy to apply the current optimal control signal, then compute new control signals starting from the next time step.

Firstly, we create a prediction horizon, we set the program to look 10 time steps ahead when choosing the best input value for the system. We also set constraints on the states and inputs, for this system the input is the participation value so it should only be between $0 \leq U \leq 1$. and the frequency in each area should not fall below the containment zone $-0.8 \leq Y_i$. For robust MPC we consider the disturbance and try and maintain optimal performance of the system considering the worst case scenario. By modelling the power loss in each area in the matrix B_w, we can set the disturbance W as $0 \leq W \leq 1$. The model will consider the worst case scenario when computing the optimal input values to choose, which is where $W = 1$.

Secondly, at each time step the algorithm computes a set of future control signals. This is done by considering the constraints put on the system and the control objective. If the constraints cannot be satisfied by any of the possible inputs then a non-feasible solution is found. For all the feasible solutions, an objective function is used to calculate the most optimal input of the group. The objective function adds weights to different control objectives. For this case study we wish to minimise Y, the frequencies of each area, and also minimise U, the input for each area. Minimising the frequency is prioritised over minimising the input so a weight of 0.01 is applied to the objective function for U. The objective O is shown in Eq. 7.7, where N is the number of areas.

$$O = \sum_{i=1}^{N} (\|Y_i\| + 0.01\|U_i\|). \tag{7.7}$$

Thirdly, By computing the objective function at the current time step, t_k, for the full prediction horizon, an optimal input value is found. This value is passed into the plant and the system evolves. The MPC algorithm now discards the previous computations and begins again this time with the current time step t_{k+1}.

In Fig. 7.15, it can be seen that the MPC algorithm provides a significant benefit for frequency response compared to the base controller. The disturbance on the system has a smaller impact than felt by the base controller. In Fig. 7.16, it can be seen that the MPC controller provides small improvements on the frequency response of all three areas of the system. Each area also satisfies the requirement of remaining outside of the containment zone. The MPC controller is robust with respect to the worst case scenario, finding an optimal solution in every time step.

This MPC controller is robust to a wide variety of disturbance scenarios. We will compute the same controller but change some of the system values—$L_1 = 0$, $L_2 = 0.3$ and $L_3 = 0$. In Fig. 7.17, a significant improvement over the base controller can be seen in this scenario. The overall contribution of the active buildings is also higher. Each area also has significant frequency response improvements over the base scenario, as seen in Fig. 7.18. It is also noteworthy, that in this scenario, area 2 has frequency transients similar to the other areas, yet for Fig. 7.16 there was a significant difference in the frequency transient of area 2 and both of the other areas.

Overall, MPC is a viable and popular control technique which has benefits when controlling systems with disturbances and uncertainty. The use of a cost function in the algorithm means MPC can provide accurate control for MIMO systems, and

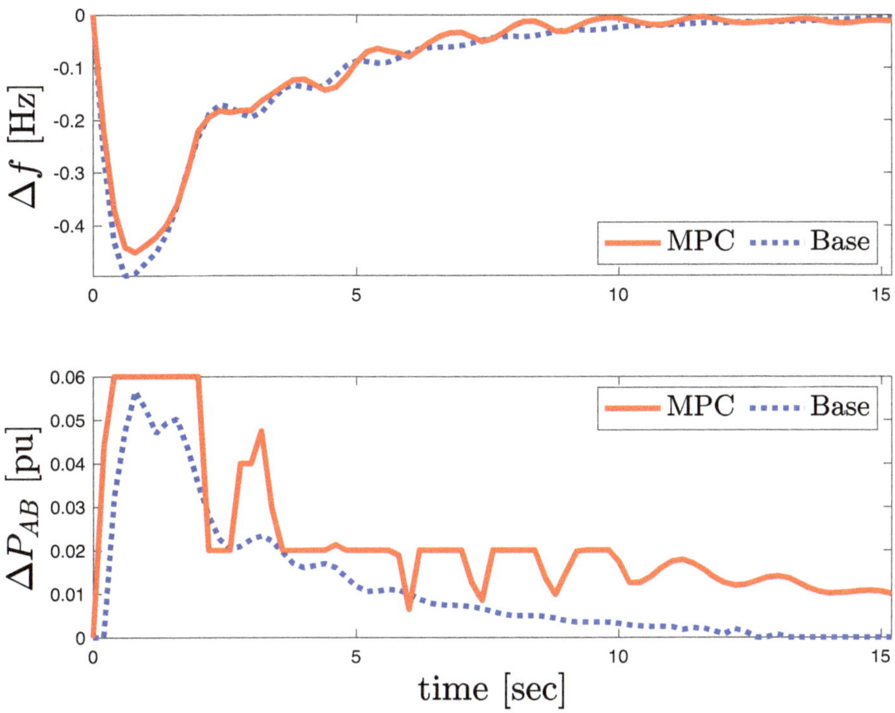

Fig. 7.15 Global frequency measurements **(top)** and participation measurements **(bottom)** for areas with $L1 = 0.3$ and $L3 = 0.3$

optimise the system despite the additional complexities, relative to the centralised scenario. This technique is beneficial for controlling ABs as seen in the case study because the algorithm was able to control the transient frequency of the smart grid to minimise the system loss while also considering AB requirements attempting to minimise their contribution.

7.4.3 Distributed

For the case study of the distributed structure, we will use the MAS control technique. We will consider a population of homogeneous residential ABs, smaller than the large ABs used in the previous two case studies, which can act as reactive or intelligent agents in the system. The buildings are connected to the smart grid and each have an ESS and PV panel. It is assumed that active buildings store energy in their ESS for potential demand-side response services which have been generated from their PV panel or stored from the grid in a previous iteration of time. Each building knows the percentage charge of the ESS it is paired too, but this is

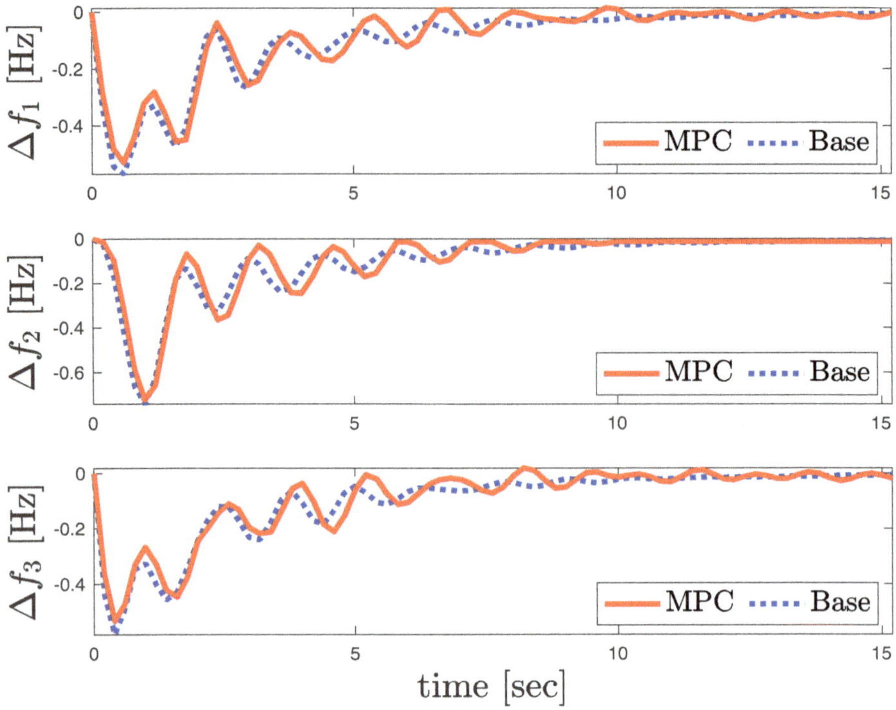

Fig. 7.16 Frequency measurements for areas 1 **(top)**, 2 **(middle)**, and 3 **(bottom)** with $L1 = 0.3$ and $L3 = 0.3$

not known by the smart grid globally. Instead, the smart grid broadcasts a signal containing the system frequency and the agents decide whether or not to respond.

Algorithm 1: Active building agent logic

running = true;
while *running* **do**
  if *f < -0.1 & c > 40* **then**
    inject power
  else
    do nothing
  end
end

The logic of each agent, as shown in Algorithm 1, is setup to be quite simple. At each time step, the agent will check the global frequency value and compute whether or not to inject power into the network. If the global frequency is below -0.1 Hz and the building has at least 40% charge in its ESS then demand-side response services are provided. By discharging the battery to the power grid the battery will have less capacity remaining. Therefore, it is assumed that, at each time step, 1% of the available capacity is lost when discharging.

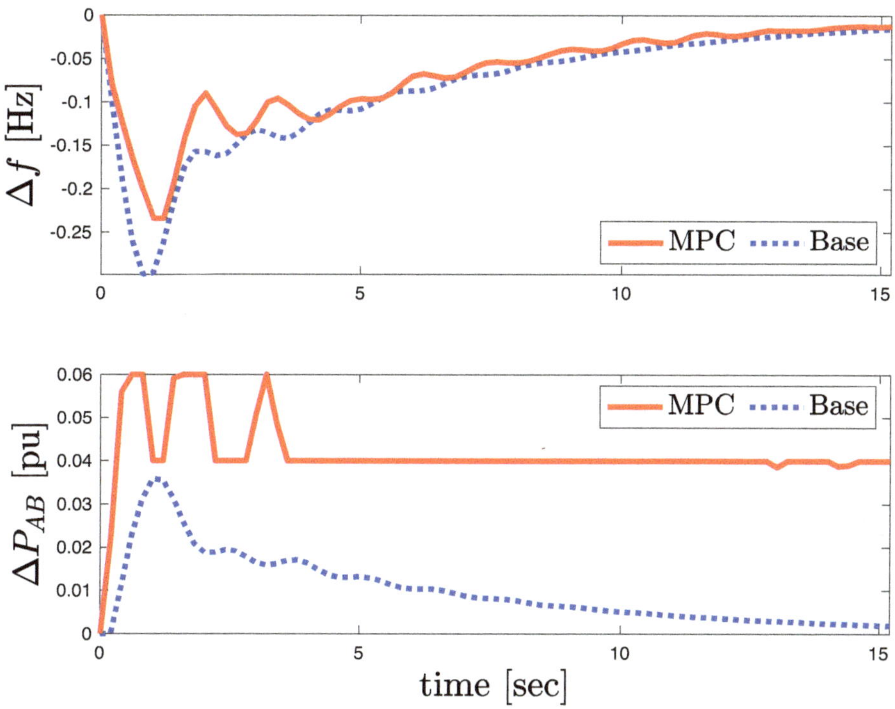

Fig. 7.17 Global frequency measurements **(top)** and participation measurements **(bottom)** for areas with $L2 = 0.3$

First, we will consider these active building agent response within the power system model described for the centralised controller. Taking the model from Fig. 7.10, and the values that produced Fig. 7.12, the model we consider is a single area model of the frequency which disconnects active buildings when the frequency drops below −0.8 Hz. We now consider that all the buildings in that network have an ESS, and that these buildings can inject their power into the power system if they decide to do so following the logic from Algorithm 1. The percentage charge of each building's ESS is chosen to be a uniformly distributed random value between 0 and 100%. The power injection per second of each homogeneous ESS is set to 28 kW. In this scenario, we will consider a population of residential buildings, $P_{AB} = 10$ kW and $N = 5000$, to show that many AB agents can collaborate for demand-side response services.

From Fig. 7.19, it is clear that having addition demand-side response from multi-agent systems can greatly enhance the frequency response of the power network and avoid LFDD schemes. Additionally, in Fig. 7.20 it is seen how different MAS logics can affect the control. In the original scenario a charge of at least 40% was required for MAS demand-side response, c value in Algorithm 1. The frequency response can be enhanced further when the charge is set to 20% or worsened when set to 60%.

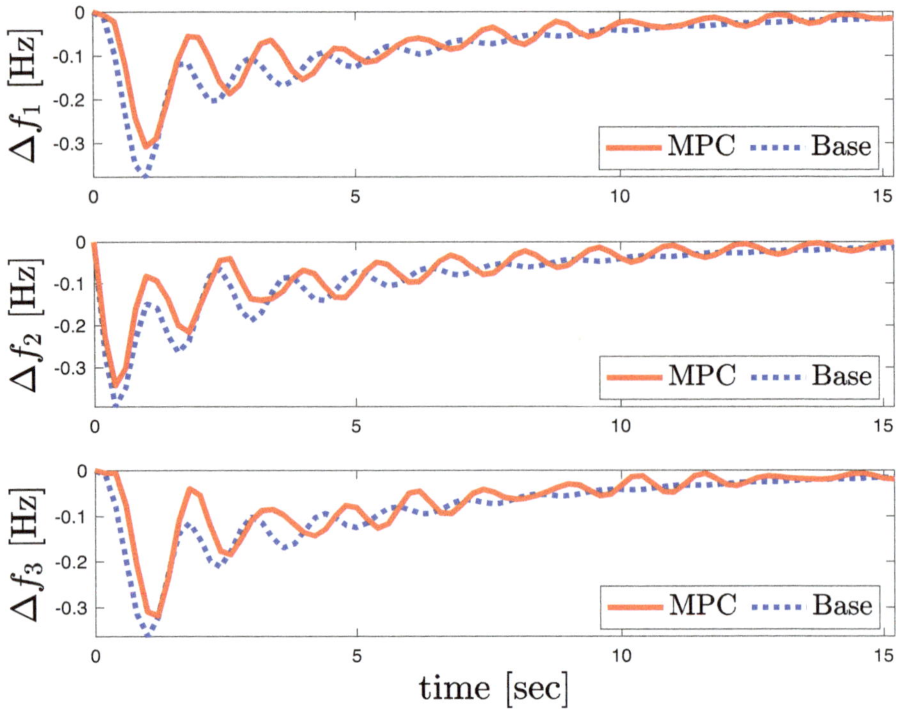

Fig. 7.18 Frequency measurements for areas 1 **(top)**, 2 **(middle)**, and 3 **(bottom)**, with $L_2 = 0.3$

In this case study, very simple agent objectives were presented; the agent wishes to contribute to demand-side response to avoid the need for LFDD. However, the agent objectives can be complex and competing objectives may arise. To solve this issue, communication algorithms are needed for the agents.

Overall, the distributed approach, such as the use of MAS, is viable for future applications because it is highly scalable. It is expected that the AB penetration in the power network will increase and so independent agents which can respond to the state of the global power network are valuable. This reduces the need for a complex centralised or decentralised control strategy that can rely on accurate models of the power network.

7.5 Status Quo, Challenges, and Outlook

In this chapter, we have looked at the control and management of active buildings. In particular, we have discussed the coordination structures and frameworks which consist of a system structure; centralised, decentralised or distributed, and an energy resource type; renewable or non-renewable. We discussed various control methods

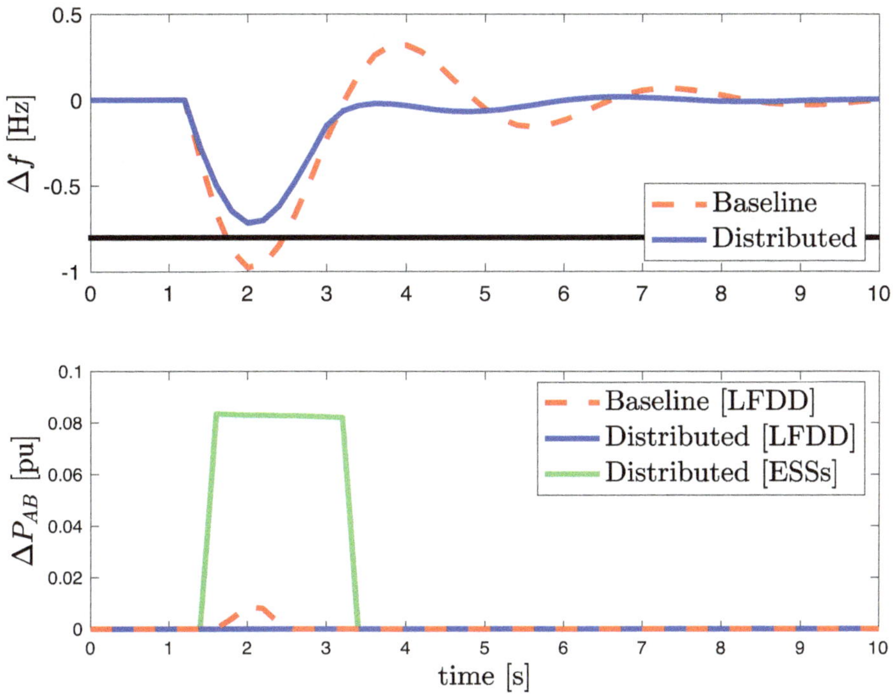

Fig. 7.19 Frequency response using distributed agent control and baseline control, the horizontal bar marks where the LFDD scheme triggers **(top)**. Power injection to system from active building agents **(bottom)**

that can be used for active building systems, the most popular being PID control and MPC, but other viable techniques are also emerging such as MAS control and DDC. Finally, we provided examples for the control and management of the active buildings at each of the centralised, decentralised and distributed system frameworks. Two future research directions are as follows:

- Firstly, the smart grid is evolving to be more distributed and electrified as more passive buildings become ABs. These new ABs and other grid components will join the power network. This increased load will add a strain to the network unless control schemes are developed to safely manage ABs while also keeping AB users satisfied. The increase in system loads leads to a larger complex network and the requirement of having scalable methods for grid control. A scalable strategy discussed in this chapter was MAS, but further research into the management of ABs and communication between competing and cooperative AB objectives needs further research.
- Secondly, as the smart grid becomes more electrified, and with the increase in ABs, it will become more complex. This complexity will create further nonlinearity within the system and be more difficult to model. Many of the current control

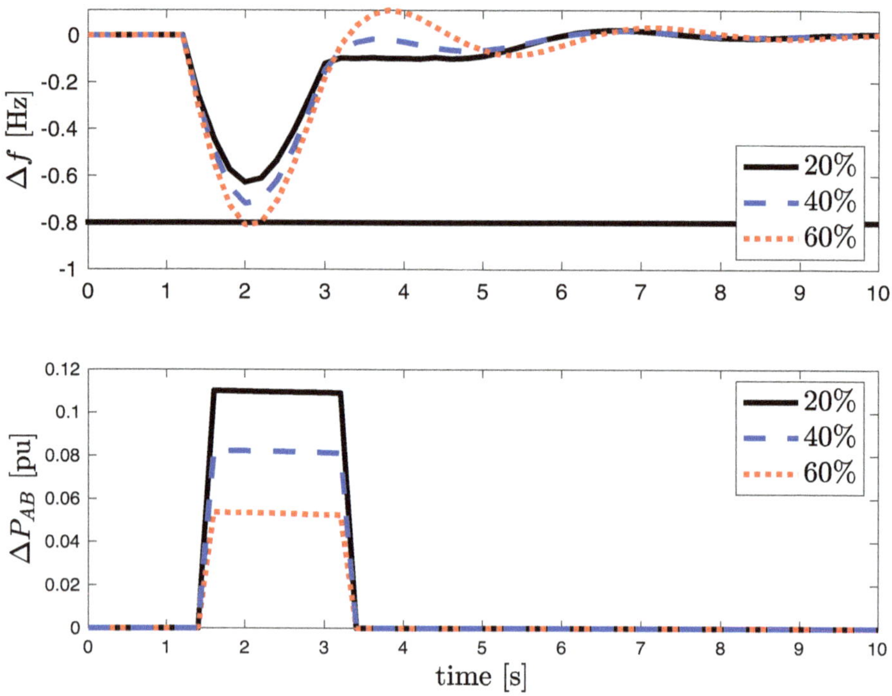

Fig. 7.20 Frequency response using distributed agent control, the horizontal bar marks where the LFDD scheme triggers **(top)**. Power injection to system from active building agents **(bottom)**

techniques are model based, such as MPC, and so rely on accurate underlying models of a system. As complexity increases a shift towards model-free or data-driven modelling as a prospective option has increased. These methods rely on large quantities of data to represent the system which can be difficult to acquire as well as potentially losing some system understanding provided by previous model's physical descriptions. Further research into efficient data-driven control schemes possibly combined or guided by simplified models are needed to manage ABs in a smart grid.

References

Annaswamy, A. M., Malekpour, A. R., & Baros, S. (2016). Emerging research topics in control for smart infrastructures. *Annual Reviews in Control, 42*, 259–270.

Arnaboldi, L., Czekster, R. M., Morisset, C., & Metere, R. (2020). Modelling load-changing attacks in cyber-physical systems. *Electronic Notes in Theoretical Computer Science, 353*, 39–60.

Başar, T., & Olsder, G. J. (1998). *Dynamic noncooperative game theory*. Philadelphia: SIAM.

Bevrani, H. (2009). *Robust power system frequency control*. Berlin: Springer.

Boyd, S., Boyd, S. P., & Vandenberghe, L. (2004) *Convex optimization*. Cambridge: Cambridge University Press.

Brunton, S. L., & Kutz, J. N. (2017). *Data driven science & engineering - machine learning, dynamical systems, and control*. Cambridge: Cambridge University Press.

Camacho, E. F., & Bordons, C. (2007). *Introduction to model predictive control*. Berlin: Springer.

Curiel, I. (2013). *Cooperative game theory and applications: Cooperative games arising from combinatorial optimization problems* (vol. 16). Berlin: Springer.

Demuth, H. (2004). Neural network toolbox documentation. NN with Matlab, p. 846.

Drgoňa, J., Arroyo, J., Cupeiro Figueroa, I., Blum, D., Arendt, K., Kim, D., Ollé, E. P., Oravec, J., Wetter, M., Vrabie, D. L., & Helsen, L. (2020). All you need to know about model predictive control for buildings. *Annual Reviews in Control, 50*, 190–232.

Energy Emergencies Executive Committee. (2019). GB power system disruption–9 August 2019. Department for Business, Energy & Industrial Strategy, Tech. Rep.

Figueiredo, J., & Sá da Costa, J. (2012). A SCADA system for energy management in intelligent buildings. *Energy and Buildings, 49*, 85–98.

Franklin, G. F., Powell, J. D., Emami-Naeini, A., & Powell, J. D. (2002). *Feedback control of dynamic systems* (vol. 4). Upper Saddle River: Prentice Hall.

Gasparyan, O. (2008). *Linear and nonlinear multivariable feedback control: A classical approach*. Hoboken: Wiley.

Hu, J., Yang, G., Kok, K., Xue, Y., & Bindner, H. W. (2017) Transactive control: A framework for operating power systems characterized by high penetration of distributed energy resources. *Journal of Modern Power Systems and Clean Energy, 5*, 451–464.

Khaled, N., & Pattel, B. (2018). Chapter 2 - theoretical foundation of MPC. In: N. Khaled & B. Pattel (Eds.) *Practical design and application of model predictive control*. Oxford: Butterworth-Heinemann.

Ławryńczuk, M. (2017). Nonlinear predictive control of a boiler-turbine unit: A state-space approach with successive on-line model linearisation and quadratic optimisation. *ISA Transactions, 67*, 476–495.

Ljung, L. (2010). Perspectives on system identification. *Annual Reviews in Control, 34*, 1–12.

Löfberg, J. (2012). Automatic robust convex programming. *Optimization Methods and Software, 27*(1), 115–129.

Miao, L., Wei, G., Fang, X., & Risheng, J. (2015). The strategy of the voltage control in smart grid based on modern control method and FPGA. In: *2015 34th Chinese Control Conference (CCC)*.

National Grid. (2020). The grid code. National Grid, Tech. Rep.

Nazir, S., Patel, S., & Patel, D. (2017). Assessing and augmenting SCADA cyber security: A survey of techniques. *Computers & Security, 70*, 436–454.

Rathore, M.M., Ahmad, A., Paul, A., & Rho, S. (2016) Urban planning and building smart cities based on the Internet of Things using big data analytics. *Computer Networks, 101*, 63–80.

Reynolds, J. (2019). *Real-time and semantic energy management across buildings in a district configuration*. Ph.D. Dissertation, Cardiff University.

Roche, R., Lauri, F., Blunier, B., Miraoui, A., & Koukam, A. (2013). *Multi-agent technology for power system control*. London: Springer.

Roughgarden, T. (2010). Algorithmic game theory. *Communications of the ACM, 53*(7):78–86.

Rumelhart, D. E., Hinton, G. E., & Williams, R. J. (1986). Learning representations by back-propagating errors. *Nature, 323*(6088), 533–536.

Saad, W., Han, Z., Poor, H. V., & Başar, T. (2012). Game-theoretic methods for the smart grid: An overview of microgrid systems, demand-side management, and smart grid communications. *IEEE Signal Processing Magazine, 29*(5), 86–105.

Silver, D., Schrittwieser, J., Simonyan, K., Antonoglou, I., Huang, A., Guez, A., Hubert, T., Baker, L., Lai, M., Bolton A., Chen, Y., Lillicrap, T., Hui, F., Sifre, L., van den Driessche, G., Graepel, T., & Hassabis, D. (2017). Mastering the game of go without human knowledge. *Nature, 550*(7676): 354–359.

Stark, C., Thompson, M., Andrew, T., Beasley, G., Bellamy, O., Budden, P., et al. (2019). Net Zero: The UK's contribution to stopping global warming.

Sutton, R. S., & Barto, A. G. (1998). *Introduction to reinforcement learning* (vol. 135). Cambridge: MIT Press.

Vezzoli, C., Ceschin, F., Osanjo, L., M'Rithaa, M. K., Moalosi, R., Nakazibwe, V., & Diehl, J. C. (2018). *Designing sustainable energy for all: sustainable product-service system design applied to distributed renewable energy*. Springer Nature.

Zhang, H., Zhou, J., Sun, Q., Guerrero, J. M., & Ma, D. (2017). Data-driven control for interlinked AC/DC microgrids via model-free adaptive control and dual-droop control. *IEEE Transactions on Smart Grid, 8*, 557–571.

Zonetti, D., Saoud, A., Girard, A., & Fribourg, L. (2019). A symbolic approach to voltage stability and power sharing in time-varying DC microgrids. In: *2019 18th European Control Conference (ECC)*.

Chapter 8
Operation and Control of a Population of Active Buildings at Network Level

Ben Wooding, Vahid Vahidinasab, and Sadegh Soudjani

8.1 Introduction to Control of Active Building Energy Systems

Buildings are seen traditionally as passive loads in the power system. Their users consume electricity, gas and water and pay for these services via some predefined tariffs. The recent evolution in smart devices and the integration of *distributed energy resources* (DERs) and storage systems enables buildings to transition toward a responsive player in the energy systems. Such buildings are called *active buildings*.

The future of active buildings is complex. Buildings are smarter, and they are coupled to complex components, known as DERs. Buildings can generate their own energy using *photovoltaic panels* (PVs) and store that energy in *energy storage systems* (ESSs) to consume later. An active building can charge an EV which is connected to and manage building temperature through TCLs. Building management systems are able to control these DERs to optimise the building's energy efficiency, even becoming self-sufficient from the power grid.

As the demand-side of the power grid contains more DERs, it also becomes smarter. The *smart grid* is able to control its frequency and voltage using *demand-side response* (DSR) services from DERs. In this way, active buildings become *prosumers*, no longer purely consuming energy from the smart grid but also generating power and injecting it into the grid when necessary or sharing it with local neighbours. Chapter 7 of this book, titled "Control and Management

B. Wooding · S. Soudjani (✉)
School of Computing, Newcastle University, Newcastle upon Tyne, UK
e-mail: b.wooding1@newcastle.ac.uk; sadegh.soudjani@newcastle.ac.uk

V. Vahidinasab
Department of Engineering, School of Science and Technology, Nottingham Trent University, Nottingham, UK
e-mail: vahid.vahidinasab@ntu.ac.uk

V. Vahidinasab, B. Mohammadi-Ivatloo (eds.), *Active Building Energy Systems*, Green Energy and Technology, https://doi.org/10.1007/978-3-030-79742-3_8

of Active Buildings", discusses coordination structures and control methods for active buildings. In this chapter, a detailed discussion of how active buildings can implement smart grid control schemes (e.g. aggregation, multi-vector control and cyber security concerns) to provide services is presented. These services include frequency response and voltage support services.

A detailed case study is also provided, the study considers a population of active buildings which are coupled to TCLs and ESSs. These devices are used to provide energy for frequency services. *Formal control synthesis* and *model predictive control* (MPC) are used to determine the optimal control strategy for the devices. These techniques are then combined for a multi-vector control method.

8.2 Control of a Population of Buildings

8.2.1 Aggregate Models for a Population of Buildings

To manage energy production and consumption and to increase stability of the network, *aggregators* and *demand-side management* (DSM) are suggested for DSR. DSR is demand-side balancing of the generation and consumption of the power network. This is a smart grid's most important function. It is vital that energy resources are carefully managed and controlled.

Currently difficulties are present with the supply-demand balance. If the consumption increases above a predicted peak value, then generation companies have to increase their energy production, or if the user consumption of energy is lower than predicted, then resources may be left unused. In both scenarios, the operator will experience unnecessary additional costs and potentially wasted resources (Anees & Chen, 2016).

Demand response can occur at a speed almost real time and can produce stabler systems with significantly reduced generation costs. In residential sectors, operators take most of the demand response benefits for themselves. Each building in a city- or district-level network has negligible negotiating power with the operators in the energy market. On top of this, the number of homes in a residential area contributes to scalability challenges, and the operator is unlikely to manage negotiations of this scale. This combination, minimal negotiating power combined with high numbers of buildings, poses a challenge to demand response schemes (Gkatzikis et al., 2013). A viable solution that has been developed is that of the aggregator.

The aggregator works as a middleman, between the operator and individual buildings in the residential region (see Fig. 8.1). The aggregator represents a group of buildings of end-users. At the energy market, the aggregator negotiates on their behalf. Since the aggregator represents a sizeable amount of demand, it is able to negotiate more effectively with the operator. This mainly happens with industrial buildings but is being transitioned into the residential sector as well.

The aggregator pays a fee to end-users to gain control over that building's appliances. By controlling the appliances of the buildings, the aggregator is able to respond to peak-demand emergencies. It can turn appliances off which are using

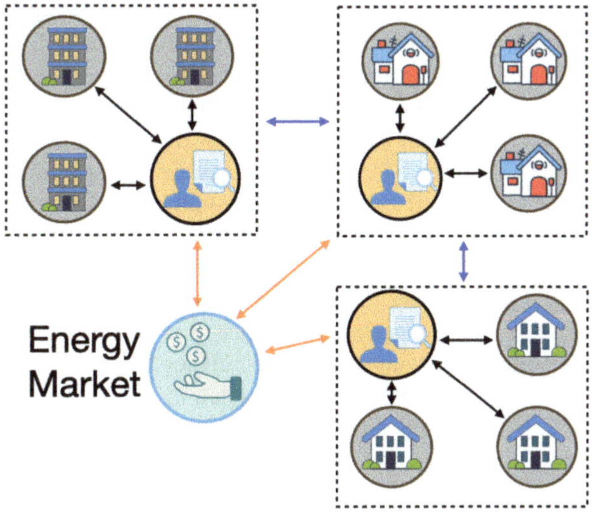

Fig. 8.1 The aggregator is able to negotiate on behalf of the end-users it represents with the energy market and with other aggregators

sizeable amounts of energy, such as air-conditioning units. This reduces the demand-side load and returns stability to the network.

By using an aggregator, individual buildings are also able to reduce their energy costs. The aggregator is aware of the total demand profile of the buildings it represents and pays them incentives to adjust their demand pattern. By scheduling loads to lower cost time periods, the operator is also able to save money on energy resources. The operator rewards the aggregator for its work in this service. By using this technique, both the aggregator and the end-users can have noteworthy financial benefits, despite the dominant position of the utility operator.

Demand response also has applications for EV charging schemes (Siano & Sarno, 2016). The aggregator can schedule when an EV charges, so as to reduce its load on the network. From a high-level perspective, the aggregator allows for collections of power units to be treated as a single power unit. This is because the aggregator manages the details of demand response within that group. By using an aggregator, it is possible to simplify and scale many models which include smart buildings, EVs and other power units.

8.2.2 Frequency Control and System Inertia

Frequency instability is the inability of the power system to maintain a global frequency value after a disturbance occurs which triggers rotor speed changes in the generators. The frequency may rise or fall out of the given operational range. Usually, instability is caused by an imbalance between the active power generation and the loads present in the system. Poor coordination, poor protective equipment, slow response times and lack of generation reserves are all possible reasons why instability would occur after a disturbance (Bevrani, 2009).

One of the control techniques to face the load and generation imbalance is *peak load shifting*. Over a 24 h period, different hours of the day experience different load values. For example, during the day, the system load is highest because people are awake and working. Large machines are also in operation which require large sums of power. At night, the opposite would be the case, people are asleep and machines are turned off. Load shifting attempts to move some of the demand from peak hours to off-peak hours, such as EVs which could charge overnight when their owners would be sleeping. By rescheduling the load away from peak hours, the amount of active power generation required at peak times decreases, providing multiple economic, environmental and efficiency benefits (Devaux & Farid, 2017).

As the number of EVs connected to the electricity grid is expected to increase. A large load will be present within the system. In Wu et al. (2019), residential homes with EVs are used to reduce the expected power system load of higher EV penetrations. A smart charging method is proposed which utilises aggregators to schedule the EV charging times. Using a method similar to MPC, the process is able to update, depending on user travel requirements and conditions of the power system. By reducing annual peak loads, the algorithm is able to reduce the charging costs for the EV owners and also defer system upgrades that would be required for the system with increased load.

Another important consideration is the *rate of change of frequency* (RoCoF). RoCoF is inversely proportional to the available inertia in the power system. When there is low inertia, RoCoF is high which means that system is more volatile and large disturbances could destabilise the system, causing load shedding (Dreidy et al., 2017).

Renewable energy sources like PVs do not provide any inertia to the system (Wang et al., 2016). This is unlike the large synchronous generators which have kinetic energy stored in their rotated mass that can be released. Therefore, with a global desire to decarbonise the energy systems and accelerate the transition to renewable-based energy systems, RoCoF will increase unless managed.

One developing solution is virtual synchronous generators (VSGs), which provides the grid with a variable virtual inertia. The VSGs consist of short-term storage with an inverter and an efficient control mechanism. The VSG reproduces the dynamic properties of real synchronous generators and keep the advantages they provided which include adjusting active and reactive power Bevrani et al. (2014).

Fixed speed wind turbines use an induction wind turbine to provide inertia response for frequency deviations, although the inertia provided is small compared to the inertia of a synchronous generator. For variable speed wind turbines, permanent magnet synchronous generators are used to provide power to the grid, but these are decoupled and so do not provide any frequency response services as well as not storing any reserve power. Techniques are developing that can release the kinetic energy stored in the rotating blades within 10 s to help with frequency response; this is known as *inertia emulation* and uses a *fast power reserve* (Dreidy et al., 2017).

For PVs, a deloading technique is used which increases the PV voltage beyond the maximum power point (MPP) voltage and allows the PV to retain some reserve power (Zarina et al., 2014). When the system frequency deviates, this reserve power can be released. However, all PVs will release the same value of power into the grid, and those with less reserves will reach MPP faster and stop contributing to the frequency response; this may cause a second frequency drop once the reserve power is used up.

Another technique to support renewable sources for frequency response is ESS devices; the ESS provides active power to try and prevent a second frequency drop in the system but also acts as a backup system to provide power if there are any deficits. Coupling ESS with wind or solar can help alleviate the risks of high RoCoF (Dreidy et al., 2017; Greenwood et al., 2017).

In Vorobev et al. (2019), it is shown that for low inertia systems, that inertia has little effect on the frequency probability distribution function for small disturbances. So, virtual inertia is insufficient on its own to keep the frequency near the nominal value. Instead, it was shown that the aggregate droop and deadband are the only parameters that have a major influence over the average frequency deviations, which suggests that energy storage solutions are viable and valuable for future smart grid frequency response services.

8.2.3 Coordinated Volt/Var Control

With the increase in renewable penetration in *transmission network*, and increased DERs in the *distribution network*, voltage control is facing significant challenges. On top of this, transmission network and distribution network are increasingly coupled, meaning both sides need to effectively be coordinated for the voltage control of the both networks. All these challenges require a greater understanding of the problem and developing advanced control techniques.

Three main control techniques exist. Restoring and maintain the system within a safe working region is known as *corrective control*. *Coordinated control* tracks a set value and mitigates disturbances. *Preventative control* tries to fix the system before any instability occurs and may use fast-response dynamic reactive power (DRP) to facilitate this. Choosing the correct control scheme may depend on response times, control cost and control effectiveness (Rogers et al., 2010).

Transmission Network

Voltage control on the transmission network has been successful for a long time; however, the power network is evolving with more renewable resources being incorporated. High penetrations of wind or solar generation have a high risk of causing voltage fluctuations. The use of HVDC lines to connect onshore and offshore wind farms to the transmission network also has the potential for cascading failures; an example cause is DC-blocking contingency. When the fault occurs, in 2–3 s the voltage will increase significantly. This may lead to curtailment of the renewable sources. For this reason, preventative control methods should be applied to wind and solar plants using DRPs for fast-response. HVDC lines have some fast-response reactive power regulation capabilities and so can also support this control scheme (Kish et al., 2015; Merlin et al., 2014).

Short-term voltage instability is usually linked to induction motor stalling and are also known as *Fault-Induced Delayed Voltage Recovery* (FIDVR) issues (Paramasivam et al., 2013). To mitigate FIDVR, dynamic reactive power sources such as *static reactive power compensators* (SVCs), *static synchronous compensators* (STATCOMs) and DERs are used. The control challenge is the optimal sizing and placement of these sources. Both aspects require the computation of the post-fault voltage trajectory which requires solving differential algebraic equations. These equations are complex due to the nonlinearity of the input-state behaviour and of the solution space, which is also nonconvex. For solving the optimal placement, selection of contingencies must be considered. Depending on the number of reactive power sources, these factors can cause a heavy computational burden. Usually solutions are found using *mixed integer programming* and *heuristic optimisation* algorithms.

For long-term voltage instability detection, the Thévenin equivalent is one method used to present a model of power supply with a fixed impedance to a voltage source for analysis. When the load impedance magnitude equals the Thévenin equivalent impedance, then the maximum power supply occurs. This is known as the impedance matching principle, and from it, the voltage stability margin can be computed. When using measurement data, there are challenges from load variations and measurement noise. SCADA and *phasor measurement unit* (PMU) data are used for these computations. There are three methods which are valuable for both online and offline estimation. These are (1) using the least mean squares to fit the data to the power-voltage curve of the Thévenin equivalent, (2) using data and a real-time algorithm to estimate the Thévenin equivalent from a given bus and (3) estimation of Thévenin reactance to compute the Thévenin voltage. Verifying and validating these measurement-based approaches is a future research area (Sun et al., 2019).

Distribution Network

The increased penetration of DERs has led to stricter grid connection requirements. The interconnection of these devices has changed the grid loading pattern and influences voltage regulation device performance. In particular, there are four challenges that high DER penetration has caused for voltage regulation in the distribution network.

Firstly, the *distribution system operator* (DSO) has to manage voltage rise. This is often triggered by the generation of solar energy from residential PVs (Karimi et al., 2016). The injection of too much power into the grid triggers a voltage rise in the system, and to mitigate the problem, the operator may impose conservative limits on PV installation. A solution is for the PV to self-regulate its voltage by reducing its active power injection or applying negative reactive power injection. Coordinating the PV inverters with demand-side management is another solution. Heat pumps,

EVs or battery storage could be coupled to the PV to increase their consumption of energy to match the PV generation injected.

Secondly, the increase in EVs in the power system could lead to overloads and large voltage drops, especially at peak times. Charging a single EV demands nearly the same energy as three houses. China, India, France and the UK have promised to phase out gas and diesel vehicles by 2040, so, with more vehicles becoming electric, the load on the grid is going to increase, and control methods such as load shifting and charging strategies will be necessary to avoid overloading the system (Wu et al., 2019). Similar to the solution for voltage rise, demand-side devices can support the network to return the voltage to its nominal value. Collections of these devices can form 'support groups' around different network buses in the system. A case study using this technique for control of the IEEE 24-bus reliability test system (RTS) is given in Rogers et al. (2010). When low-voltage buses are discovered in the system, the support group would be tasked with returning the system voltage to its expected region.

Thirdly, the distribution network is designed for unidirectional power flow. *Line voltage regulators* (LVRs) are included in the network for voltage control from the load side. If a DER increases the voltage where it is situated, the LVR will try to reduce this from a load side. However, it is possible the voltage at the DER connection point would remain high. Even if the LVR could participate in bidirectional power flow, the DER and/or local voltage correction devices would be needed to stabilise the network. In Agalgaonkar et al. (2014), reactive power control options are coordinated to avoid the continuous operation of devices such as LVR. This avoids both device deterioration and operation of devices at their control limit.

Finally, cloud cover variations affect PVs and can lead to voltage fluctuations and lower power quality across the distribution network. Higher deadband and slope values would ensure system stability of the volt-var curves but also lead to a reduced range of the voltage margin. Other solutions, such as coupling the DERs for coordinated control could be successful (Mahmud & Zahedi, 2016).

Transmission/Distributed System Operator Coordination

For voltage control, it is also necessary to coordinate the control of the *transmission system operator* (TSO) and the DSO. This is to prevent reactive power exchange when the reactive power is low in the transmission network. A distribution network with coordinated DERs can flexibly adjust its reactive power consumption to provide power reserves and improve the transmission network voltage. There are two types of coordination of the TSO and DSO, these are rule-based methods and distributed optimisation. Both methods involved exchanging boundary voltage and power with one another to stabilise both the transmission network and the distribution network. The distribution network can help by providing fast-response power injection into the transmission network or by reducing their demands and increasing local generation. If the distribution network control scheme ignores the transmission network conditions, it is possible long-term instability will result. To analyse the transmission-distribution reactive power support, a co-simulation of the transmission-distribution is needed. Effective coupling of the transmission networks and distribution networks leads to economic benefits, as a transmission network does not need to invest in voltage control devices and can leverage the DERs of the distribution networks (Sun et al., 2019).

8.2.4 Coordinated Control of Buildings as a Multi-vector Nano Energy Hub

Since the buildings are a junction for the interconnection of multiple energy vectors including electricity, gas, water, renewable energy resources and transportation, they should be modelled and studied as a multi-vector nano energy hub. This is the reason that buildings will play an important role in the energy context. Considering the increasing global awareness about the holistic approach to energy issues (Elliott et al., 2020), control and operation of the buildings of the future need to be revised in such a way that they can be efficiently managed and controlled; this will accelerate the transition to decarbonisation. Due to the different dynamics and specifications in each energy vector, control of integrated coupled energy systems inside a building is a tough task.

District energy systems attempt to manage and control multiple vectors, this makes them complicated and they require a detailed understanding of both modelling and optimisation. Multi-vector control is the requirement to control modern energy networks that consist of coupled vectors. Previously, heat, electricity, water and gas have been controlled independently, but this is no longer suitable for the future smart grid. For example, a heat pump may use electricity to create heat or a combined heat and power unit (CHP) creates electricity from heat and gas. Controlling just a single vector may actually provide lesser control strategies than the combined approach. In one example, a framework is developed that assesses technical, economic and environmental (TEE) benefits of integrating gas and electricity distributed networks with storage devices and discussed how a vector coupling storage system is able to increase whole energy systems efficiency (Reza Hosseini et al., 2021).

Understanding the multi-vector components provides increased accuracy in the control. Solar and wind energy are both very uncertain. Solar energy generation uncertainty comes from its proportionality to solar irradiance, but wind speed is the hardest energy source for modelling and prediction. When accurately modelled, a component such as CHP, which uses a heat byproduct from the electricity generation of natural gas, can increase its energy efficiency. In the case of CHP, this can be an increase from 30–40% to 80–90% (Reynolds et al., 2018).

Power-to-gas (P2G) is an interesting developing technology. It is used to store excess electricity from stochastic renewable power as either hydrogen or methane gas when it cannot be utilised. These gases can then be used in other areas such as fuel for hydrogen vehicles or by injection into the gas network. P2G is still largely being tested, and there is some worry about high costs and a low conversion efficiency. But a major benefit of this technology will be the large amounts of wind energy storage it can provide.

Buildings also require modelling as they will be integral to future smart grids and district energy systems. The buildings are prosumers and take part in demand-side control.

- **White box model**—based on the physical principles of the building;
- **Black box model**—based solely on data;
- **Grey box model**—hybrid approach of white and grey box modelling.

Grey box models and black box models are effective for modelling many building variables. But white box models of buildings are less effective; this is particularly for the cases of real-time optimisation with timesteps less than an hour.

In Strbac et al. (2020), an Integrated Whole Energy System (IWES) model is used to quantify the benefits of using a multi-vector approach with regard to active buildings. Using a whole system modelling approach shows significant economic saving opportunities. The combined flexibility can increase the proportion of electricity production from renewables and reduce the reliance on low carbon generation like nuclear power. This flexibility will also work for decarbonisation and reaching carbon emission targets. As efficiency of the active building improves, the total system costs reduce. Modelling active buildings as multi-vector systems allows for system complementing behaviours to be recognised. An example of this in an active building could be *thermal energy storage* (TES) which would charge thermal energy when the carbon intensity of electric heating is low and discharge it when it is high. These can improve short-term operational costs as well as long-term investment costs and reducing carbon emissions. In the report, cost savings were doubled when considering a multi-vector approach.

8.2.5 Security Aspects of Coordinated Control of Active Buildings as a Cyber-Physical System

From a high-level power system perspective, the major danger of a cyber-attack is if it can permeate through the network. Localised instability, or failure, should be prevented from affecting other areas of the network. Should this mitigation fail, it might trigger cascading failures across the whole network, which would have certain economic consequences as well as possible consequences to transportation (e.g. airports), healthcare (e.g. hospitals) and/or education (e.g. universities/schools).

Sensors and actuators are critical resources for power system control. These devices may connect to an Internet of Things (IoT). Devices of the IoT are known to have vulnerabilities, either on the device directly or through applications that connect to them, e.g. a linked smartphone app. One example attack could be the *Manipulation of Demand via IoT* (MadIoT) attack. An attack intending to deliberately cause load shedding in the system, which would have huge repercussions. In

Huang et al. (2019), a discussion is given about how the attack could be resisted. The solution describes embedding protections into the operation of the transmission grid. Another proposed solution, from (Szymanski, 2017), regards deterministic virtual networks (DVNs)—a lightweight encryption which would provide security, privacy, performance and energy efficiency to the IoT.

In 2017, a cyber security researcher proposed a cyber-attack known as the Horus scenario which targets PV panel inverters. Consider thousands of PV panels on the rooftops of European residential buildings, an attacker might send a signal which would be picked up by these PV panels and cause them to stop storing energy. The aggregate loss of energy across the power network would then lead to load shedding schemes across the continent.

In a response from SMA Solar Technology, the low likelihood of this attack was shown. Three factors are given that show cyber-attacks of this kind require significantly large efforts when attempting to destabilise the grid network. They are distributed regenerative power generation, decentralized production and the heterogeneity of the PV devices and manufacturers. In essence, it was said that diversity in the grid was the important feature for its safety. An attack would need to simultaneously affect multiple distinct types of devices. Even the large-scale use of bots would have limited success as each system would require an individually configured attack profile (SMA Solar Technology AG, 2017).

Ultimately, the danger of any cyber-attack in the smart grid is related to its communication channels. As a smart grid relies on sending information, communication-specific security requirements need to be evaluated. From known security standards, the key parameters that relate to communication are as follows:

- **Confidentiality**—only those with permission should be able to read communicated information;
- **Authentication**—the true sender should be known to the receiver of the communication;
- **Integrity**—information should arrive as it was sent, without any tampering;
- **Access Control**—access to the communication network should only be available to those with the correct clearance;
- **Non-repudiation**—a sender or receiver should not be able to deny their part in a transaction;
- **Availability**—communication channels should not fail, communication should always be possible.

An attacker would benefit should they exploit any one of these principles. Providing authentication and integrity are particularly applicable to a distributed network. Distributed network nodes are fluid and may connect and disconnect from the network at any time (e.g. EV that disconnects to drive). Sender-receiver pairs are unpredictable and depend on the current network state, as well as which nodes are active. Exploiting these could lead to man-in-the-middle attacks, impersonation, message editing or forgery (Rogers et al., 2010). One defensive authentication technique is digital signatures via hashing and decryption. The receiver compares the sender's signature with a known signature of that sender. If the signatures match, the receiver can be confident in the sender's authenticity and that the message had not been tampered with.

Another dangerous cyber-attack is the replay attack, which would listen and copy a message as its being broadcast. The attacker then sends this message again, some time in the future, attempting to use it maliciously. If a harmful message is replayed, it could cause quite problematic consequences. To address this problem, the receiver needs to know that it has received the message before. Timestamps or random number sequences embedded in the message can provide a good solution.

8.3 Case Studies

In this section, we provide a case study for frequency control. In Fig. 8.2, a population of active buildings is shown. Each building contains either a TCL or an ESS. ESSs and TCLs are combined to provide a demand-side energy response. A centralised controller sends a signal to these buildings to use their respective building components when required. When required, the ESSs is able to charge or discharge energy to provide a response service. The TCLs can also provide energy for response services by changing their thermal setpoint, reducing or increasing their consumption of energy from the grid. This case study is broken into four subsections; in the first two subsections, we discuss formal control synthesis and use it for frequency control using ESSs. In the third subsection, we use MPC power tracking to control the TCLs to a desired setpoint value. In the final section, both of

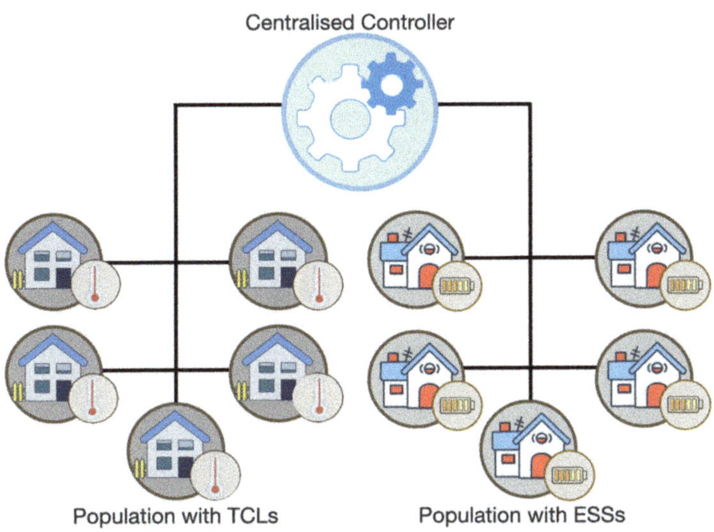

Fig. 8.2 The case study consists of a population of buildings with connected ESSs and TCLs

these approaches are combined for a joint ESS and TCL demand-side response for frequency regulation.

8.3.1 *Emerging* Formal Control Synthesis *Approaches*

A system can be described as a mathematical model of a dynamical phenomenon. Different models of the same phenomenon can be used in different tasks, and relationships between those systems can be described (Tabuada, 2009).

Describing Systems
Infinite-state systems are described by difference or differential equations, depending on whether they are described in discrete or continuous time. The system can be called *deterministic* if for any state x in the system, and for any input u, there is at **most** one successor state. This can be written more formally as in Eq. (8.1).

$$(x \xrightarrow{u} x') \cup (x \xrightarrow{u} x'') \Rightarrow (x' = x''), \forall x, u \rightarrow x \in X, u \in U \tag{8.1}$$

If for an input u at state x, it is possible to have two or more distinct successor states, then the system is *non-deterministic*. Different control approaches are needed dependent on the system's determinism. To describe these principles graphically, the most common approach is to use circles for system states with a transition between states represented by an arrow. A notation is added on top of the arrows which shows the input required for the state to undergo the transition. In Fig. 8.3, a deterministic system is shown, and Fig. 8.4 shows a non-deterministic system.

Equivalence Property

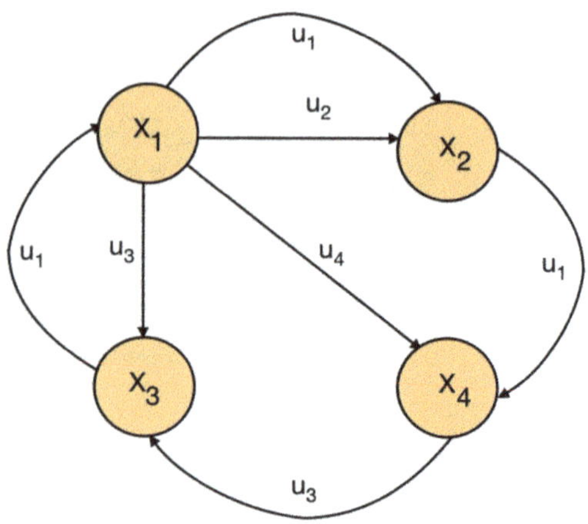

Fig. 8.3 Determinism—for each input u_i applied to a state X_i, a state transition is represented by an arrow. There is at **most** one successor state for each input-state pair

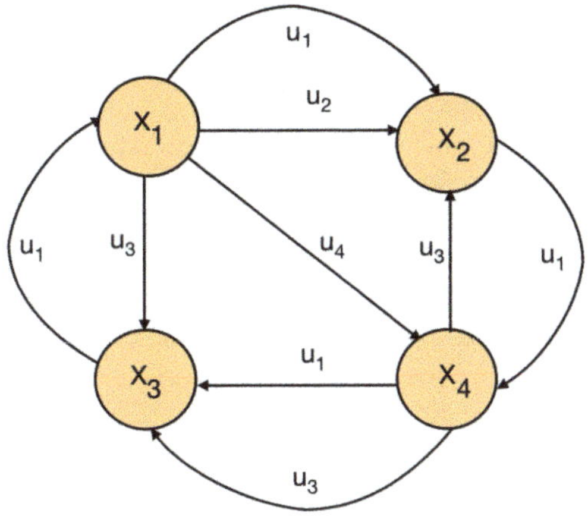

Fig. 8.4 Non-determinism—for each input u_i applied to a state X_i, a state transition is represented by an arrow. There may be two or more successor states for an input-state pair

For infinite-state systems, controlling the system's behaviour is difficult. Mostly this is due to the complexity of the system and the computational time and effort required to do any calculations. One tactic is to transform the system to a second, simpler, potentially finite, state system which satisfies the desired *specification* of the system and show it is equivalent to the known infinite-state system (see Eq. (8.2)).

$$S_a \cong S_b \tag{8.2}$$

After the transformation, further computations can be completed on this second simpler system. Additionally, the system outcomes will be as if they were computed on the more complex system. Even only an approximate equivalence, inside a margin of precision, can provide useful guarantees.

Designing System Specifications

When discussing the equivalence property, it was mentioned that the second system was one that satisfied a desired specification. If a system fails, the economic repercussions will be high. Whether that is purely in lost revenue or repair costs and other aspects affected by a system failure. By designing a rigorous specification that the system holds, it is possible to provide guarantees on system behaviour.

Model checking verifies the system operation against a given specification. The two main specifications for systems relate to safety and reachability. A *safety specification* attempts to control a system within a safe interval, while *reachability specifications* check if a system will transition and 'reach' a certain state or set of states. Specifications can also be combined, such as a *reach-and-stay specification* to reach a safe interval and then remain within it.

Special logic is used to define the system specifications; the two main ones are *Linear Temporal Logic (LTL)* which is used for specifying linear time properties and *Computational Time Logic (CTL)* used for branching time properties. In this chapter, we consider LTL specifications, but further details on both can be found in Baier and Katoen (2008).

Behaviour

Formal control synthesis leverages off the previous discussions on specifications and the equivalence properties. For two systems S_a and S_b, an equivalence is wanted. Consider that S_a relates to an infinite-state system model and S_b relates to a system which satisfies the desired specification for that system. If the behaviour of S_a is encapsulated inside the system S_b, then S_a is behaviourally included in S_b (see Eq. (8.3)).

$$B^w(S_a) \subseteq B^w(S_b) \implies S_a \preceq S_b \tag{8.3}$$

If both $S_a \preceq S_b$ and $S_b \preceq S_a$ hold, then the two systems are behaviourally equivalent, acting in the same manner. Extending this notion, it is also proven that under the same actions w, both systems will behave the same (see Tabuada (2009)).

$$B(S_a) = B(S_b) \implies B^w(S_a) = B^w(S_b) \tag{8.4}$$

$$B^w(S_a) = B^w(S_b) \implies B(S_a) = B(S_b) \tag{8.5}$$

Formal Control Synthesis

For a system S_a, by finding an equivalent simpler system, S_b, computations for control can be done in less time. The shift to a simpler system is done via an abstraction. The system S_a is partitioned, and each partition is treated as a single *symbolic state* in a *symbolic model*. The symbolic model becomes the equivalent system S_b, and this was shown mathematically in the previous sections.

The desired specification is written in formal language and given along with S_b to a formal methods software tool. The tool attempts to create a formal controller. One guaranteed to satisfy the desired specification for system S_b. If the tool is successful, then the new formal controller can be attached to system S_a. As the controller provided guarantees for S_b, it must also provide the same guarantees for S_a, as it is in Eqs. (8.4) and (8.5). Tools used for this type of controller synthesis when the system is under the influence of uncertainty are MASCOT (Majumdar et al., 2020, 2021), FAUST^2 (Soudjani et al., 2015), SReachTools (Vinod et al., 2019), AMYTISS (Lavaei et al., 2020), StocHy (Cauchi & Abate, 2019), PRISM Kwiatkowska et al. (2011), among others (Abate et al., 2020).

8.3.2 *Formal Control of Fast-Response Elements for Frequency Regulation*

ESSs, such as EVs, can connect to the grid through homes or buildings and are used in aggregation for primary frequency response. *Primary frequency response* is the fastest response that the system can provide when a large disturbance event occurs. The ideal scenario returns the frequency to a stable region in as short of a time horizon as possible. In Wooding et al. (2020), formal control synthesis techniques are used, as discussed in Sect. 8.3.1. Formal controller design requires three main steps:

1. **Define the formal specification**—using LTL, the desired system specification is written in a temporal logic. Choosing this specification can be tricky as a good understanding of the system is required in order to define wanted behaviour and unwanted behaviour;
2. **Create the symbolic model**—here the original system dynamics are converted to a symbolic representation by representing partitions of the state space as single symbolic states;
3. **Synthesise symbolic controller**—the symbolic model and the specification are combined to work out legal transitions around the system. Transitions which do not lead to the satisfaction of the specification are removed from the symbolic model. Once this process has finished, any remaining state transitions form the symbolic controller. From the mathematical properties of the system such as bisimulations, the symbolic controller can be applied to the original system with formal guarantees that the specification will be satisfied (Tabuada, 2009).

To define the formal specification, the GB grid code is used for reference (National Grid, 2020). Two key specifications are determined; these are that the frequency should remain in the safe interval (I_1) of 50 ± 0.5 Hz or for large disturbances only leave the interval for 60 s. The frequency should never fall into a containment zone 49.2 Hz. To show the strength of the formal methodology, an additional constraint was added; when the frequency returns to I_1, a second controller should step in and return the system to an even smaller region (I_2). The formal LTL description is shown in Eq. (8.6), where I_1 is the larger safe interval, C_{zone} is the containment zone frequency and I_2 is the smaller safe interval.

$$\begin{aligned}\psi := \Box(f \geq C_{zone}) \wedge [\neg(f \in I_1) \implies \Diamond(f \in I_1)] \wedge \\ [(f \in I_1 \wedge f \notin I_2) \implies \Diamond(f \in I_2)]\,. \end{aligned} \tag{8.6}$$

To provide a proof of concept, a simulation based on simplified power system dynamics (see Eqs. (8.7)–(8.9)) is used with an expected infrequent infeed loss of 2000 MW. The system is linear and has four dimensions, including the frequency which is the system output. For the system input, the percentage of ESSs which participate in demand-side response is used. Variable values can be found in the original paper (Wooding et al., 2020).

$$\dot{x}(t) = Ax(t) + Bu(t) + B_w w(t), \tag{8.7}$$

$$A = \begin{bmatrix} \frac{-D}{2H} & 0 & 0 & \frac{1}{2H} \\ \frac{1}{T_g R_{eq}} & \frac{-1}{T_g} & 0 & 0 \\ \frac{T_1}{T_2 T_g R_{eq}} & \frac{T_g - T_1}{T_g T_2} & \frac{-1}{T_2} & 0 \\ 0 & 0 & \frac{1}{T_t} & \frac{-1}{T_t} \end{bmatrix} \tag{8.8}$$

$$B = \begin{bmatrix} \frac{p_{av} \times N_{ev}}{2H} \\ 0 \\ 0 \\ 0 \end{bmatrix}, \quad B_w = \begin{bmatrix} \frac{-1}{2H} \\ 0 \\ 0 \\ 0 \end{bmatrix}. \tag{8.9}$$

The formal synthesis is completed inside the SCOTS software tool (Rungger & Zamani, 2016). Passing the formal specification, system dynamics, maximum and minimum values of the states and size of the partitions into the SCOTS tool will enable it to attempt the formal controller synthesis. As the technique involves removing states from the state space that do not lead to satisfaction of the specification, it is possible for too many transitions to be removed and dead states to exist. In this case, the synthesis will have failed and the SCOTS tool returns no controller to the user. From this, a guarantee is given that should the SCOTS tool return a controller to the user, then a guarantee is provided that the returned controller will satisfy the specification.

After successfully synthesising the controller for the system, simulations to compare the performance to a baseline controller are devised. The baseline controller accepts the system frequency, Δf, as an input and consists of a frequency deadband, a transfer function and a saturation block to determine the percentage of ESSs required to participate in frequency regulation, Δu. This participation value is multiplied by the aggregate ESS power, P_{ESS} and number of ESSs, N_{ESS}, to calculate a total demand response from ESSs for frequency regulation at that time instance, as shown in Fig. 8.5.

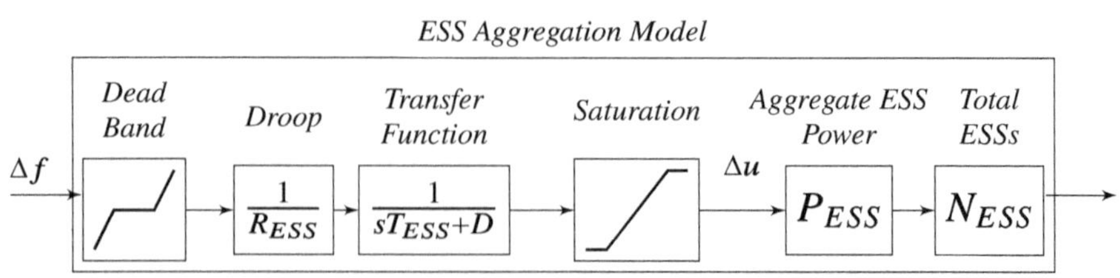

Fig. 8.5 Baseline controller of ESS frequency response services when a large power loss occurs, adapted from Mu et al. (2018), Izadkhast et al. (2015)

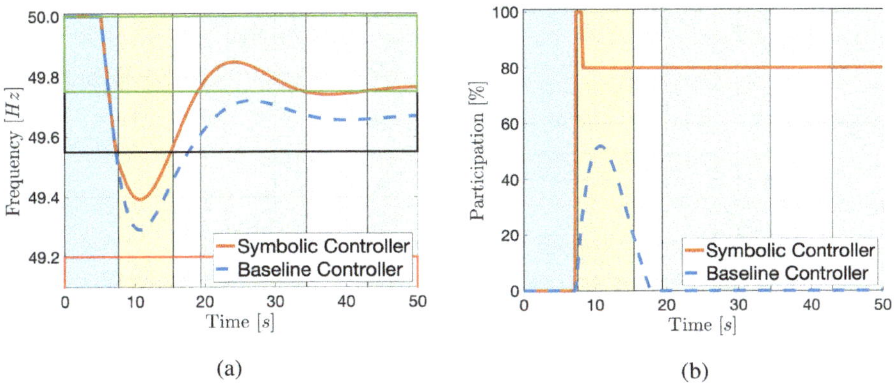

Fig. 8.6 Symbolic control of the frequency using ESSs **(a)** and the participation of ESSs over time **(b)**

The control strategy implemented involved several phases which are each represented in the further figures by a shaded colour in the background. In the first phase, there is no initial ESS participation as the frequency is inside the safe interval (I_1), phase shown in light blue. When the frequency leaves the safe interval, the first controller C_1 attempts to return the frequency to I_1 again, phase shown in yellow. Once achieved, controller C_2 takes over and tries to transition the frequency to smaller region I_2, phase shown in white. Upon entering I_2, the ESS participation value is fixed, unless the frequency oscillates outside the region or normal system operation is restored, phase shown in green. In the same manner, the baseline controller provides no participation when the frequency is inside the larger safe zone I_1, and this is managed by the deadband.

In Fig. 8.6 and the following figures, the region I_1 is marked in black, I_2 in green and C_{zone} in red. This model updated every 0.2 seconds. The frequency under the synthesised symbolic controller satisfies the formal specification ψ in (8.6) with $I_1 = [49.55, 50]$ and $I_2 = [49.75, 50]$ Hz, but the baseline controller is unable to shape the frequency with respect to ψ.

To test the robustness of the controller, uncertainty is added to ESS response. Adding a uniformly distributed random uncertainty of up to 10% participation was found to still satisfy the specification as shown in Fig. 8.7 and outperform the equivalent baseline controller. In both controller scenarios, the minimum frequency value shifts slightly lower; this is because the uncertainty in the participation reduces the amount of energy used for response. This leads to a larger RoCoF, and so, for the same time horizon, lower frequency values are reached.

This case study approach can also be applied to more complex and nonlinear systems, but with increased computations. Figure 8.7 showed the results of the system with an uncertain participation value, but extensions can also be made to include uncertainty of the plant values or increasing the participation uncertainty. To continue to provide guarantees in the cases of uncertainty, one approach uses

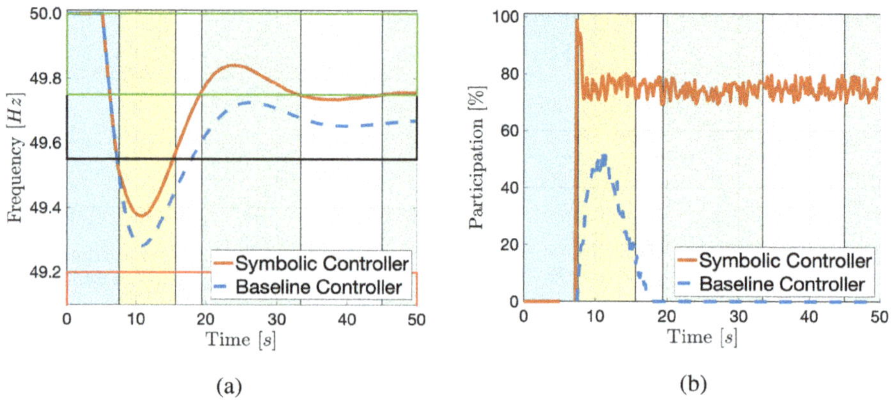

Fig. 8.7 Symbolic control of the frequency using uncertain ESSs **(a)** and the participation of uncertain ESSs over time **(b)**

formal methods for stochastic systems. The guarantees are to a certain confidence level; this can be a % confidence that the guarantees are provided or a full guarantee within a particular range.

8.3.3 MPC-Enabled Heating/Cooling Systems for Flexibility Provision

A component of uncertainty in the power system is renewable generation sources which depend on highly unpredictable factors such as wind speed and solar irradiance. Power tracking is a tool which can be used to control the changes surrounding such uncertainties. TCLs are an example component than can be used for power tracking. By using a temperature setpoint as an input, it is possible to control the energy consumption of the TCLs and use the difference in energy consumption for demand-side response services.

In this section, we will discuss a two-step formal approach for TCL power tracking using a model based on the probabilistic evolution of the TCL temperature. Firstly, the continuous space model is converted to a finite state model, which obtains a population of Markov chains. Secondly, a cross product of the Markov chain models is done to find the coarsest probabilistically bisimular Markov chain. This abstraction then has guaranteed error bounds. At the end of the section, a numerical example will be given using a modified Kalman filter to estimate the system state space and a one-step MPC for total power consumption.

For the power tracking, an accurate TCL model is required. We consider a population of homogeneous TCLs. For a single TCL, the temperature evolution can be given by Eq. (8.10), where θ_a is the ambient temperature, C is thermal capacitance, R is thermal resistance, P_{rate} is the rate of energy transfer, $w(t),\ t \in \mathbb{Z}$ is process noise and $a = e^{-h/RC}$ with discretisation step h.

$$\theta(t+1) = a\theta(t) + (1-a)(\theta_a \pm m(t)RP_{rate}) + w(t) \tag{8.10}$$

For TCLs which heat, + is used, while for TCLs which cool − is used. $m(t)$ signifies the state of the TCL; if 0, then the TCL is in the OFF state, and if 1, then the TCL is in the ON state. The temperature dynamics are regulated by

$$f(m,\theta) = \begin{cases} 0, & \theta > \theta_s - \frac{\delta}{2} \doteq \theta_- \\ 1, & \theta > \theta_s + \frac{\delta}{2} \doteq \theta_+ \\ m, & else, \end{cases}$$

where θ_s is the temperature setpoint and δ is the deadband. Together, these describe the temperature range. The power consumption of the TCL when in the ON position is given as

$$P_{rate,ON} = \frac{1}{\eta} P_{rate} \tag{8.11}$$

where η is the *coefficient of performance* (COP). Therefore, the power consumption at time t is

$$t = m(t)P_{rate,ON} \tag{8.12}$$

From the abstraction given, a single TCL can be considered a *stochastic hybrid system* (SHS). The TCL is a discrete-time Markov process evolving over a hybrid state space characterised by a variable $s = (m,\theta) \in \mathbb{Z}_1 \times \mathbb{R}$. By interpreting the TCL as an SHS, an abstraction technique from Abate et al. (2010) can be used to reduce the uncountable state space Markov process to a finite state Markov chain.

We focus on a population of cooling devices, but heating devices can be similarly derived. We will represent the homogeneous Markov chain by Ξ. Ξ has the state

$$\mathbf{z} = [z_1, z_2, \ldots, z_{n_p}]^T \in \mathcal{Z} = \mathbb{N}_{2n}^{n_p}, \tag{8.13}$$

where $z_j \in \mathbb{N}_{2n}$ represents the state of the jth Markov chain.

For the second step of the abstraction, we wish to further aggregate the model and use a notion of exact probabilistic bisimulation (Baier & Katoen, 2008). Given the labelling function, L, an equivalence relation, $\mathcal{R}$, of the state space, $\mathcal{Z}$, is defined such that

$$\forall(\mathbf{z},\mathbf{z}') \in \mathcal{R} \Leftrightarrow L(\mathbf{z}) = L(\mathbf{z}') \tag{8.14}$$

The equivalence relation provides a partition of the state space of $\mathcal{Z}$ into equivalent classes belonging to a quotient set $\mathcal{Z}/\mathcal{R}$, where each class is uniquely specified by the label of its elements. A proof is given as to why $\mathcal{R}$ is an exact probabilistic bisimulation relation on Ξ in Soudjani and Abate (2015), Soudjani and Abate (2013). From this proof, we now have an abstract model of the evolution of the TCL population based on linear stochastic equations. We refer the reader to the papers (Kamgarpour et al., 2013; Soudjani et al., 2014) for closely related formal treatment of aggregate energy modelling and regulation.

The power consumption of the aggregate model is therefore

$$y_{total}(t) = \sum_{i=1}^{n_p} m_i(t) P_{rate,ON} \tag{8.15}$$

For the abstract model, this is

$$y_a(t) = H\mathbf{X}(t), \quad H = n_p P_{rate,ON}[0_n, 1_n] \tag{8.16}$$

where the variable $\mathbf{X}$ is the normalised value of the labels $\mathbf{x}$ and $\mathbf{x}(t)$ is a bisimulation of $\mathbf{z}(t)$ of Ξ,

$$\mathbf{z}(t) \rightarrow \mathbf{x}(t)$$

and $[0_n, 1_n]$ are row vectors with all entries equal to 0 and 1, respectively. Finding the difference between the expected values of $y_{total}(t)$ and $y_a(t)$ allows for tuning the error made in estimating the total power consumption of the population from the abstraction.

Finally we will present a numerical simulation of the power tracking. The simulation uses a TCL population size of 500, temperature set point of 20 °C and a time horizon of 1 h. Other simulation values can be found in Soudjani and Abate (2015).

We set the control input to be the temperature setpoint, θ_s, of the TCL. We apply a control input to all TCLs uniformly, to obtain a homogeneous population of TCLs since this represents no prior knowledge of the state of a single TCL. A prior knowledge of the states in $\mathbf{X}(t)$ is not necessary. Instead, an online measurement of the total power consumption of the TCL population is used which allows estimating the states in $\mathbf{X}$(t) and using θ_s to track any reference signal based on a one-step output prediction.

We consider the total power consumption to be

$$y_m(t) = H\mathbf{X}(t) + v(t) \tag{8.17}$$

where v(t) is measurement noise. The state of the system is then measured using a modified classical Kalman filter. This state estimation is included in a one-step MPC

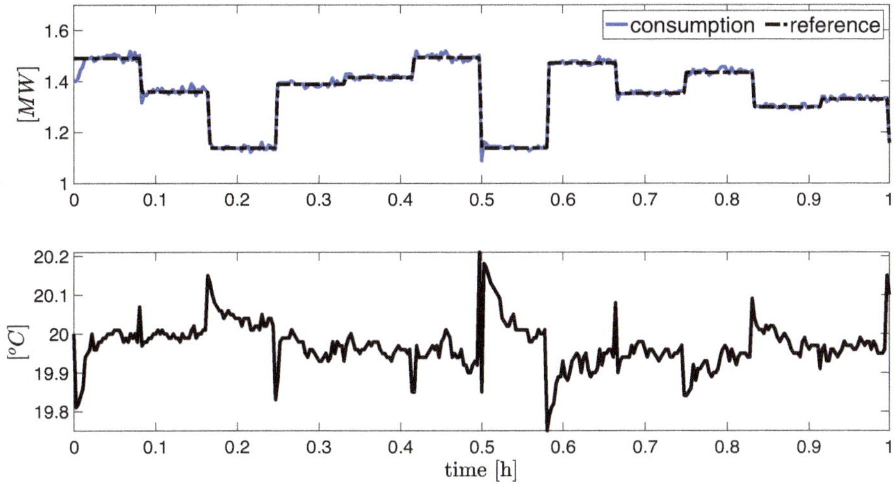

Fig. 8.8 TCL tracking outcome **(top)** and temperature setpoint θ_s **(bottom)**

scheme to synthesise the next control input. The obtained optimal value $\theta_s(t+1)$ is applied to the following iteration of the TCL population.

In Fig. 8.8, the outcomes of the TCL tracking of a reference signal is shown for a time horizon of 1 hour. Here it can be seen that a reference value can be set to control the population of TCLs; when the reference is increased, then more power is required and a percentage of TCLs switch to the ON state, and when the reference is reduced, then a percentage of TCLs switch to the OFF state. From the formal abstraction, we have guarantees over this behaviour and can exactly quantify the error added by the abstraction.

8.3.4 Coordinated Control for Provision of Frequency Response Services

In the previous two sections, we discussed a formal control approach to provide demand-side response for frequency regulation using ESSs and power tracking TCLs which were able to accurately follow a reference signal to consume a desired level of energy. In this section, we will combine the two approaches to provide a frequency response that consists of both ESSs and TCLs, in a multi-vector control scenario. We decide that the energy used in responding to the frequency loss event will come from both ESSs and TCLs, with a ESS-to-TCL ratio of $3:1$.

As described in the first section, the symbolic controller chooses a participation value, u_{ESS}, based on the current frequency. Using this participation, we can determine the amount of power that the controller is asking to be used for frequency response. By quartering that value, a desired reference, y_{des} is found that can be

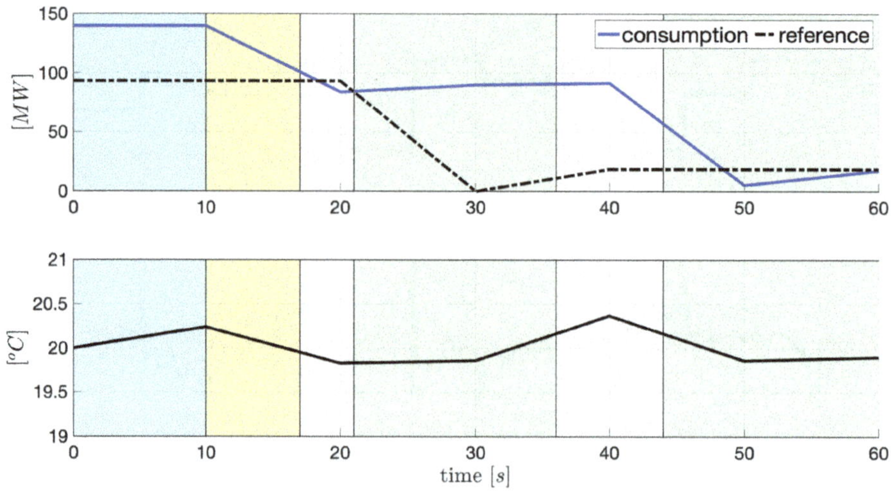

Fig. 8.9 **Top**—TCL reference tracking to provide demand response. **Bottom**—TCL temperature setpoint changes used to track the reference

fed into the power tracking algorithm of the TCLs; see Eq. (8.18) where $N_{ESS} = 25,000$, $N_{TCL} = 50,000$, $P_{ESS} = 0.028$ MW and $P_{TCL} = 0.014$ MW. The desired reference is used to select the setpoint temperature, θ_s, which reduces TCL energy consumption to be used for demand response.

$$y_{des} = (1 - u_{ESS}) \times \frac{N_{ESS} P_{ESS}}{4 N_{TCL} P_{TCL}} \tag{8.18}$$

TCLs respond slower than ESSs as there is a delay for the components to heat or cool to the desired setpoint. This can be seen in Fig. 8.9 where there is a clear delay before the reference value and consumption value align. This consumption value is what will be used in our frequency response model of the TCLs as the delay provides a realistic response of the TCL behaviour. **Note:** the decrease in consumption is equivalent to an increase in energy generation as it will be shown in Fig. 8.10b.

Unlike ESSs, TCLs cannot respond to frequency changes in under a second, so the timestep of the model will be set to 1 s rather than 0.2 s as in the previous section.

In Fig. 8.10, it is clear that ESSs and TCLs can combine to perform a valuable frequency response. The performance of the combined approach is slower than for 100% ESS scenario, and the frequency reaches to a lower minimum frequency value than that scenario. However, the combined demand-response still fulfils the requirements given in the formal specification from Eq. (8.6) and provides an improved frequency response compared to the baseline controller. On top of this, the slow response requires less initial energy from the responding components than for the purely ESS scenario, where 100% ESSs were asked to participate.

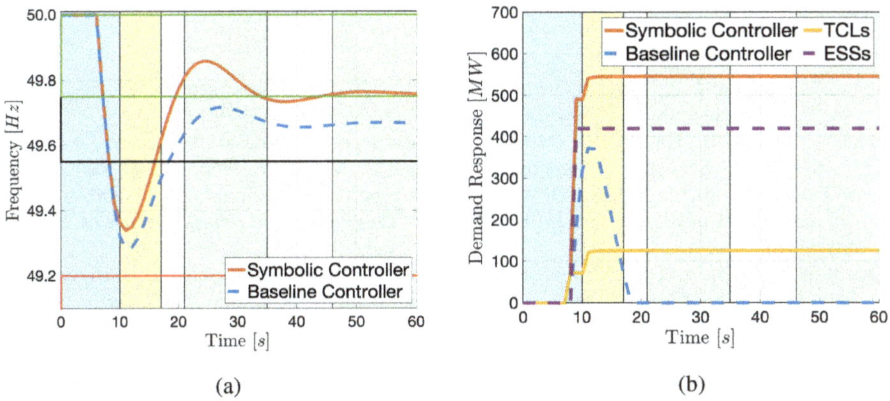

Fig. 8.10 Symbolic control of the frequency using ESSs and TCLs **(a)** and the combined demand response over time **(b)**

8.4 Status Quo, Challenges and Outlook

To conclude, this chapter has discussed the control and management of a community of active buildings. Specifically, we looked at the control of frequency and voltage and discussed the controls used in multi-vector energy and aggregators problem, as well as cyber security aspects. Finally, a case study using a symbolic controller for demand-side frequency response of ESSs and TCLs was shown to provide beneficial controls, particularly that the control technique has guarantees on performance.

There are three future directions for researchers to consider having read this chapter.

- Firstly, the smart grid is evolving to be more distributed and electrified as more components join the power network. This increased load will add a strain to the network unless control schemes are developed to safely manage these new components while also keeping users satisfied.
- Secondly, as the smart grid becomes more distributed, the need for multi-vector energy controls will keep growing; this requires accurate models of the vector components and also inclusion of distributed resources such as buildings and EVs. The importance of accurate models is a recurring theme of this chapter and will benefit all aspects of control.
- Finally, as more uncertain renewable sources are incorporated into the grid, the amount of system inertia decreases and the RoCoF increases. VSGs and ESSs will become increasingly important to the future of the smart energy systems.

References

Abate, A., Katoen, J.-P., Lygeros, J., & Prandini, M. (2010). Approximate model checking of stochastic hybrid systems. *European Journal of Control, 16*(6), 624–641.

Abate, A., Blom, H., Cauchi, N., Delicaris, J., Hartmanns, A., Khaled, M., Lavaei, A., Pilch, C., Remke, A., Schupp, S., Shmarov, F., Soudjani, S., Vinod, A., Wooding, B., Zamani, M., & Zuliani, P. (2020). ARCH-COMP20 category report: Stochastic models. In *ARCH20. 7th International Workshop on Applied Verification of Continuous and Hybrid Systems (ARCH20)*. EasyChair.

Agalgaonkar, Y. P., Pal, B. C., & Jabr, R. A. (2014). Distribution voltage control considering the impact of PV generation on tap changers and autonomous regulators. *IEEE Transactions on Power Systems, 29*(1), 182–192.

Anees, A., & Chen, Y. P. P. (2016). True real time pricing and combined power scheduling of electric appliances in residential energy management system. *Applied Energy, 165*, 592–600.

Baier, C., & Katoen, J.-P. (2008). *Principles of Model Checking*. MIT Press.

Bevrani, H. (2009). *Robust power system frequency control*. Springer.

Bevrani, H., Ise, T., & Miura, Y. (2014). Virtual synchronous generators: A survey and new perspectives. *International Journal of Electrical Power & Energy Systems, 54*, 244–254.

Cauchi, N., & Abate, A. (2019). StocHy: automated verification and synthesis of stochastic processes. In *25th International Conference on Tools and Algorithms for the Construction and Analysis of Systems (TACAS)* (2019)

Devaux, P., & Farid, M. M. (2017). Benefits of PCM underfloor heating with PCM wallboards for space heating in winter. *Applied Energy, 191*, 593–602.

Dreidy, M., Mokhlis, H., & Mekhilef, S. (2017). Inertia response and frequency control techniques for renewable energy sources: a review. *Renewable and Sustainable Energy Reviews, 69*, 144–155.

Elliott, T., Geske, J., & Green, R. (2020). *White Paper: Can You Make Money from Active Buildings? Challenges Facing Business Models*. Active Buildings Centre Research Programme, Tech. Rep.

Gkatzikis, L., Koutsopoulos, I., & Salonidis, T. (2013). The role of aggregators in smart grid demand response markets. *IEEE Journal on Selected Areas in Communications, 31*(7), 1247–1257.

Greenwood, D. M., Lim, K. Y., Patsios, C., Lyons, P. F., Lim, Y. S., & Taylor, P. C. (2017). Frequency response services designed for energy storage. *Applied Energy, 203*, 115–127.

Huang, B., Cardenas, A. A., & Baldick, R. (2019). Not everything is dark and gloomy: Power grid protections against IoT demand attacks. *Proceedings of the 28th USENIX Security Symposium*, 1115–1132.

Izadkhast, S., Garcia-Gonzalez, P., & Frías, P. (2015). An aggregate model of plug-in electric vehicles for primary frequency control. *IEEE Transactions on Power Systems, 30*(3), 1475–1482.

Kamgarpour, M., Ellen, C., Soudjani, S., Gerwinn, S., Mathieu, J., Mullner, N., Abate, A., Callaway, D., Fränzle, M., & Lygeros, J. (2013). Modeling options for demand side participation of thermostatically controlled loads. In *International Conference on Bulk Power System Dynamics and Control (IREP)*, August 2013 (pp. 1–15).

Karimi, M., Mokhlis, H., Naidu, K., Uddin, S., & Bakar, A. H. A. (2016). Photovoltaic penetration issues and impacts in distribution network – A review. *Renewable and Sustainable Energy Reviews, 53*, 594–605.

Kish, G. J., Ranjram, M., & Lehn, P. W. (2015). A modular multilevel DC/DC converter with fault blocking capability for HVDC interconnects. *IEEE Transactions on Power Electronics, 30*(1), 148–162.

Kwiatkowska, M., Norman, G., & Parker, D. (2011). PRISM 4.0: Verification of probabilistic real-time systems. In G. Gopalakrishnan & S. Qadeer, (Eds.) *Proceedings of the 23rd International Conference on Computer Aided Verification (CAV'11), LNCS*(Vol. 6806, pp. 585–591). Springer.

Lavaei, A., Khaled, M., Soudjani, S., & Zamani, M. (2020). AMYTISS: Parallelized automated controller synthesis for large-scale stochastic systems. In *Proceedings of the 32nd International Conference on Computer Aided Verification (CAV), LNCS*. Springer.

Mahmud, N., & Zahedi, A. (2016). Review of control strategies for voltage regulation of the smart distribution network with high penetration of renewable distributed generation. *Renewable and Sustainable Energy Reviews, 64*, 582–595.

Majumdar, R., Mallik, K., & Soudjani, S. (2020). Symbolic controller synthesis for büchi specifications on stochastic systems. In *Proceedings of the 23rd International Conference on Hybrid Systems: Computation and Control. Series. HSCC '20*. New York, NY, USA: Association for Computing Machinery.

Majumdar, R., Mallik, K., Schmuck, A.-K., & Soudjani, S. (2021). *Symbolic control for stochastic systems via parity games*. arXiv:2101.00834.

Merlin, M. M. C., Green, T. C., Mitcheson, P. D., Trainer, D. R., Critchley, R., Crookes, W., & Hassan, F. (2014). The alternate arm converter: A new hybrid multilevel converter with dc-fault blocking capability. *IEEE Transactions on Power Delivery, 29*(1), 310–317.

Mu, C., Liu, W., & Xu, W. (2018). Hierarchically adaptive frequency control for an EV-integrated smart grid with renewable energy. *IEEE Transactions on Industrial Informatics, 14*(9), 4254–4263.

National Grid (2020). The Grid Code. National Grid, Tech. Rep.

Paramasivam, M., Salloum, A., Ajjarapu, V., Vittal, V., Bhatt, N. B., & Liu, S. (2013). Dynamic optimization based reactive power planning to mitigate slow voltage recovery and short term voltage instability. *IEEE Transactions on Power Systems, 28*(4), 3865–3873.

Reynolds, J., Ahmad, M. W., & Rezgui, Y. (2018). Holistic modelling techniques for the operational optimisation of multi-vector energy systems. *Energy and Buildings, 169*, 397–416.

Reza Hosseini, S. H., Allahham, A., Vahidinasab, V., Walker, S. L., & Taylor, P. (2021). Techno-economic-environmental evaluation framework for integrated gas and electricity distribution networks considering impact of different storage configurations. *International Journal of Electrical Power & Energy Systems, 125*, 106481.

Rogers, K. M., Klump, R., Khurana, H., Aquino-Lugo, A. A., & Overbye, T. J. (2010). An authenticated control framework for distributed voltage support on the smart grid. *IEEE Transactions on Smart Grid, 1*(1), 40–47.

Rungger, M., Zamani, M. (2016). SCOTS: A tool for the synthesis of symbolic controllers. In *Hybrid Systems: Computation and Control (HSCC)*. ACM.

Siano, P., & Sarno, D. (2016). Assessing the benefits of residential demand response in a real time distribution energy market. *Applied Energy, 161*, 533–551.

SMA Solar Technology AG (2017). *Statement by SMA Solar Technology AG on the Cyber Security of PV Inverters (Horus Scenario)*. White Paper.

Soudjani, S., & Abate, A. (2013). Aggregation of thermostatically controlled loads by formal abstractions. In *European Control Conference*, Zurich, Switzerland, July 2013 (pp. 4232–4237).

Soudjani, S., & Abate, A. (2015). Aggregation and control of populations of thermostatically controlled loads by formal abstractions. *IEEE Transactions on Control Systems Technology, 23*(3), 975–990.

Soudjani, S., Gerwinn, S., Ellen, C., Fränzle, M., & Abate, A. (2014). Formal synthesis and validation of inhomogeneous thermostatically controlled loads. In *International Conference on Quantitative Evaluation of Systems* (pp. 57–73). Springer.

Soudjani, S., Gevaerts, C., & Abate, A. (2015). FAUST^2: formal abstractions of uncountable-state stochastic processes. In *TACAS. LNCS* (Vol. 9035, pp. 272–286). Springer

Strbac, G., Woolf, M., Pudjianto, D., Zhang, X., Walker, S., & Vahidinasab, V. (2020). *White Paper: The Role of Active Buildings in the Transition to a Net Zero Energy System*. Active Buildings Centre Research Programme, Tech. Rep.

Sun, H., Guo, Q., Qi, J., Ajjarapu, V., Bravo, R., Chow, J., Li, Z., Moghe, R., Nasr-Azadani, E., Tamrakar, U., Taranto, G. N., Tonkoski, R., Valverde, G., Wu, Q., & Yang, G. (2019). Review of challenges and research opportunities for voltage control in smart grids. *IEEE Transactions on Power Systems, 34*, 2790–2801.

Szymanski, T. H. (2017). Security and privacy for a green internet of things. *IT Professional, 19*(5), 34–41.

Tabuada, P. (2009). *Verification and control of hybrid systems: A symbolic approach*. Springer US.

Vinod, A. P., Gleason, J. D., & Oishi, M. M. K. (2019). SReachTools: A MATLAB stochastic reachability toolbox. In *Proceedings of the 22nd ACM International Conference on Hybrid Systems: Computation and Control. HSCC '19* (pp. 33–38). New York, NY, USA: Association for Computing Machinery.

Vorobev, P., Greenwood, D. M., Bell, J. H., Bialek, J. W., Taylor, P. C., & Turitsyn, K. (2019). Deadbands, droop, and inertia impact on power system frequency distribution. *IEEE Transactions on Power Systems, 34*(4), 3098–3108.

Wang, Y., Silva, V., & Lopez-Botet-Zulueta, M. (2016). Impact of high penetration of variable renewable generation on frequency dynamics in the continental Europe interconnected system. *IET Renewable Power Generation, 10*(1), 10–16.

Wooding, B., Vahidinasab, V., & Soudjani, S. (2020). Formal controller synthesis for frequency regulation utilising electric vehicles. In *2020 International Conference on Smart Energy Systems and Technologies (SEST)* (pp. 1–6)

Wu, D., Radhakrishnan, N., & Huang, S. (2019). A hierarchical charging control of plug-in electric vehicles with simple flexibility model. *Applied Energy, 253*, 1-1.

Zarina, P. P., Mishra, S., & Sekhar, P. C. (2014). Exploring frequency control capability of a PV system in a hybrid PV-rotating machine-without storage system. *International Journal of Electrical Power & Energy Systems, 60*, 258–267.

Chapter 9
Cybersecurity Roadmap for Active Buildings

Ricardo M. Czekster, Charles Morisset, Aad van Moorsel, John C. Mace, Walter A. Bassage, and John A. Clark

9.1 Introduction

Strict cyber-physical security requirements in modern smart infrastructures (Mo et al., 2011) shield users and customers from adversaries committing malicious interventions. The smart grid (SG) (Greer et al., 2014; Amin & Wollenberg, 2005; Fang et al., 2011) is a cyber-physical system (CPS) (Gunes et al., 2014; Humayed et al., 2017) considered a critical infrastructure. The SG comprises buildings in residential, commercial, and industrial settings. Buildings are targets for performance improvements because they are the primary pollution drivers responsible for 40% of global energy use according to the International Energy Agency (IEA).[1] The UK government's ambition is to curtail carbon emissions by 80% until 2050 (baseline of 1990) according to the Future Energy Scenarios[2]; thus, urgent action is necessary.

[1] IEA publishes energy datasets to support secure and sustainable energy with global scope. Link (for buildings): https://www.iea.org/topics/buildings. All links here were accessed in January/2021.

[2] For the UK's Future Energy Scenarios (2020 edition), please refer the following link: https://www.nationalgrideso.com/future-energy/future-energy-scenarios/fes-2020-documents.

R. M. Czekster (✉) · C. Morisset · A. van Moorsel · J. C. Mace
Newcastle University, Newcastle, UK
e-mail: ricardo.melo-czekster@ncl.ac.uk; charles.morisset@ncl.ac.uk; aad.vanmoorsel@ncl.ac.uk; john.mace@ncl.ac.uk

W. A. Bassage · J. A. Clark
University of Sheffield, Sheffield, UK
e-mail: w.bassage@sheffield.ac.uk; john.clark@sheffield.ac.uk

V. Vahidinasab, B. Mohammadi-Ivatloo (eds.), *Active Building Energy Systems*, Green Energy and Technology, https://doi.org/10.1007/978-3-030-79742-3_9

Active buildings (AB) (Clarke et al., 2020) are a novel approach built on top of the SG functioning as a grid-connected microgrid.[3] They retain connectivity to the conventional infrastructure and are able to operate in islanded mode if required (Hatziargyriou et al., 2007). The idea is to extend functionality by three axes (Strbac et al., 2020): (i) increased adoption of renewable energy resources (RER) such as solar panels using photovoltaic (PV) technologies or wind turbine power converter systems (WTPCS); (ii) distributed energy storage systems (ESS) such as static batteries attached to buildings or mobile ones in electric vehicles (EV); (iii) thermal energy storage systems (TES) for balanced building thermodynamics within its envelope, energy storage, and comfort level adjustments while heating or cooling. This decentralised infrastructure operates in low-voltage distribution grids (LVDG) promoting decarbonisation and grid stability and congregating customers that may produce and consume power as so called *prosumers* (Greer et al., 2014; Zafar et al., 2018) that participate in dynamic energy markets. RER, ESS and TES (among others) are considered key power-based assets referred as distributed energy resources (DER) (Driesen & Katiraei, 2008) that are managed and controlled by a diversified number of systems and stakeholders.

AB aim to shift passive energy users towards active behaviour enacting '*buildings as power stations*' (Bankovskis, 2017; Nikolaidou et al., 2020). Under these contexts, bi-directional energy flows add stability, reliability and trading capabilities across peer-to-peer (P2P) agents. It improves ancillary services such as frequency regulation and the management of power reserves for continuous energy provision (Ma et al., 2013). AB operate on the edge of conventional grids as nearly net-zero energy buildings (nZEB) (Attia, 2018; Kurnitski et al., 2011). nZEB approximate energy generation to consumption over a fixed duration.[4] The buildings in the P2P network trade the energy surplus in the market or with their neighbours and make operational decisions about planning and management.

AB encompass a plethora of sub-systems and devices, notably Internet of Things (IoT) capable of sensing and communicating data. It relies on the pervasive use of information and communication technologies (ICT) and information systems (IS). These elements relay the operational status of power and telecommunication networks to an operation control centre (OCC) that provide situational awareness, enact timely response coordination and make planning decisions to address supply and demand prognostics.

[3] According to NISTIR 7628 Rev. 1, microgrid is *'an implied hierarchy in availability and resilience eliminates potential peer-to-peer negotiations between microgrids. Its models suggest that availability starts in a local microgrid and that resilience is gained by aggregating and interconnecting those microgrids. They are intended to operate either as islands or interconnected; islands are key where critical operations need to be maintained'*..

[4] This definition is sanctioned by the European Commission on Directive 2010/31/EC, stating that *''nearly zero-energy building' means a building that has a very high energy performance, as determined in accordance with Annex I. The nearly zero or very low amount of energy required should be covered to a very significant extent by energy from renewable sources, including energy from renewable sources produced on-site or nearby;.'*

In such highly connected and data-centric settings, it is essential to provide *trustworthiness*, i.e. security, privacy, safety, reliability and resilience[5] (Sridhar et al., 2012; He & Yan, 2016; Humayed et al., 2017; Giraldo et al., 2017) to withstand the harmful effects of cyber-attacks or abnormal situations. Organisations usually require cybersecurity officers and managers to perform risk management[6] processes throughout system's life cycle. One way is to employ a risk assessment (RA) methodology to map vulnerabilities and threats to compute risk and exposure level (Abercrombie et al., 2013; Wangen, 2017; Leszczyna, 2018). Organisations devise RA as an ongoing effort to increase the trustworthiness of their underlying systems and infrastructure. Over the years, authoritative bodies have defined standards, assessment methodologies, recommendations, and guidelines in the USA and Europe (Leszczyna, 2018). For example, Ruland et al. (2017) discussed security standards, whereas Gritzalis et al. (2018) addressed how to select suitable RA methodologies. In terms of NIST overview on cybersecurity and guidelines for RA, we mention NISTIR 7628 (Pillitteri & Brewer, 2014) and most notably NIST.SP 800-30 (Initiative, 2012). The level of detail, effort, rigour and quantitative/qualitative aspects to consider are left for the organisation to decide in line with its objectives.

We present here a roadmap for tackling cyber-physical security in AB. Our aim is to focus on RA, security metrics, intrusion detection, threat modelling and simulation. The audience are cybersecurity officers and building managers overviewing the vulnerability and threat landscape posed by AB. In terms of RA, we describe major methodologies from accredited institutions and complement the research with established cybersecurity risk factors behind AB's designs. The novelty of our approach concerns infrastructures with active energy agents under coupled power and telecommunications networks. These architectures have sizeable DER attached in LVDG, so one must observe the cyber-physical security implications and the potential attack surface. Our work points out the overlap of AB with usual SG designs with respect to cybersecurity as we survey related mechanisms and technologies to address effective RA.

The chapter is organised as follows. Section 9.2 describes active buildings, and Sect. 9.3 details cyber-physical security of SG and CPS discussing its application to AB. Section 9.4 presents a research roadmap and suggestions to tackle cybersecurity across AB domains, and Sect. 9.5 concludes the work with our final considerations. Figure 9.1 serves as a visual guide to readers outlining the topics covered in our chapter.

---

[5] *Resilience* in the context of this work is the ability of a system to remain in operation and continue to service in the presence of incidents or events that attempt to disrupt or shut down assets.

[6] The US National Institute of Standards and Technology (NIST) has published a wealth of standards for cybersecurity and risk management, including a comprehensive framework to help managers. Link: https://csrc.nist.gov/projects/risk-management/.

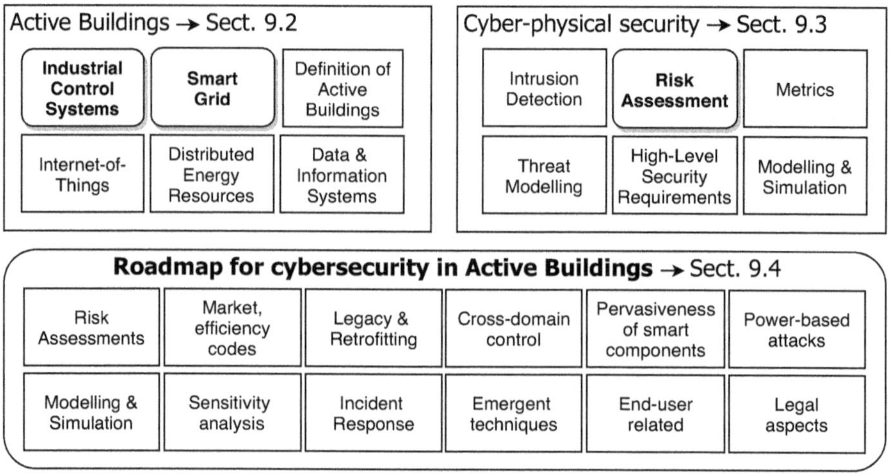

Fig. 9.1 Overview of topics covered by this chapter

9.2 Active Buildings

Electrical power grids (EPG) are responsible for delivering high-quality energy to consumers in power generation, transmission and distribution (GT&D). AB are *'buildings to sustain a country's energy infrastructure'* (Nikolaidou et al., 2020) on top of EPG. They operate in a grid-connected microgrid handling bi-directional energy flows with respect to conservation, generation, storage and release of power. AB contrast with traditional (centralised) approaches in GT&D by presenting a decentralised infrastructure operating in smaller contexts for increased control.

The core idea behind AB is *flexibility* (Nikolaidou et al., 2020; Strbac et al., 2020). It shifts from passive energy consumers to active entities that efficiently respond to supply and demand for optimal grid stability. As millions of potential energy subscribers attach their RER into the infrastructure, buildings may act as agents for managing resilience while reducing dependence on high-pollutant conventional power generators. As a direct consequence, the carbon footprint generated by buildings diminishes considerably and cascades across national and international boundaries.[7]

Energy operators achieve frequency regulation (He & Yan, 2016) in the grid by employing a number of strategies, namely, (i) toggling electricity in power plants; (ii) shaping user behaviours thus curtailing electricity through pricing incentives; (iii) importing or exporting electricity with close geographical neighbours; (iv)

[7] The UK's National Grid Electricity System Operator (ESO) has published a document entitled *Operability Strategy Report 2021* in December/2020 with comments on existing frequency response services, voltage requirements, and grid stability. Link: https://www.nationalgrideso.com/research-publications/system-operability-framework-sof.

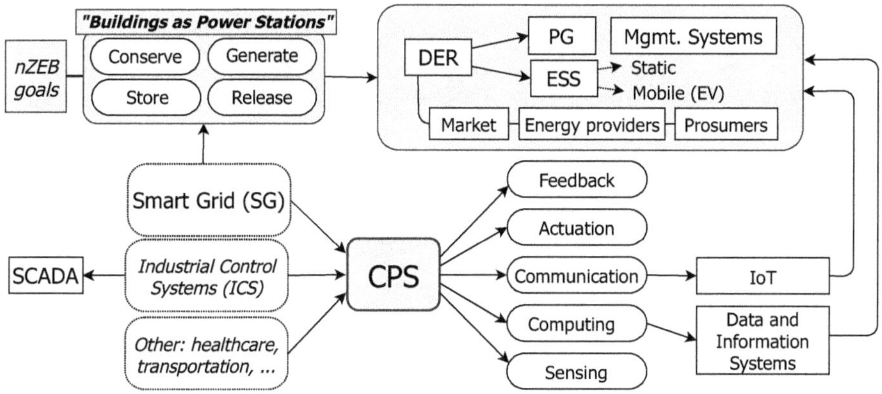

Fig. 9.2 Architectural overview of AB and CPS

employing *load shedding* mechanisms, i.e. partially disconnecting a region from the grid; and (v) increasing electricity storage capacity with batteries (static or EV) (Greenwood et al., 2017). Figure 9.2 depicts the major components of AB, i.e. CPS and IoT, ICS (Stouffer et al., 2015) and SCADA, where IS aggregate data from different sources for timely decision-making (Humayed et al., 2017).

AB attach and manage DER across locations where installed ESS capacity helps tackling renewable energy intermittency to provide 'around the clock' electricity to customers (Izadkhast et al., 2015). They leverage power release and storage employing optimisation strategies that consider season, weather forecasts, time of the day and energy profiles. AB interface with the energy market in bidding and committing energy contracts. Ubiquitous telecommunication guarantees near real-time state estimation for control and response, while cybersecurity measures in place have the potential to isolate the effects of incidents and avoid propagation.

The major stakeholders of AB are as follows: (i) building managers: people working in OCC assess data provided by building management systems (BMS), building energy management systems (BEMS), advanced metering infrastructure (AMI) (Greer et al., 2014) specialists and SCADA operators; (ii) decision-makers and procurement officers acquiring power equipment or supporting ICT systems and compliance officers overseeing regulations dictated by the ESO; (iii) security officers addressing conformance in safety and security, accountability and digital forensics, privacy protections, Network Intrusion Detection Systems (NIDS) and network administrators; (iv) incident response teams in attack remediation; and (v) aggregators, energy suppliers in wholesale and retail, prosumers and application developers implementing solutions on top of the infrastructure.

Figure 9.3 shows a schematic for AB where the power and the telecommunication are dissimilar, and power may be dispatched or aggregated by the set of DER.

The AB's infrastructure is not immune against malicious incursions. We detail next a few notable cybersecurity concerns (Sridhar et al., 2012; He & Yan, 2016;

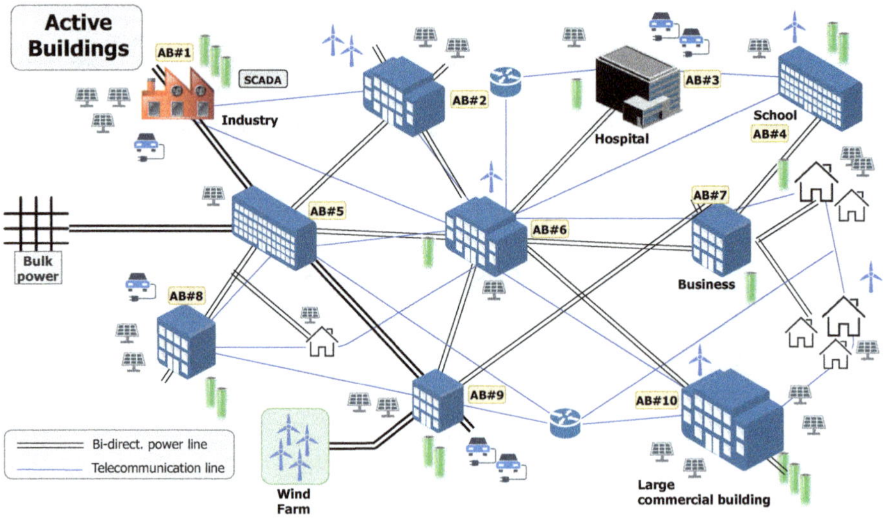

Fig. 9.3 AB schematic for power and telecommunication, components and systems

Ding et al., 2018) and NIST 7628 recommendations (Pillitteri & Brewer, 2014) particular to IS and cybersecurity requirements for the SG.

- *Business proposition*: trading energy, procurement (suppliers), generation and consumption, impacting market fairness.
- *Data and communication*: invalid readings/measurements from devices. Adversaries may corrupt or delay data traversed in the telecommunication network, jamming exchanges or inserting spurious packets.
- *Components and IS*: attackers inputting corrupted data in CPS, IoT, SCADA elements, BMS/BEMS.
- *Optimality*: the AB environment is supported by several systems, so attacks aiming at the control algorithms executing on top of the infrastructure may cause critical responses to be wrongly assigned with catastrophic consequences.
- *Synchronised incursions*: attackers may install malicious firmware in high-wattage devices in attempts to imbalance frequency through synchronised actions, i.e. simultaneously turning on or off (refer to Sect. 9.3.1).
- *Legacy systems*: retrofitting buildings may increase vulnerabilities and pose new threats as customers attach DER or smart appliances. They may disregard basic cybersecurity measures (e.g. plugging-in unsigned devices), lax systems maintenance (patching, updates) or inadvertently installing malicious firmware or as recipients of phishing attacks (Sridhar et al., 2012).

Other sources of concern are advanced persistent threats (APT) (Kim & Tong, 2013; Camana Acosta et al., 2020) and load redistribution attacks (LRA) (Yuan et al., 2011) that are hard to track and pinpoint their sources, often requiring long duration tracking and historic datasets. APT have devastating consequences to EPG (Kim & Tong, 2013; Kumar et al., 2019), as attackers may patiently gather data on grid responses in long-term surveillance and hostile reconnaissance. They are more destructive than usual attacks as it may involve large organisations or state-sponsored agents exploiting backdoors in CPS or other vulnerable assets (Gunduz & Das, 2020). Honeypots and continuous auditing are known countermeasures that may identify and contain APT.

9.2.1 Attack Surface

AB bring together energy and ICT infrastructure connecting and upgrading smart buildings (SB) to co-exist in an active P2P environment. SB are a data-centric approach where cyber-physical structures are intertwined with sensing capabilities, intelligent systems and feedback loops. They allow for the remote management of assets ensuring inhabitants' comfort considering thermodynamics and automatically adapting to weather conditions and luminosity, among other features.

Under AB, one must distinguish security as the set of contingencies in place in the event of faults from cybersecurity incidents on system's vulnerabilities and threats. Standardisation institutes such as the North American Electric Reliability Corporation (NERC) in the USA and the European Programme for Critical Infrastructure Protection (EPCIP) defined security measures and contingencies for power systems. For example, EPG are designed to sustain N-1 single contingency criterion[8] to meet reliability constraints (there are other contingencies to address, out of the scope of this work). This means that the system admits a single failure to remain operating.

The AB infrastructure must withstand cybersecurity incidents by implementing and enforcing security requirements (Pillitteri & Brewer, 2014). Adversaries[9] target cyber-physical elements, exploiting vulnerabilities. The reasons behind attacks are usually related to competitive advantage, industrial espionage (or state sponsored agent), energy theft (monetary incentives or ransomware), technical challenge or terrorism. Adversaries attempt to destabilise operation, cause components to fail, exploit vulnerabilities, increase downtime or impair communications (Humayed et al., 2017).

AB present a large attack surface for adversaries due to the sheer size of their infrastructure, as shown in Fig. 9.4. Telecommunication is pervasive across AB and attackers attempt to maliciously influence major AB's objectives on power conservation, generation, storage, release, or a combined approach for greater

[8] Please, refer to NERC Standard 51—*Transmission System Adequacy and Security* (2005). Link: http://www.nerc.com.

[9] The 'adversary' or 'attacker' definition in use here is based on the *Categories of Adversaries to IS* described in NISTIR 7628r1: careless/poorly trained employees, malicious customers, insiders, organised crime, nation states, disgruntled employees, terrorists, script kiddies, hacktivists and black/white hat hackers.

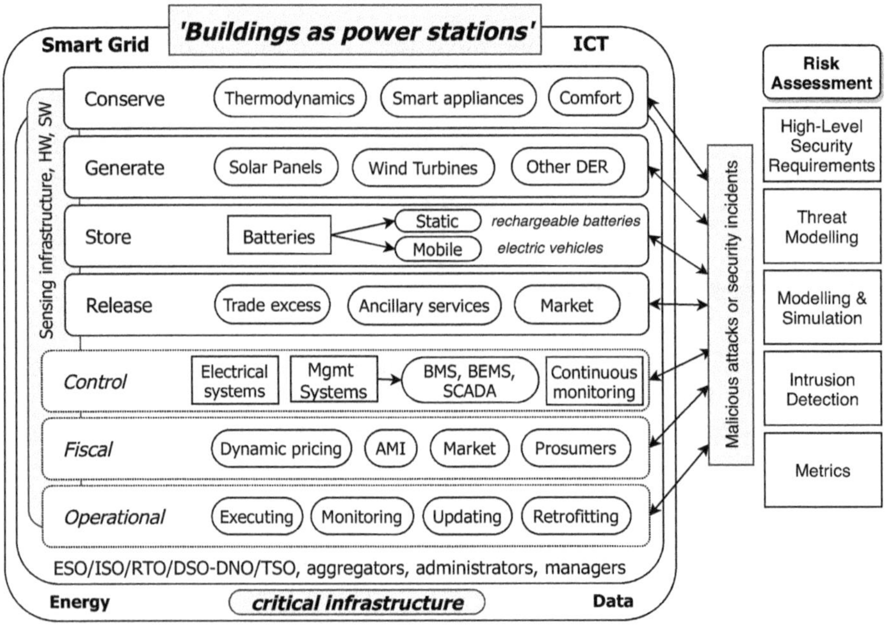

Fig. 9.4 Attack surface of AB and cybersecurity

damage. The peers interface with the energy market, so adversaries have financial incentives (or other motives) as they try to artificially alter electricity prices.

9.2.2 Interfacing with the Transactive Energy Market

The AB are connected to the energy market as prices fluctuate according to supply and demand. Transactive Energy (TE) aims to coordinate DER to react to energy prices and system conditions that could be used in frequency regulation (Huang et al., 2019b). The idea is to use the market to shape demand through dynamic pricing mechanisms. The ESO interacts with the systems using middleware capabilities offered by a layer known as the virtual power plant (VPP) to communicate with deployed field devices.

The VPP also coordinates interactions with distribution system operators (DSO), transmission system operators (TSO), independent system operators (ISO) and regional transmission operators (RTO) when submitting energy bids and commitments (Pudjianto et al., 2007). The VPP triggers actions to access DER scattered across locations, turning buildings into actual power plants (Royapoor et al., 2020). From a cybersecurity perspective, the VPP is a high-valued target due to the amount of IS it encompasses and the responsibilities to timely command DER.

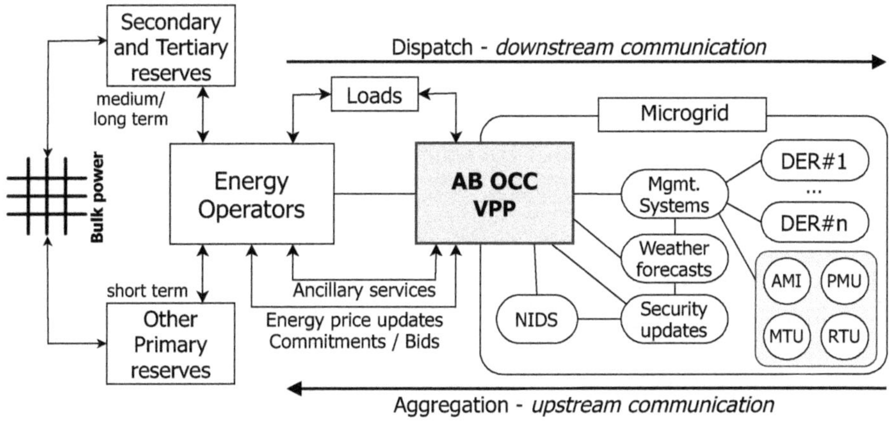

Fig. 9.5 OCC for AB managing power loads with the VPP across DER and energy operators in a grid-connected microgrid

Figure 9.5 shows a typical OCC for AB serving as an interface for the market and energy operators and the VPP coordinating DER (Zajc et al., 2019; Kolenc et al., 2017). The OCC for AB monitors the infrastructure and coordinates responses in the event of failures.

This architecture is flexible enough to accommodate changes as the stakeholders see fit to mitigate attacks, prevent vulnerabilities, and harden the infrastructure against attacks. The VPP is fed by IS installed within the AB's infrastructure. Energy operators keep nominal frequency around 50Hz (the UK's National ESO admits a 1% variation) to avoid disconnecting parts of the grid, used as a last resort. Control systems in EPG perform load frequency control (LFC) responsible for the maintenance of load over generation ratio in the power grid (He & Yan, 2016). The cybersecurity modules in the VPP update their systems with recent vulnerabilities and weather forecasts. SCADA devices are responsible for providing data from devices deployed away from the OCC (Stouffer et al., 2015). Phasor measurement units (PMU) monitor frequency, whereas remote terminal units (RTU) relay data in the network to master terminal units (MTU).

9.3 Cyber-Physical Security

Cyber-physical security encompasses a wealth of techniques to protect systems and customers. Next, we discuss an overview of cybersecurity in the SG, RA, intrusion detection, threat modelling, security metrics and co-simulation of incidents in AB.

9.3.1 Overview of Smart Grid Cybersecurity

We present here a cybersecurity overview and list related work on CPS and ICS (Stouffer et al., 2015) that could be borrowed for adaptation and application in AB. Authors compartmentalise the complex infrastructure of the SG and focus effort on more manageable parts, e.g. CPS, IoT, smart homes, IS and synchronised attacks.

Usual attacks perpetrated at network components in EPG are Man-in-The-Middle (MiTM), Denial-of-Service (DoS or its distributed variation—DDoS), eavesdropping, jamming, spoofing and packet flooding, to mention a few (He & Yan, 2016; Kumar et al., 2019). Another class of malicious events are called *zero-day attacks* where adversaries may choose targets not yet patched due to some recently discovered security vulnerability, which may compromise resources and quickly propagate across the network. Radhakrishnan et al. (2019) compiled a list of zero-day and malware incursions over past years, discussing detection and mitigation.

Over recent past security breaches targeting energy have caused major disruption and blackouts. Attacks targeting energy infrastructure caused severe financial losses and even physical damage to components. One example is the Stuxnet worm that attacked SCADA systems in late 2015 significantly interrupting the progress of the Iranian nuclear program (Langner, 2011; Farwell & Rohozinski, 2011). Another malicious incursion with massive proportions was the attack at the Ukrainian power grid that blacked-out large portions of the infrastructure with interruptions that lasted for months. Examples include Duqu, Red October, Dragonfly 2.0 and Black Energy (Kimani et al., 2019; Gunduz & Das, 2020).

SG has a broad scope so we will focus on CPS, IoT, data and smart homes.

- *CPS:* We highlight the work of Humayed et al. (2017) that discussed security of CPS with examples in ICS, SG, healthcare and EV. For each CPS, it breaks down the cyber and the physical part showing the propensity of attacks by each component, discussing major threats and listing vulnerabilities. It concludes the work by presenting real-world attacks with targets, impacts, preconditions with mitigation strategies and methods used by adversaries.

  He and Yan (2016) commented on attacks and defences in the SG, listing vulnerabilities and security issues in energy generation and transmission. The work has focused on electrical issues arising in the SG, describing energy management systems in detail, LRA, and switching attacks in the market, and on PMU. For defence, it listed countermeasures such as protection, detection and coordinated mitigation. Ashibani and Mahmoud (2017) explained the differences between CPS and IoT as well as shared characteristics. They have also addressed security in three layers common in CPS, namely, (i) perception layer; (ii) transmission layer; and (iii) application layer. For the most common types of attacks, they related security with countermeasures for combating malicious events.
- *IoT security:* IoT research, protocols, technologies, privacy, safety and security are the topic of many surveys and literature reviews with a significant increase in

recent years (Al-Fuqaha et al., 2015; Alaba et al., 2017; Yang et al., 2017) across CPS (transportation systems, smart cities, SG, ICS and SB). Emmanuel and Rayudu (2016) addressed communication issues aligned with the IEEE Guide for SG Interoperability and NIST standards, commenting on protocols for media access and governing issues.

Lin et al. (2017) discussed the integration of IoT in edge/fog computing and its relationship with CPS. Also, they presented the enabling technologies behind IoT to realise a truthful cyber-physical world with opportunities for stakeholders to address cybersecurity, privacy, resilience and research challenges. Gunduz and Das (2020) surveyed cybersecurity in SG's IoT. The authors evaluated cyber-attacks in the communication layer and explained defence strategies. They listed a wealth of research on attacks directed at SG and detailed threats and countermeasures.

- *Smart Homes:* Heartfield et al. (2018) discussed attack vectors using IoT technologies in smart homes, major CPS threats and impact. Other security aspects are of interest in smart homes such as the discussion in Lin and Bergmann (2016) that tackled privacy of IoT and compared enabling technologies.
- *Data:* Authors described how to deal with false data injection (FDI) attacks (Deng et al., 2016; Liang et al., 2016; Musleh et al., 2019) to prevent storing data with outliers or skewed measurements (Huang et al., 2013). For mitigation, alternatives are deep learning (He et al., 2017) or Kalman filters (Manandhar et al., 2014). Paté-Cornell et al. (2018) presented a risk analysis using statistical data for quantitative assessments in critical infrastructures. They discussed three case studies: (i) a Bayesian Network for high-impact attack scenarios; (ii) risk analysis of connectivity; and (iii) software upgrade decisions to thwart attacks, countermeasures and anticipation when dealing with malicious events.
- *Synchronised attacks in the SG:* Attackers may compromise low security or outdated software of smart appliances attached to high-wattage devices and promote synchronised attacks. Adversaries choose most susceptible scheduling (e.g. peak hours) to switching on or off a massive number of heating/cooling units, water heaters, or pumps simultaneously. They try to imbalance frequency as systems will not be able to cope with the extra energy load in such short notice. Examples covered in the literature include BlackIoT (Soltan et al., 2018), Load Changing Attack (LCA) (Dabrowski et al., 2017; Yankson & Ghamkhari, 2020), Manipulation of Demand (MAD) (Soltan et al., 2019), flash attacks (Kumar & Bhama, 2019) or switching attacks (He & Yan, 2016). Arnaboldi et al. (2020) used Markov Chains to model LCA and investigate appropriate mix of power generation units. Grid-connected microgrids are susceptible of LCA as they could imbalance frequency; however, Huang et at. (2019a) discussed that current contingencies may in fact thwart attacks.
- *Surveys in security:* A wealth of surveys was published over the years. We highlight Giraldo et al. (2017) tackling privacy issues and a survey based on a control perspective for attack detection in ICS (Ding et al., 2018). There were also surveys combining the strengths of SG and smart homes' cyberse-

curity (Komninos et al., 2014) and a comprehensive review on NIDS research in CPS (Mitchell & Chen, 2014).

9.3.2 Risk Assessments

This topic has been discussed over the years as standards, methodologies, frameworks and tools emerged. Next, we present related work on RA for these dimensions.

- *Standards:* Over the years, analysts, managers and researchers have placed security as a significant and crucial aspect of future EPG. They have defined a significant number of standards to help stakeholders conduct RA in organisations tailoring their approach according to the objective. Leszczyna (2018) surveyed standards and discussed their relationships, advantages and drawbacks highlighting 35 relevant publications over standards and guidelines published by NIST, ISO/IEC and NERC.
- *RA methodologies:* Gritzalis et al. (2018) selected popular frameworks such as *Expression des Besoins et Identification des Objectifs de Sécurité* (EBIOS), MEthod for Harmonized Analysis of RIsk (MEHARI), Operationally Critical Threat and Vulnerability Evaluation (OCTAVE), IT-Grundschutz, *Metodología de Análisis y Gestión de Riesgos de los Sistemas de Información* (MAGERIT), Central Computing and Telecommunications Agency Risk Analysis and Management Method (CRAMM), Harmonized Threat Risk Assessment (HTRA), NIST.SP 800, RiskSafe, and CORAS for a numerical comparative analysis using multi-criteria decision methods. Out of those, EBIOS, MEHARI, HTRA, NIST.SP 800-30 and CORAS are free, whereas other methodologies demand a fixed or variable fee.

The authors employed risk calculation classes, e.g. high-level formulas to address quantitative indices according to a methodology or standard, following previous work by Zambon et al. (2011). For instance, the operator $\otimes$ defines a combination between two factors, as shown in Table 9.1.

Table 9.1 Risk classes and risk calculation formulas according to cybersecurity characteristics

Class	*Risk calculation*
A	Risk (τ, α) = Likelihood (τ) $\otimes$ Vulnerability (τ, α) $\otimes$ Impact (τ, α)
B	Risk (τ, α, ρ) = Vulnerability (τ, α) $\otimes$ Impact (τ, ρ)
C	Risk (τ, α) = Annual Loss Expectancy (τ, α) = Likelihood (τ, α) $\otimes$ Average Loss (τ, α)
D	Risk $(\tau$, Critical $\alpha)$ = Vulnerability (Critical α) $\otimes$ Impact $(\tau$, Critical $\alpha)$
E	Risk (ι, α) = Likelihood (ι) $\otimes$ Consequences (ι, α)
Where τ is *Threat*, α is *Asset*, ρ is *Requirement*, ι is *Incident*	

Class A uses threats and assets over likelihood, vulnerability and impact, whereas Class B includes security requirements to the formula. Class C adds costs and losses, and Class D considers only critical assets. Class E includes incidents in the infrastructure and its consequences. For the authors, MEHARI, MAGERIT, CRAMM, HTRA, NIST SP800 and RiskSafe are Class A, whereas EBIOS is Class B, OCTAVE is Class D and IT-Grundshutz is Class E.

Information Security Risk Assessment (ISRA) deals with threats, vulnerabilities and risks associated with IS. Throughout the years, several approaches emerged for organisations (Shamala et al., 2013; Shameli-Sendi et al., 2016; Wangen, 2017), such as NIST.SP 800-30, ISO/IEC 27005, CRAMM, Facilitated Risk Assessment Process (FRAP), Consultative, Objective and Bi-functional Risk Analysis (COBRA), CORAS, Microsoft's Risk Assessment model and OCTAVE/OCTAVE Allegro.

The European Network and Information Security Agency (ENISA) has compiled an inventory of risk assessment tools.[10] The inventory has 12 tools (January/2021) with 22 attributes describing major characteristics and templates. More specifically to NIST 7628, Abercrombie et al. (2013) have proposed an RA for the SG.

- *Dependability issues:* Chemweno et al. (2018) relating dependability modelling using failure mode and effect analysis, fault tree analysis, stochastic Petri nets, Bayesian networks and Monte Carlo methods. Nagaraju et al. (2017) discussed benefits and limitations in risk-based modelling using fault and attack trees.
- *RA in CPS/IoT:* Cherdantseva et al. (2016) discussed SCADA in a literature review. They divided case studies into formula and model-based approaches. For risks in the SG, part of the European Community for research called Security for Smart Electricity Grids (SEGRID) initiative, aiming to increase protection against attacks. A framework emerged from comparing pre-existing methodologies and as the SEGRID Risk Management Methodology (SRMM) (Rossebø et al., 2017). SRMM is based on the European Telecommunications Standards Institute's (ETSI's) Threat Vulnerability and Risk Analysis (TVRA) method. Teixeira et al. (2015) discussed networked control systems in coupled systems using NIST's *Risk Management Framework*, whereas Mace et al. (2020) investigated a real-world case study performing a RA in a SB detailing challenges and lessons learned.

With respect to RA in IoT, Nurse et al. (2017) discussed emerging issues such as connectivity and pervasiveness of devices. Radanliev et al. (2018) studied the economic impact of risks using different approaches, and Casola et al. (2019) investigated automated threat assessment and risk modelling.

---

[10] Link: https://www.enisa.europa.eu/topics/threat-risk-management/risk-management/current-risk/risk-management-inventory/rm-ra-tools.

9.3.3 Security Metrics

The approaches in Table 9.1 do not quantify the actual security of a building. For building designers, this is usually unsatisfactory, since they need to determine if considered security techniques are sufficient, or they need to trade-off cybersecurity with other system properties, such as reliability and performance. In this section, we review a few security metrics proposed in the literature.

A key metrics-based tool for computer systems security professionals is the Common Vulnerability Scoring System (CVSS) (Mell et al., 2006), in conjunction with the US National Vulnerability Database (NVD).[11] The NVD keeps track of all known vulnerabilities, reported by industry and individuals, and scores each of the vulnerabilities using CVSS. Such scoring is important for practitioners, since the numeric score provides an immediate appreciation of severity and potential impact of the vulnerability. In cooperation with the NIST and the NVD, MITRE Corporation maintains the Common Vulnerability and Exposures (CVE) database.[12]

Cybersecurity practitioners use the periodic list of 'top' vulnerabilities to guide their activities, allowing them to prioritise the most urgent threats according to the CVSS scores. Details of CVSS can be found in the standard, now at version 3.1,[13] but the main thrust is that scores are derived from expert opinions about elements such as the complexity to exploit the vulnerability or the impact it may have on confidentiality. Vulnerabilities with higher scores are of more immediate concern to the cybersecurity team that manages IS, for instance, in SB.

A quite different approach, also of practical importance, is that of scoring cybersecurity activities carried out. In so doing, one scores how well protected is the ICT system. For instance, in a SB, one can count the percentage of devices for which firmware or software has been updated in the last month. The NIST Performance Measurement Guide for Information Security[14] provides a useful introduction into such approaches, albeit not tailored to SB. As the authors point out, such performance quantification is particularly useful if one can integrate it with risk management. However, establishing this relation between security activities and their impact on risk is usually far from straightforward and requires further research.

Thus far, the security metrics in this section provide a static measure for the state of the security, either by considering the vulnerabilities that are present or the mediation actions that have been taken. This emphasis on static metrics implies that there still is a gap in identifying measures that can help in decision-making that relate to the dynamics of the system, e.g., to make technology trade-offs between security and system efficiency or security and reliability.

[11] National Vulnerability Database. Link: https://nvd.nist.gov/.

[12] Common Vulnerability and Exposures. Link: https://cve.mitre.org/.

[13] CVSS v3.1: Specification Document. Link: https://www.first.org/cvss/specification-document.

[14] NIST.SP 800-55r1, 2008. Link: https://csrc.nist.gov/publications/detail/sp/800-55/rev-1/final.

To alleviate this shortcoming, researchers have proposed metric frameworks to incorporate security considerations in the design of networked IS, such as these for SB. Particularly Ramos et al. (2017) provide an extensive and exhaustive survey and discussion of metrics that consider security within dynamic systems, including the notion of Quality of Protection. Metrics such as Quality of Protection typically assume model-based assessment approaches, executed in the design phases, to configure and dimension ICT systems. In Nicol et al. (2004) and in Ramos et al. (2017), a survey of model techniques is provided, both with and without representation of time. Security breaches and incidents are represented in the model as system artefacts, similar to software failures and system malfunctions (Avizienis et al., 2004). In so doing, the impact of security on traditional performance and dependability metrics can be assessed.

In conclusion, there is no 'silver bullet' in terms of a metric that meaningfully capture system security for all situations. Instead, the literature either proposes static metrics that quantify the level of vulnerability or mediating actions or refers back to traditional performance and dependability metrics. In the latter case, security techniques and attack patterns are represented within the model, not in the metric. Clearly, significant research challenges remain in connecting up the various approaches, for instance, relating vulnerability metrics with SB risk management equations and security management investments in buildings with their impact on the system's combined Quality of Service and Quality of Protection.

9.3.4 Intrusion Detection

With the rapidly growing market for IoT devices and the significant advancements in CPS, we have found that NIDS have been playing a major role in cybersecurity solutions (Radoglou-Grammatikis & Sarigiannidis, 2019; Mitchell & Chen, 2014). They are designed to safeguard computer systems from a diverse range of malicious activities and attacks.

Literature divides designs into three classifications: (i) anomaly-based (Garcia-Teodoro et al., 2009), (ii) signature-based systems (Challa et al., 2017) and (iii) NIDS (Raghunath & Mahadeo, 2008). One of the main issues is detecting new or known threats that have been slightly modified and how they have an altered effect on a network traffic. For example, Shenfield et al. (2018) used artificial intelligence (AI) algorithms trained to detect malicious network traffic with the use of artificial neural networks in deep packet inspection.

Research focuses primarily on comparing NIDS rather than evaluation methods to identify the advantages of the methodology and the disadvantages for ensuring whether or not it meets requirements. Next, we present a list of evaluation criteria to measure NIDS performance and identify open research questions.

- *Identification of threats:* IoT and CPS environments encompass a range of devices and sensors connected by in a wide range of networks. There is a challenge on how to account for all types of intrusions, bringing to the focus of a hybrid NIDS to cover both network and devices. Reporting every threat detected is impractical, and it would be advised to conduct classification rankings. For instance, Snort[15] (Premathilaka et al., 2013) uses a grading system from 1 to 10, with 1 ranking in as a low interest threat that need no intervention and can be dealt with by the NIDS whereas 10 represents a major threat requiring user intervention.

- *Scalability and adaptability:* As infrastructures scale over time, so the amount of control and monitoring to keep systems updated with new occurrences to NIDS. That is the main reason as to why it should be able to adapt to changing attack strategies and the ability to scale over time to meet the additional devices and networks. This would include to work over multiple NIDS pertaining a wide range of sub-systems as well as being able to combine reports for decision-making and traceability efforts. Ensuring that the NIDS is capable of adaptability can prove useful when having to adapt to additional components or changes in the environment, allowing it to be customised to meet new requirements or to carry a stricter level of security on networks.
- *Known vulnerabilities:* It is the core responsibility and purpose of a NIDS to prevent known exploits and vulnerabilities. Unfortunately, recent research has highlighted that many commercial NIDS fail to meet this responsibility, with the given reason being the failure to swiftly update systems with recent vulnerabilities as and when they are discovered (Debar & Morin, 2002). As discussed in Sect. 9.3.3, some of these vulnerabilities are well known to the security community. The cybersecurity module in the VPP must constantly update the NIDS to ensure that periodic reviews are conducted on recent and past exploits. One example is the zero-day attack where adversaries explore recent discovered vulnerabilities to trigger attacks before administrators have the chance to update their systems.
- *Dynamic signature updating:* A NIDS is highly dependent on its ability to detect signatures to identify intrusions. However, they are not always able to efficiently detect most recently developed intrusions or in fact even slightly modified ones (Alhanahnah et al., 2018). For a NIDS to be effective, one will need to address the capability in which the system can be updated with recent signatures as and when administrators discover new exploits and vulnerabilities.
- *Third-party support:* Understanding the support one can receive from third parties and vendors can prove useful when developing a NIDS. In addition, it may highlight areas in which the system does not meet pre-approved standards of threat detection.

The AB environment pose several research opportunities on NIDS adapting efforts to cope with security requirements as they start interchanging data or interlinking components. One key issue is to investigate how NIDS deal with IoT

[15] Snort—Network Intrusion Detection and Prevention System. Link: https://www.snort.org/.

devices and CPS that are no longer supported or are simply outdated. There is potential research on how to best integrate AI and machine learning to negate vulnerabilities by leveraging the computational tractability of the algorithms.

9.3.5 Threat Modelling

One significant technique used in cybersecurity is known as threat modelling (TM) (Humayed et al., 2017; Rizvi et al., 2020; Shevchenko et al., 2018). The idea is to devise abstractions of systems and consider possible attackers with aims, goals and methods. As output, it generates a list of threats that the system has and should be addressed. Methods used in this approach vary in objectives and scope; as an example, we highlight STRIDE, i.e. a mnemonic for security concerns accounting for Spoofing, Tampering, Repudiation, Information Disclosure, Denial of Service, and Elevation of Privilege. Another example is to model and devise *attack trees* where paths of attacks and defences are built to mitigate incidents (Camana Acosta et al., 2020). In terms of TM research, Sion et al. (2018) enriched threat models with risk analysis information helping prioritisation and triaging. On a similar approach, Marksteiner et al. (2019) used TM in LVDG combining the approach with RA over legacy and newly added devices in the SG. Best practices for TM can be broken up into five steps: (i) asset identification; (ii) threat reconnaissance; (iii) risk assessment; (iv) threat mapping; and (v) mitigation capabilities.

TM can be quite complex and therefore not always limited to a singular methodology. It is not uncommon for organisations in industry to adopt multiple methodologies to insure all threats are covered within their given environment. For this reason, some industries and academic researchers might broaden their choices by including methodologies such as Process for Attack Simulation and Threat Analysis (PASTA), Hybrid Threat Modelling Method (hTMM), or Security Cards focusing mostly on brainstorming threats.

PASTA is a TM methodology designed as an integrated application threat analysis. It works on a risk- or asset-based approach, making it ideal for business focus environments such as active buildings. It is designed as an approach to dynamic threat detection, enumeration and producing a scoring process. Once the threat model is completed, it allows security experts to analyse vulnerabilities and other identified threats to the system environment, highlighting any areas in which security controls need to be added (UcedaVelez & Morana, 2015). PASTA methodology works around an attacker-focused view of the system and over all environment, in which the developers can design an asset-focused defence.

The hTMM methodology has been recently introduced threat models by Security Equipment Inc. (SEI) in 2018 with widespread application in CPS (Mead et al., 2018). It consists of two other methodologies, *Security Quality Requirements Engineering* (SQUARE) designed to extract, categorise and prioritise security requirements and *Persona non Grata*, used to identify ways in which a system can be attacked according to adversaries' goals.

The Security Card methodology approach to TM is designed around brainstorming and creative thinking, unlike the other methodologies we have seen so far which focus on structured approaches. This methodology is designed to help identify less common or novel attacks. It incorporates the use of 42 cards covering Human Impact (9 cards), Adversary's Motivations (13 cards), Adversary Resources (11 cards) and Adversary's Methods (9 cards) (Shevchenko et al., 2018).

9.3.6 Modelling and Simulation

Modelling and simulation help the design of virtually any system by artificially creating approximated versions of a target infrastructure. Modellers then introduce controlled incidents in components devising 'what-if' scenarios for making most likely assumptions on the system and evaluate yielded output (Li & Zhang, 2014) for analysis. There is a substantial integration of MATLAB/Simulink with real-time simulators such as OPAL-RT (Poudel et al., 2017) to investigate power-related problems with built-in primitives, integrated toolboxes and shared libraries. Modellers may use Hardware-in-the-Loop (HIL) simulation (Ren et al., 2008) to assess designs by toggling the level-of-detail according to desired feature representations.

Cybersecurity-oriented simulation focuses on the *consequence* of attacks (Nicol et al., 2004), i.e. what happens after an attack or breach took place (data corruption in IS, stolen credentials, or increased privileges). Modellers tackle issues arising in both the power network and in the telecommunication network (Mets et al., 2014; Müller et al., 2016) with co-simulation (Gomes et al., 2018).

A plethora of frameworks exists for co-simulating the SG such as GridLAB-D (Chassin et al., 2014) or OpenDSS (Dugan & McDermott, 2011) combined with the Network Simulator (ns) (Riley & Henderson, 2010) or OMNeT++/INET (Varga, 2010). In terms of platforms or tool-chains specific for co-simulation, we mention the Framework for Network Co-simulation (FNCS) (Ciraci et al., 2014) gradually being replaced by the Hierarchical Engine for Large-Scale Infrastructure Co-simulation (HELICS) (Palmintier et al., 2017), Mosaik (Schütte et al., 2011), and pandapower (Thurner et al., 2018).

For modelling buildings and energy efficiency across buildings, analysts usually employ EnergyPlus[16] (Crawley et al., 2001), broadly adopted due to validation issues and trustful simulation results, thermal analysis and other building related features. It is yearly updated where new versions provide fixes to software defects and backwards compatibility (working with older model versions). Modellers use EnergyPlus in conjunction the graphical interface offered by OpenStudio,[17] and both tools are free.

[16] Link: https://energyplus.net/.

[17] Link: https://openstudio.net/.

There are co-simulation tools that offer modelling of DER in power distribution networks. For instance, Open Platform for Energy Networks (OPEN) (Morstyn et al., 2020) used pandapower (Thurner et al., 2018) for an integrated simulation. The authors presented a case study consisting of a BEMS with PV and energy trading, and another of an EV fleet with an unbalanced three-phase distribution network. The SCEPTRE toolchain (Johnson et al., 2020) (Sandia National Laboratories, US) is used to model networks of DER and cybersecurity defences where latency may impact grid performance. In its current version, the platform is online (live) and uses virtualised servers for the co-simulation engine.

Some authors applied those frameworks to model cybersecurity aspects such as Souza et al. (2020) that combined OpenDSS, Mosaik and ns-3 into a platform and Le et al. (2019) that used FNCS to model threats in the SG.

9.3.7 Discussion

There is substantial push for efficient planning (short, medium, and long term), preparedness and accurate state estimation as the SG is a critical infrastructure prone to unintended or malicious cybersecurity incidents. Measures in place address rigorous access control, real-time tracking of people while sensing the buildings' infrastructure, and collecting data.

From a cybersecurity perspective, officers across responsibilities should balance trade-offs against functionality, usability, control, and privacy while quickly acting to respond and prevent incidents. In AB, managers tackle these concerns by overseeing the whole infrastructure involving the DER, the set of SB, the VPP controlling CPS and IoT, SCADA sub-systems (if present), AMI and the TE. Historic energy yields throughout seasons should account for scale, i.e. new devices are attached to the grid as incentives are put in place over time. The OCC for AB integrates with ESO, energy providers, aggregators, consumers and building managers, continuously gathering data for accurate state estimation (an explanation on how the state estimator works is provided by He and Yan (2016)), while enabling trustworthiness. It accounts for short-, medium- and long-term planning and prepare for adverse weather conditions, avoiding the reliance on conventional power to meet nZEB requirements.

One notices a significant increase in market penetration of rooftop PV, WTPCS, or other DER. Akram et al. (2017) researched on sizing the RER mix in grid-connected microgrids and compared reliability over minimum costs. Attackers may target energy theft as customers will start behaving as small generators trading in the market as *prosumers*. It is a substantial shift from conventional interactions with the energy market, from passive consumers to active players that could even help balancing frequency if required by the management system or controller. These concerns will increase with wide adoption of EV and smart chargers participating in the network as they may also regulate frequency (Izadkhast et al., 2015; Greenwood et al., 2017).

Business stakeholders and cybersecurity experts must convene and use the RA for insights when addressing threats and vulnerabilities. Establishing clear and quantifiable measures could help understanding the level of exposure faced within AB and what actions to take to diminish risk and exploitation opportunities. They must discuss which cybersecurity metrics they will consider, whether related to quality properties, system status (in terms of updating), number of breaches or attack attempts per time unit or a customised approach that address quantitative measures. Within the context of SB, there is research to conduct in defining the important, easily measurable, and Quality of Protection metrics. Moreover, it is crucial to research the relationship between different metrics, e.g. between software updates and risk, so that different stakeholders can make use of each other's metrics.

In AB settings, it is essential to deal with uncertainty, i.e. imperfect or incomplete knowledge of threats and new vulnerabilities. The intermittent aspect of renewable generation contributes to raise uncertainty; however, one could leverage it with the use of ESS in buildings and EV. Borrowing N-1 security ideas from the EPG, AB could address how to cope with more localised contingencies to avoid load shedding or energy blackouts in their vicinity. One example is to employ redundancy or signal the market and customers on prices so power frequency may be even out during crisis.

9.4 Research Roadmap for AB Cybersecurity

AB and nZEB are expected to permeate future power grid across residential, commercial and industrial counterparts. Their advantages will be weighed against the current model where customers will balance convenience, trustworthiness, trade-offs in investments *versus* billing and security and privacy guarantees to make a decision. As potentially millions of IoT devices attached to DER proliferate into buildings and households, continuously updated RA across system life cycles will be compulsory. We outline next cybersecurity-oriented research roadmap for AB:

1. ***Risk assessments***: research on timely, bespoke, and automated RA throughout (any) system life cycle. Address how to cope with emergent threats and vulnerabilities and requirement changes. RA will be tailored to meet organisation's objectives and will require quick adapting to cope with increased levels of attack sophistication, involving cybersecurity experts in the process (Mace et al., 2020). The RA could help TM and vulnerability assessment through combinations with security metrics. For instance, Rizvi et al. (2020) modelled weaknesses of IoT in home, commerce and healthcare facilities and used NIST's CVSS to compute vulnerability scores for the devices.
2. ***Efficiency codes***: just as energy efficiency labels grade appliances for power consumption, modern infrastructure could address other concerns such as emissions, amount of energy generation and storage capability. Nikolaidou et al. (2020) devised a code for AB; however, it could be extended to consider trustworthiness.

It could help customers evaluating trade-offs when purchasing estate or adjusting security in their premises. With respect to security metrics, a significant practical open problem is to establish relationships on the metrics used by different stakeholders, for instance, between the technology metrics used by ICT professionals and risk metrics used in business continuity processes.

3. ***Market features***:

    - *New players*: the number of prosumers in AB will increase as potential thousands of new power generators are attached to the grid. This is highly motivated by low acquisition prices and governmental incentives. It promises lower electric bills (among other advantages), and authors have studied how they manage and share energy in the SG (Zafar et al., 2018). The envisioned massive penetration of rooftop solar panels and EV will change the energy landscape by adding new DER and helping the flexibility of power grids; however, it will enlarge the attack surface considerably.
    - *Transactive energy*: TE may help prevent power cuts and blackouts, for instance, Yankson and Ghamkhari (2020) used it for thwarting load-based attacks and flexible loads to balance frequency. Huang et al. (2019b) used co-simulation to study the valuation in responsive loads for comparing TE and non-TE agents.

4. ***Smart grid components***:

    - *Cross-domain control*: more research is required to understand the required multidisciplinary approach involving a plethora of systems and components bringing together experts from different domains.
    - *Pervasiveness of smart components*: the SG infrastructure allows potential millions of smart appliances to be easily attached. In a cybersecurity point of view, it could open backdoors for malicious interventions as devices may eavesdrop or divert communications, influence automated incident responses (or preventing them from happening) and postpone (or prevent) events from reaching the OCC.
    - *Component life cycle*: in AB, the infrastructure is composed by a plethora of components, and each one has their own life cycle from pre-acquisition and procurement to deployment, i.e. when they start servicing customers until decommission. Cybersecurity officers must address RA throughout these phases.
    - *Legacy systems and retrofitting*: one must consider the effect on cybersecurity aspects as customers retain outdated equipment in their buildings.
    - *Heterogeneity and interoperability*: cybersecurity officers employ device heterogeneity to increase protection as it difficults attacks. However, under these circumstances, they will have to account for other issues such as intra-communication, timely maintenance and interoperability.

5. ***Power-based attacks***:
   - *Synchronised cyber incidents*: these attacks occur when malicious actors inject malware and coordinate events that may impact frequency, e.g. switching on or off a massive number of high-wattage devices. There is a need to better understand LCA, MAD or LRA (Dabrowski et al., 2017; Soltan et al., 2019; Kumar & Bhama, 2019), in detection and countermeasures to avoid its propagation.
   - *APT*: measures consist on training systems and personnel on identifying long-term incursions, monitoring data, energy use based on previous consumption or generation and building energy profiling over prolonged time scales (Skopik et al., 2014; Friedberg et al., 2015). Also, investigating FDI attacks (Musleh et al., 2019) and comparing multiple datasets to unveil patterns that could hint diversions or inconsistent loads across the infrastructure to detect, assess, confirm, mitigate and cope with APT.
6. ***Modelling & Simulation***:
   - *Modelling scale*: as the advantages of AB become visible to customers, companies and the government, other buildings and households will desire to participate and even trade energy, so the solution must account for scaling the infrastructure to withstand the demand.
   - *Testbeds*: tackle realistic modelling and co-simulation efforts when designing and experimenting with varying scale EPG, addressing power supply and demand in the presence of security incidents, physical destruction or service interruption, to name a few (Cintuglu et al., 2016). Address preparedness, proactive responses and accurate detection and confirmation of incidents.
   - *Cyber-physical counterparts*: one idea to test effectiveness of responses is to artificially inject attacks to see how countermeasures in place react. Current research tackles replicas of physical and virtual counterparts into *Digital Twins*, aimed to facilitate analysis by juxtapositioning these elements.
   - *Prediction under uncertainty*: user behaviour and weather intermittency may drive planning for short, medium, and long term. The idea is to iterate over multiple scenarios considering computational tractability of running a massive number of simulations. It is an open research question how to model the stochasticity of DER, ESS and EV.
   - *Controlled stress testing*: In order to avoid actual attacks on the physical infrastructure, simulation could be used to add controlled failures to components in single or multiple sources. It could address susceptible scenarios under different energy loads to drive design decisions on how to best tackle electricity-based shortcomings.
   - *HIL*: it allows for the modelling of hardware and its internal structures as they increasingly demand modelling on varying level-of-detail. Kochanneck et al. (2018) considered a HIL co-simulation in a LVDG comparing a multi-building design with a real-world building setting.

- *Sensitivity analysis*: this effort involves multiple parametrisations for comparing the impact of characteristics and discover scenarios deemed deficient (according to metrics) for in-depth analysis.

7. ***Incident response***: fast reaction to malicious incursions and recovery from security incidents or unanticipated breaches, learning from the situations and documenting details, actions, and reactions for future reference. Incident response could be of crucial use in the event of catastrophes or natural disasters impairing the infrastructure.
8. ***Emergent techniques***: increased use of machine learning and AI algorithms for processing Big Data from multiple sources (e.g. IoT measurements, fine-grained weather forecasts, logging and so on) and distributed ledgers for smart contracts in TE systems (Siano et al., 2019).
9. ***End-user related***: the inhabitants, managers, maintenance personnel and officers will have to adapt, learn and point out inconsistent behaviours, anomalies, improper use, invalid situations or user mishandling of technology throughout the premises of AB. The interaction with AB's systems will have to leverage functionality over cybersecurity, coping with easiness of operation with different user levels, e.g. children, the elderly or non-savvy customers.
10. ***Legal aspects***: accountability, ownership, forensics, i.e. questions as to the magnitude of the attacks, impacted systems and who are responsible (and liable) for the AB infrastructure. These matter in both new and retrofitted buildings as well as assigning unique individuals to attacks with evidence (forensics). As data pours in into IS, security requirements (Pillitteri & Brewer, 2014) and data protection mechanisms will ensure better managerial actions in conjunction with life-cycle RA, General Data Protection Regulation (GDPR) adherence, staff training and health and safety measures.

AB will revolutionise the way people handle power in buildings. It will support future energy systems evolving current paradigms towards more active roles instead of just passive consumption. The advantages brought forth by AB will improve balancing the power grid during peak demand while enabling prosumers to interact in the market in P2P trading. The greener and sustainable approach offered by AB may also reduce electric bills as more conscious customers will shift their behaviour according to electricity prices or other incentives.

Figure 9.6 shows an overview of the proposed cybersecurity roadmap for AB, showing areas of concern and most susceptible targets. It covers the attack surface with highest likelihood of adversarial interventions, accounting major issues for current and future architectures and technologies.

Bespoke RA aligns business objectives with cybersecurity trade-offs. As mentioned, the organisation leverages the rigour, formality and depth of the RA accounting business proposition, optimality and potential threats that could undermine service provision. The sheer complexity of AB calls for strict cybersecurity management as officers and stakeholders ensure the quality of data that are fed into the systems for accurate system state estimation and situational awareness. These

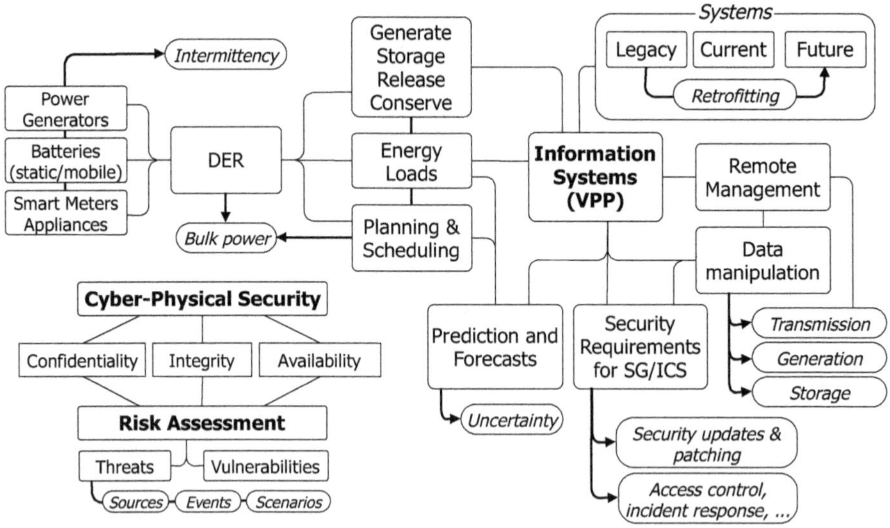

Fig. 9.6 Cyber-physical security, risk assessment, and potential targets in AB settings

protections safeguard IS and prevent the infrastructure to fall short on cyber-physical security.

As stated throughout this work, adversaries may target the AB business proposition (energy generation, storage, release and conservation), corrupt data in IS, skew predictions or transmit erroneous status of field devices. It is the responsibility of the cybersecurity officers to quickly enact, identify, confirm and respond to incidents in a timely fashion, fast track exposures and precisely reporting on occurrences to account for minimum level of false positives, i.e. teams responding to false events.

Finally, it is worth mentioning that adversaries may surveil the infrastructure and monitor signals and responses to learn how to inflict more damage in future malicious incursions. To mitigate these situations, research is needed for modelling attackers' actions to establish effective countermeasures and thwart cyber-physical security incidents, thus preventing them from re-occurring.

9.5 Conclusion

Security and safety encompass the proposition of AB and the intertwined nature of P2P energy provision. The AB is defined as buildings that are active energy agents sustaining a nation's power infrastructure. So, any adversarial incursion against the infrastructure and all that it entails are attacks targeted at AB as well. Under these settings, one must consider the harmful effects that attacks have that could promote undesired load shedding or frequency imbalance, not to mention the consequences to inhabitants and managers. Security, after all, is a trade-off between adhering to requirements and user/customer/stakeholder interaction/experience. Riskier outcomes must be prioritised before unimportant ones as cybersecurity officers continuously assess threats and vulnerabilities.

For these ideas to become reality, actions are required in a trans-national level, for example, a joint effort on how to best decommissioning current power plants with few impacts to the environment. Renewable energy is known for its uncertainty, as WTPCS and PV panels are not used 100% of the time (only a fraction) due to climatic conditions. Besides this, another problem is the required surface when commissioning PV panels since it demands a large area for a given energy yield whereas wind turbines may impact wildlife and vegetation. Nevertheless, despite those shortcomings, cross-domain action must be taken to meet energy objectives for the near future, enshrining the solution with cyber-physical security and safety.

AB adoption undoubtedly poses significant opportunities aligned with the decarbonisation strategy set forth by the government. We have proposed here a roadmap for tackling cyber-physical security using RA, modelling and intrusion detection over the AB's life cycle, as well as future research opportunities. We have raised awareness to cybersecurity in AB by commenting how to adapt existing mechanisms in CPS, IoT and ICS. Our work has presented an overview on the required cross-fertilisation effort across domains for increased preparedness to assess and respond to cyber-physical incidents.

9.5.1 Status Quo, Challenges and Outlook

Cybersecurity is vital for any critical system design as it sustains operational capabilities in the presence of adversarial incursions. Power and building managers as well as security officers are expected to employ previous research on infrastructure, learning from previous experiences and lessons learned as their cyber-physical systems become increasingly hardened against attacks. As shown throughout the chapter, several challenges still persist in modern (and smart) infrastructures as attackers devise creative means to criminally access systems and thwart operations. As adversaries will always have different incentives to thwart the SG due to its capabilities, managers should always consider cybersecurity throughout the AB proposition and recruiting, whenever possible, the help of its underlying stakeholders (e.g. reporting anomalies, phishing attempts, uncommon behaviour and so on).

Our work has pointed out key aspects to observe in cyber-physical security from CPS research (SG and ICS), showing related work on SCADA vulnerabilities and mitigations, risk assessments, threat modelling, and modelling & simulation. Next efforts are expected to follow this path of building up from previous ideas and adapting systems, techniques and methods to cope with frequently changing cyber-physical topologies to accommodate emergent technologies seamlessly to customers and administrators. The cybersecurity challenge in AB should take into account attacker dynamics as the defences adapt in near real time, detect malicious advances and prevent more serious consequences from propagating across the infrastructure.

Acknowledgments The authors wish to acknowledge funding from the Industrial Strategy Challenge Fund and Engineering and Physical Sciences Research Council, EP/V012053/1, for The Active Building Centre Research Programme (ABC RP).

References

Abercrombie, R. K., Sheldon, F. T., Hauser, K. R., Lantz, M. W., & Mili, A. (2013). Risk assessment methodology based on the NISTIR 7628 guidelines. In *2013 46th Hawaii International Conference on System Sciences* (pp. 1802–1811). IEEE.

Akram, U., Khalid, M., & Shafiq, S. (2017). Optimal sizing of a wind/solar/battery hybrid grid-connected microgrid system. *IET Renewable Power Generation, 12*(1), 72–80.

Al-Fuqaha, A., Guizani, M., Mohammadi, M., Aledhari, M., & Ayyash, M. (2015). Internet of things: A survey on enabling technologies, protocols, and applications. *IEEE Communications Surveys and Tutorials, 17*(4), 2347–2376.

Alaba, F. A., Othman, M., Hashem, I. A. T., & Alotaibi, F. (2017). Internet of Things security: A survey. *Journal of Network and Computer Applications, 88*, 10–28.

Alhanahnah, M., Lin, Q., Yan, Q., Zhang, N., & Chen, Z. (2018). Efficient Signature Generation for Classifying Cross-Architecture IoT Malware (pp. 1–9).

Amin, S. M., & Wollenberg, B. F. (2005). Toward a smart grid: power delivery for the 21st century. *IEEE Power Energy Magazine, 3*(5), 34–41.

Arnaboldi, L., Czekster, R. M., Morisset, C., & Metere, R. (2020). Modelling Load-Changing Attacks in Cyber-Physical Systems. *Electronic Notes in Theoretical Computer Science, 353C*, 39–60.

Ashibani, Y., & Mahmoud, Q. H. (2017). Cyber physical systems security: Analysis, challenges and solutions. *Computer Security, 68*, 81–97.

Attia, S. (2018). *Net Zero Energy Buildings (NZEB): Concepts, frameworks and roadmap for project analysis and implementation*. Butterworth-Heinemann.

Avizienis, A., Laprie, J.-C., Randell, B., & Landwehr, C. (2004). Basic concepts and taxonomy of dependable and secure computing. *IEEE Transactions on Dependable and Secure Computing, 1*(1), 11–33.

Bankovskis, A. (2017). One million homes constructed as 'buildings as power stations' – report of indicative benefits. *SPECIFIC online report*.

Camana Acosta, M. R., Ahmed, S., Garcia, C. E., & Koo, I. (2020). Extremely randomized trees-based scheme for stealthy cyber-attack detection in smart grid networks. *IEEE Access, 8*, 19921–19933.

Casola, V., De Benedictis, A., Rak, M., & Villano, U. (2019). Toward the automation of threat modeling and risk assessment in IoT systems. *Internet of Things, 7*, 100056.

Challa, S., Wazid, M., Das, A. K., Kumar, N., Reddy, A. G., Yoon, E.-J., & Yoo, K.-Y. (2017). Secure signature-based authenticated key establishment scheme for future IoT applications. *IEEE Access, 5*, 3028–3043.

Chassin, D. P., Fuller, J. C., & Djilali, N. (2014). GridLAB-D: An agent-based simulation framework for smart grids. *Journal of Applied Mathematics, 2014*, 1–12.

Chemweno, P., Pintelon, L., Muchiri, P. N., & Van Horenbeek, A. (2018). Risk assessment methodologies in maintenance decision making: A review of dependability modelling approaches. *Reliability Engineering & System Safety, 173*, 64–77.

Cherdantseva, Y., Burnap, P., Blyth, A., Eden, P., Jones, K., Soulsby, H., & Stoddart, K. (2016). A review of cyber security risk assessment methods for SCADA systems. *Computer Security, 56*, 1–27.

Cintuglu, M. H., Mohammed, O. A., Akkaya, K., & Uluagac, A. S. (2016). A survey on smart grid cyber-physical system testbeds. *IEEE Communications Surveys and Tutorials, 19*(1), 446–464.

Ciraci, S., Daily, J., Fuller, J., Fisher, A., Marinovici, L., & Agarwal, K. (2014). FNCS: a framework for power system and communication networks co-simulation. In *Proceedings of the Symposium on Theory of Modeling & Simulation-DEVS integrative* (pp. 1–8).

Clarke, J., Jones, P., Littlewood, J., & Worsley, D. (2020). Active buildings in practice. In *Sustainability in Energy and Buildings* (pp. 555–564). Springer.

Crawley, D. B., Lawrie, L. K., Winkelmann, F. C., Buhl, W. F., Huang, Y. J., Pedersen, C. O., Strand, R. K., Liesen, R. J., Fisher, D. E., Witte, M. J., et al. (2001). EnergyPlus: creating a new-generation building energy simulation program. *Energy and Buildings, 33*(4), 319–331.

Dabrowski, A., Ullrich, J., & Weippl, E. R. (2017). Grid shock: Coordinated load-changing attacks on power grids: The non-smart power grid is vulnerable to cyber attacks as well. In *Proceedings of the 33rd Annual Computer Security Applications Conference* (pp. 303–314).

de Souza, E., Ardakanian, O., & Nikolaidis, I. (2020). A co-simulation platform for evaluating cyber security and control applications in the smart grid. In *IEEE International Conference on Communications (ICC)* (pp. 1–7). IEEE.

Debar, H., & Morin, B. (2002). Evaluation of the diagnostic capabilities of commercial intrusion detection systems. In Wespi, A., Vigna, G., & Deri, L., editors, *Recent Advances in Intrusion Detection* (pp. 177–198). Springer, Berlin, Heidelberg.

Deng, R., Xiao, G., Lu, R., Liang, H., & Vasilakos, A. V. (2016). False data injection on state estimation in power systems - Attacks, impacts, and defense: A survey. *IEEE Transactions on Industrial Informatics, 13*(2), 411–423.

Ding, D., Han, Q.-L., Xiang, Y., Ge, X., & Zhang, X.-M. (2018). A survey on security control and attack detection for industrial cyber-physical systems. *Neurocomputing, 275*, 1674–1683.

Driesen, J., & Katiraei, F. (2008). Design for distributed energy resources. *IEEE Power Energy Magazine, 6*(3), 30–40.

Dugan, R. C., & McDermott, T. E. (2011). An open source platform for collaborating on smart grid research. In *2011 IEEE Power and Energy Society General Meeting* (pp. 1–7). IEEE.

Emmanuel, M., & Rayudu, R. (2016). Communication technologies for smart grid applications: A survey. *Journal of Network and Computer Applications, 74*, 133–148.

Fang, X., Misra, S., Xue, G., & Yang, D. (2011). Smart grid – the new and improved power grid: A survey. *IEEE Communications Surveys and Tutorials*, 14(4), 944–980.

Farwell, J. P., & Rohozinski, R. (2011). Stuxnet and the future of cyber war. *Survival, 53*(1), 23–40.

Friedberg, I., Skopik, F., Settanni, G., & Fiedler, R. (2015). Combating advanced persistent threats: From network event correlation to incident detection. *Computer Security, 48*, 35–57.

Garcia-Teodoro, P., Diaz-Verdejo, J., Maciá-Fernández, G., & Vázquez, E. (2009). Anomaly-based network intrusion detection: Techniques, systems and challenges. *Computer Security, 28*(1-2), 18–28.

Giraldo, J., Sarkar, E., Cardenas, A. A., Maniatakos, M., & Kantarcioglu, M. (2017). Security and privacy in cyber-physical systems: A survey of surveys. *IEEE Design & Test, 34*(4), 7–17.

Gomes, C., Thule, C., Broman, D., Larsen, P. G., & Vangheluwe, H. (2018). Co-simulation: a survey. *ACM Computing Surveys (CSUR), 51*(3), 49.

Greenwood, D., Lim, K. Y., Patsios, C., Lyons, P., Lim, Y. S., & Taylor, P. (2017). Frequency response services designed for energy storage. *Applied Energy, 203*, 115–127.

Greer, C., Wollman, D. A., Prochaska, D. E., Boynton, P. A., Mazer, J. A., Nguyen, C. T., FitzPatrick, G. J., Nelson, T. L., Koepke, G. H., Hefner Jr, A. R., et al. (2014). NIST framework and roadmap for smart grid interoperability standards, release 3.0. Technical report.

Gritzalis, D., Iseppi, G., Mylonas, A., & Stavrou, V. (2018). Exiting the Risk Assessment maze: A meta-survey. *ACM Computing Surveys (CSUR), 51*(1), 1–30.

Gunduz, M. Z., & Das, R. (2020). Cyber-security on smart grid: Threats and potential solutions. *Computer Networks, 169*, 107094.

Gunes, V., Peter, S., Givargis, T., & Vahid, F. (2014). A survey on concepts, applications, and challenges in cyber-physical systems. *KSII Transactions on Internet & Information Systems, 8*(12), 4242–4268.

Hatziargyriou, N., Asano, H., Iravani, R., & Marnay, C. (2007). Microgrids. *IEEE Power Energy Magazine, 5*(4), 78–94.

He, H., & Yan, J. (2016). Cyber-physical attacks and defences in the smart grid: a survey. *IET Cyber-Physical Systems: Theory & Applications, 1*(1), 13–27.

He, Y., Mendis, G. J., & Wei, J. (2017). Real-time detection of false data injection attacks in smart grid: A deep learning-based intelligent mechanism. *IEEE Transactions on Smart Grid, 8*(5), 2505–2516.

Heartfield, R., Loukas, G., Budimir, S., Bezemskij, A., Fontaine, J. R., Filippoupolitis, A., & Roesch, E. (2018). A taxonomy of cyber-physical threats and impact in the smart home. *Computer Security, 78*, 398–428.

Huang, Y., Esmalifalak, M., Nguyen, H., Zheng, R., Han, Z., Li, H., & Song, L. (2013). Bad data injection in smart grid: attack and defense mechanisms. *IEEE Communications Magazine, 51*(1), 27–33.

Huang, B., Cardenas, A. A., & Baldick, R. (2019a). Not everything is dark and gloomy: Power grid protections against IoT demand attacks. In *28th {USENIX} Security Symposium* (pp. 1115–1132).

Huang, Q., McDermott, T. E., Tang, Y., Makhmalbaf, A., Hammerstrom, D. J., Fisher, A. R., Marinovici, L. D., & Hardy, T. (2019b). Simulation-based valuation of transactive energy systems. *IEEE Transactions on Power Systems, 34*(5), 4138–4147.

Humayed, A., Lin, J., Li, F., & Luo, B. (2017). Cyber-physical systems security – a survey. *IEEE Internet of Things Journal, 4*(6), 1802–1831.

Initiative, J. T. F. T. (2012). Guide for conducting risk assessments (nist. sp 800-30rev1). *The National Institute of Standards and Technology (NIST), Gaithersburg.*

Izadkhast, S., Garcia-Gonzalez, P., & Frías, P. (2015). An aggregate model of plug-in electric vehicles for primary frequency control. *IEEE Transactions on Power Systems, 30*(3), 1475–1482.

Johnson, J., Onunkwo, I., Cordeiro, P., Wright, B. J., Jacobs, N., & Lai, C. (2020). Assessing DER network cybersecurity defences in a power-communication co-simulation environment. *IET Cyber-Physical Systems: Theory & Applications, 5*(3), 274–282.

Kim, J., & Tong, L. (2013). On topology attack of a smart grid: Undetectable attacks and countermeasures. *IEEE Journal on Selected Areas in Communications, 31*(7), 1294–1305.

Kimani, K., Oduol, V., & Langat, K. (2019). Cyber security challenges for IoT-based smart grid networks. *International Journal of Critical Infrastructure Protection, 25*, 36–49.

Kochanneck, S., Mauser, I., Phipps, K., & Schmeck, H. (2018). Hardware-in-the-Loop Co-simulation of a Smart Building in a Low-voltage Distribution Grid. In *IEEE Innovative Smart Grid Technologies Conference Europe* (pp. 1–6). IEEE Power & Energy Society.

Kolenc, M., Nemček, P., Gutschi, C., Suljanović, N., & Zajc, M. (2017). Performance evaluation of a virtual power plant communication system providing ancillary services. *Electric Power Systems Research, 149*, 46–54.

Komninos, N., Philippou, E., & Pitsillides, A. (2014). Survey in smart grid and smart home security: Issues, challenges and countermeasures. *IEEE Communications Surveys and Tutorials, 16*(4), 1933–1954.

Kumar, C. O., & Bhama, P. R. S. (2019). Detecting and confronting flash attacks from IoT botnets. *The Journal of Supercomputing, 75*(12), 8312–8338.

Kumar, P., Lin, Y., Bai, G., Paverd, A., Dong, J. S., & Martin, A. (2019). Smart grid metering networks: A survey on security, privacy and open research issues. *IEEE Communications Surveys and Tutorials, 21*(3), 2886–2927.

Kurnitski, J., Allard, F., Braham, D., Goeders, G., Heiselberg, P., Jagemar, L., Kosonen, R., Lebrun, J., Mazzarella, L., Railio, J., et al. (2011). How to define nearly net zero energy buildings nZEB. *Rehva Journal, 48*(3), 6–12.

Langner, R. (2011). Stuxnet: Dissecting a cyberwarfare weapon. *IEEE Security & Privacy, 9*(3), 49–51.

Le, T. D., Anwar, A., Beuran, R., & Loke, S. W. (2019). Smart Grid Co-Simulation Tools: Review and Cybersecurity Case Study. In *7th International Conference on Smart Grid (icSmartGrid)* (pp. 39–45). IEEE.

Leszczyna, R. (2018). Standards on cyber security assessment of smart grid. *International Journal of Critical Infrastructure Protection, 22*, 70–89.

Li, W., & Zhang, X. (2014). Simulation of the smart grid communications: Challenges, techniques, and future trends. *Computers & Electrical Engineering, 40*(1), 270–288.

Liang, G., Zhao, J., Luo, F., Weller, S. R., & Dong, Z. Y. (2016). A review of false data injection attacks against modern power systems. *IEEE Transactions on Smart Grid, 8*(4), 1630–1638.

Lin, H., & Bergmann, N. W. (2016). IoT privacy and security challenges for smart home environments. *Information, 7*(3), 44.

Lin, J., Yu, W., Zhang, N., Yang, X., Zhang, H., & Zhao, W. (2017). A survey on internet of things: Architecture, enabling technologies, security and privacy, and applications. *IEEE Internet of Things Journal, 4*(5), 1125–1142.

Ma, O., Alkadi, N., Cappers, P., Denholm, P., Dudley, J., Goli, S., Hummon, M., Kiliccote, S., MacDonald, J., Matson, N., Olsen, D., Rose, C., Sohn, M. D., Starke, M., Kirby, B., & O'Malley, M. (2013). Demand response for ancillary services. *IEEE Transactions on Smart Grid, 4*(4), 1988–1995.

Mace, J. C., Czekster, R. M., Morisset, C., & Maple, C. (2020). Smart Building Risk Assessment Case Study: Challenges, Deficiencies and Recommendations. In *16th European Dependable Computing Conference (EDCC)* (pp. 59–64). IEEE.

Manandhar, K., Cao, X., Hu, F., & Liu, Y. (2014). Detection of faults and attacks including false data injection attack in smart grid using Kalman filter. *IEEE Transactions on Control of Network Systems, 1*(4), 370–379.

Marksteiner, S., Vallant, H., & Nahrgang, K. (2019). Cyber security requirements engineering for low-voltage distribution smart grid architectures using threat modeling. *Journal of Information Security and Applications, 49*, 102389.

Mead, N. R., Shull, F., Vemuru, K., & Villadsen, O. (2018). A hybrid threat modeling method. Technical report, CMU – Software Engineering Institute.

Mell, P., Scarfone, K., & Romanosky, S. (2006). Common vulnerability scoring system. *IEEE Security & Privacy, 4*(6), 85–89.

Mets, K., Ojea, J. A., & Develder, C. (2014). Combining power and communication network simulation for cost-effective smart grid analysis. *IEEE Communications Surveys and Tutorials, 16*(3), 1771–1796.

Mitchell, R., & Chen, I.-R. (2014). A survey of intrusion detection techniques for cyber-physical systems. *ACM Computing Surveys (CSUR), 46*(4), 1–29.

Mo, Y., Kim, T. H.-J., Brancik, K., Dickinson, D., Lee, H., Perrig, A., & Sinopoli, B. (2011). Cyber-physical security of a smart grid infrastructure. *Proceedings of the IEEE, 100*(1), 195–209.

Morstyn, T., Collett, K. A., Vijay, A., Deakin, M., Wheeler, S., Bhagavathy, S. M., Fele, F., & McCulloch, M. D. (2020). OPEN: An open-source platform for developing smart local energy system applications. *Applied Energy, 275*, 115397.

Müller, S. C., Georg, H., Nutaro, J. J., Widl, E., Deng, Y., Palensky, P., Awais, M. U., Chenine, M., Küch, M., Stifter, M., et al. (2016). Interfacing power system and ICT simulators: Challenges, state-of-the-art, and case studies. *IEEE Transactions on Smart Grid, 9*(1), 14–24.

Musleh, A. S., Chen, G., & Dong, Z. Y. (2019). A survey on the detection algorithms for false data injection attacks in smart grids. *IEEE Transactions on Smart Grid, 11*(3), 2218–2234.

Nagaraju, V., Fiondella, L., & Wandji, T. (2017). A survey of fault and attack tree modeling and analysis for cyber risk management. In *2017 IEEE International Symposium on Technologies for Homeland Security (HST)* (pp. 1–6). IEEE.

Nicol, D. M., Sanders, W. H., & Trivedi, K. S. (2004). Model-based evaluation: from dependability to security. *IEEE Transactions on Dependable and Secure Computing, 1*(1), 48–65.

Nikolaidou, E., Fosas, D., Roberts, M., Allen, S., Walker, I., & Coley, D. (2020). Buildings as Energy Infrastructure, not Passive Consumers. *University of Bath.*

Nurse, J. R., Creese, S., & De Roure, D. (2017). Security risk assessment in Internet of Things systems. *IT Professional, 19*(5), 20–26.

Palmintier, B., Krishnamurthy, D., Top, P., Smith, S., Daily, J., & Fuller, J. (2017). Design of the HELICS high-performance transmission-distribution-communication-market co-simulation framework. In *2017 Workshop on Modeling and Simulation of Cyber-Physical Energy Systems (MSCPES)* (pp. 1–6). IEEE.

Paté-Cornell, M.-E., Kuypers, M., Smith, M., & Keller, P. (2018). Cyber risk management for critical infrastructure: a risk analysis model and three case studies. *Risk Analysis, 38*(2), 226–241.

Pillitteri, V. Y. and Brewer, T. L. (2014). Guidelines for smart grid cybersecurity (NIST 7628). Technical report.

Poudel, S., Ni, Z., & Malla, N. (2017). Real-time cyber physical system testbed for power system security and control. *International Journal of Electrical Power & Energy Systems, 90*, 124–133.

Premathilaka, N. A., Aponso, A. C., & Krishnarajah, N. (2013). Review on state of art intrusion detection systems designed for the cloud computing paradigm. In *2013 47th International Carnahan Conference on Security Technology (ICCST)* (pp. 1–6). IEEE.

Pudjianto, D., Ramsay, C., & Strbac, G. (2007). Virtual power plant and system integration of distributed energy resources. *IET Renewable Power Generation, 1*(1), 10–16.

Radanliev, P., De Roure, D. C., Nicolescu, R., Huth, M., Montalvo, R. M., Cannady, S., & Burnap, P. (2018). Future developments in cyber risk assessment for the internet of things. *Computers in Industry, 102*, 14–22.

Radhakrishnan, K., Menon, R. R., & Nath, H. V. (2019). A survey of zero-day malware attacks and its detection methodology. In *TENCON 2019-2019 IEEE Region 10 Conference (TENCON)* (pp. 533–539). IEEE.

Radoglou-Grammatikis, P. I., & Sarigiannidis, P. G. (2019). Securing the smart grid: A comprehensive compilation of intrusion detection and prevention systems. *IEEE Access, 7*, 46595–46620.

Raghunath, B. R., & Mahadeo, S. N. (2008). Network intrusion detection system (NIDS). In *1st International Conference on Emerging Trends in Eng. and Tech.* (pp. 1272–1277). IEEE.

Ramos, A., Lazar, M., Holanda Filho, R., & Rodrigues, J. J. (2017). Model-based quantitative network security metrics: A survey. *IEEE Communications Surveys and Tutorials, 19*(4), 2704–2734.

Ren, W., Steurer, M., & Baldwin, T. L. (2008). Improve the stability and the accuracy of power hardware-in-the-loop simulation by selecting appropriate interface algorithms. *IEEE Transactions on Industry Applications, 44*(4), 1286–1294.

Riley, G. F., & Henderson, T. R. (2010). The ns-3 network simulator. In *Modeling and Tools for Network Simulation* (pp. 15–34). Springer.

Rizvi, S., Pipetti, R., McIntyre, N., Todd, J., & Williams, I. (2020). Threat model for securing internet of things (IoT) network at device-level. *Internet of Things, 11*, 100240.

Rossebø, J. E., Wolthuis, R., Fransen, F., Björkman, G., & Medeiros, N. (2017). An enhanced risk-assessment methodology for smart grids. *Computer, 50*(4), 62–71.

Royapoor, M., Pazhoohesh, M., Davison, P. J., Patsios, C., & Walker, S. (2020). Building as a virtual power plant, magnitude and persistence of deferrable loads and human comfort implications. *Energy and Buildings, 213*, 109794.

Ruland, K. C., Sassmannshausen, J., Waedt, K., & Zivic, N. (2017). Smart grid security – an overview of standards and guidelines. *e&i Elektrotech Inf Tech, 134*(1), 19–25.

Schütte, S., Scherfke, S., & Tröschel, M. (2011). Mosaik: A framework for modular simulation of active components in Smart Grids. In *2011 IEEE First International Workshop on Smart Grid Modeling and Simulation (SGMS)* (pp. 55–60). IEEE.

Shamala, P., Ahmad, R., & Yusoff, M. (2013). A conceptual framework of info structure for information security risk assessment (ISRA). *Journal of Information Security and Applications, 18*(1), 45–52.

Shameli-Sendi, A., Aghababaei-Barzegar, R., & Cheriet, M. (2016). Taxonomy of information security risk assessment (ISRA). *Computer Security, 57*, 14–30.

Shenfield, A., Day, D., & Ayesh, A. (2018). Intelligent intrusion detection systems using artificial neural networks. *ICT Express*, 4.

Shevchenko, N., Chick, T. A., O'Riordan, P., Scanlon, T. P., & Woody, C. (2018). Threat modeling: a summary of available methods. Technical report, CMU – Software Engineering Institute.

Siano, P., De Marco, G., Rolán, A., & Loia, V. (2019). A survey and evaluation of the potentials of distributed ledger technology for peer-to-peer transactive energy exchanges in local energy markets. *IEEE Systems Journal, 13*(3), 3454–3466.

Sion, L., Yskout, K., Van Landuyt, D., & Joosen, W. (2018). Risk-based design security analysis. In *Proceedings of the 1st International Workshop on Security Awareness from Design to Deployment* (pp. 11–18).

Skopik, F., Friedberg, I., & Fiedler, R. (2014). Dealing with advanced persistent threats in smart grid ICT networks. In *Innovative Smart Grid Technologies (ISGT)* (pp. 1–5). IEEE Power & Energy Society.

Soltan, S., Mittal, P., & Poor, H. V. (2018). BlackIoT: IoT botnet of high wattage devices can disrupt the power grid. In *27th {USENIX} Security Symposium* (pp. 15–32).

Soltan, S., Mittal, P., & Poor, V. (2019). Protecting the Grid against MAD Attacks. *IEEE Transactions on Network Science and Engineering*.

Sridhar, S., Hahn, A., & Govindarasu, M. (2012). Cyber-physical system security for the electric power grid. *Proceedings of the IEEE, 100*(1), 210–224.

Stouffer, K., Falco, J., & Scarfone, K. (2015). Guide to industrial control systems (ICS) security – Rev. 2. *NIST Special Publication, 800*(82).

Strbac, G., Woolf, M., Walker, S., & Vahidinasab, V. (2020). The Role of Active Buildings in the Transition to a Net Zero Energy System. *Imperial College, Newcastle University.*

Teixeira, A., Sou, K. C., Sandberg, H., & Johansson, K. H. (2015). Secure control systems: A quantitative risk management approach. *IEEE Control Systems Magazine, 35*(1), 24–45.

Thurner, L., Scheidler, A., Schäfer, F., Menke, J.-H., Dollichon, J., Meier, F., Meinecke, S., & Braun, M. (2018). pandapower – an open-source python tool for convenient modeling, analysis, and optimization of electric power systems. *IEEE Transactions on Power Systems, 33*(6), 6510–6521.

UcedaVelez, T. and Morana, M. M. (2015). *Risk centric threat modeling*. Wiley Online Library.

Varga, A. (2010). OMNeT++. In *Modeling and Tools for Network Simulation* (pp. 35–59). Springer.

Wangen, G. (2017). Information security risk assessment: A method comparison. *Computer, 50*(4), 52–61.

Yang, Y., Wu, L., Yin, G., Li, L., & Zhao, H. (2017). A survey on security and privacy issues in Internet-of-Things. *IEEE Internet of Things Journal, 4*(5), 1250–1258.

Yankson, S., & Ghamkhari, M. (2020). Transactive energy to thwart load altering attacks on power distribution systems. *Future Internet, 12*(1), 4.

Yuan, Y., Li, Z., & Ren, K. (2011). Modeling load redistribution attacks in power systems. *IEEE Transactions on Smart Grid, 2*(2), 382–390.

Zafar, R., Mahmood, A., Razzaq, S., Ali, W., Naeem, U., & Shehzad, K. (2018). Prosumer based energy management and sharing in smart grid. *Renewable and Sustainable Energy Reviews, 82*, 1675–1684.

Zajc, M., Kolenc, M., & Suljanović, N. (2019). Virtual power plant communication system architecture. In *Smart Power Distribution Systems* (pp. 231–250). Elsevier.

Zambon, E., Etalle, S., Wieringa, R. J., & Hartel, P. (2011). Model-based qualitative risk assessment for availability of IT infrastructures. *Software & Systems Modeling, 10*(4), 553–580.

Chapter 10
Energy Management Systems of Grid-Connected Active Buildings

Mohammad Shaterabadi, Hosna Khajeh, Hooman Firoozi, and Hannu Laaksonen

Status Quo, Challenges, and Outlook
Energy management is a well-known method to deal with the cost of operation and energy efficiency either in generation-side or demand-side systems. However, due to the increasing number of active end-users on the demand side, the need for energy management has become more significant, especially in active buildings. In order to implement the energy management approaches, both software and hardware solutions are required, which could be challenging. Moreover, the future of energy management must consider the new requirements of electricity networks at all levels of the implementation. Therefore, we need to answer the question that how energy management could be applied in active buildings? What are the requirements and constraints? And what techniques could be deployed to make energy management more efficient and user-friendly?

10.1 Overview

Increasing energy consumption in the world due to the industrialization of countries, population growth, and the need for providing the required energy for consumers have had negative environmental impacts such as global warming due to increases in CO_2 emission or other greenhouse gases stemming from fossil fuels (Beaudin & Zareipour, 2017; Gholamzadehmir et al., 2020). Therefore, in recent years,

M. Shaterabadi
Department of Electrical Engineering, Kermanshah University of Technology, Kermanshah, Iran

H. Khajeh · H. Firoozi (✉) · H. Laaksonen
Flexible Energy Resources, School of Technology and Innovations, University of Vaasa, Vaasa, Finland
e-mail: hooman.firoozi@uwasa.fi

V. Vahidinasab, B. Mohammadi-Ivatloo (eds.), *Active Building Energy Systems*, Green Energy and Technology, https://doi.org/10.1007/978-3-030-79742-3_10

energy management issues have become more important in most countries. Energy management can be considered both in the generation and demand sides (Tabar et al., 2017). Initially, to overcome these problems, electrical network stakeholders were obsessed with demand-side management, in which their main focus was on energy management as well as increasing the utilization of renewable energy resources located on the generation side. However, since all the activities of the power system aim to provide the required power and energy for the consumers, and a great portion of power is consumed at the consumer level by different kind of buildings, including residential, commercial, industrial, etc. Most of the waste in energy occurs in this side of power systems. This attracted the attention of systems' operators to spend more time and money on the energy management studies in the downstream networks (Shaterabadi & Jirdehi, 2020). Accordingly, energy management at the building level entered a new phase of the study. Energy consumption in buildings for cooling, heating, lighting, etc. accounts for a significant part of energy consumption. According to available statistics, this amount is about 40 % of total generation capacity which is also responsible for about 36 % of the total emission (Rathor & Saxena, 2020). The share of energy consumption in buildings is about 20.1% of the total delivered energy production consumed in the world that is expected to have an annual growth of 1.5% by 2040 (Badar & Anvari-Moghaddam, 2020; Jamil & Mittal, 2018). The negative effects of this staggering increase in energy demand in communities are not limited to rising costs and pollution. It might also result in system-wide blackouts due to high energy demand during peak hours and the consequent reliability problems in the whole system. Hence, the energy management systems (EMSs) and their related considerations on the demand side especially in active (smart) buildings have become a crucial concept (Saad Al-Sumaiti et al., 2014).

Grid-connected ABs are those that not only utilize several renewable and non-renewable generation resources within themselves but also perform energy management levels in conjunction with a variety of automation systems. Figure 10.1 shows an overview of a grid-connected AB. As shown in the figure, these buildings are also interacting and cooperating with the upstream power network. This interaction and cooperation will be done for specific purposes such as reducing the operating costs of the building and, as result, reducing the costs imposed on the network due to an increase in energy consumption. Moreover, other benefits of such buildings could be reducing pollution, peak shaving, increasing the reliability of the power network, improving the power quality and robustness of the power system, etc. (Al Dakheel et al., 2020; Zhou et al., 2016). Energy management in AB allows users to monitor and control the amount of energy consumed in different sectors at any time and find out where they can reduce their costs and prevent energy waste. The automation system in these buildings should be designed and implemented in such a way that it leads to the minimum need for the presence of the occupant and the impacts of his/her behavior on decisions regarding energy-saving and management's issues. Building energy management system (BEMS) typically controls the heating, cooling, ventilation, air conditioning, lightning, etc. of the buildings (Sun et al., 2013). Control, monitoring, and optimization of these systems

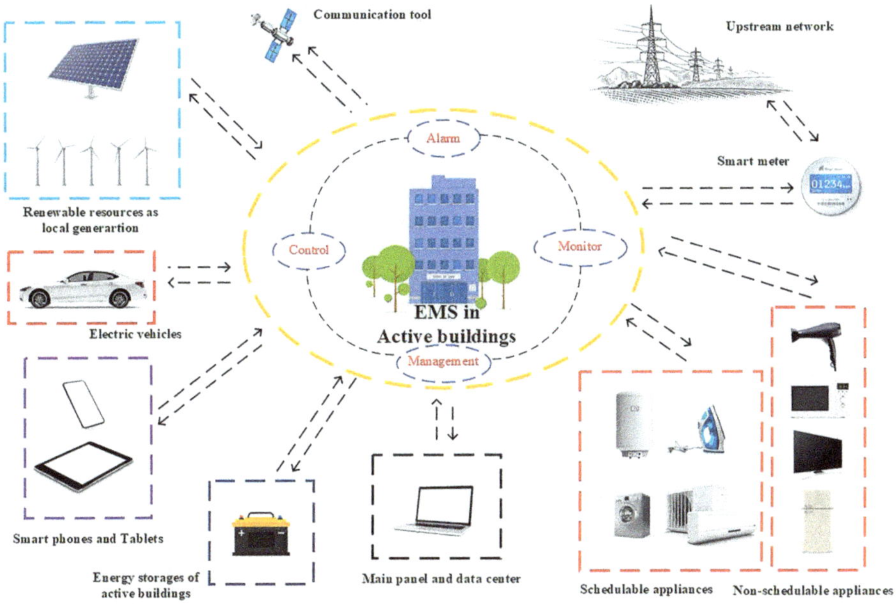

Fig. 10.1 General scheme of energy management in grid-connected active buildings (ABs)

in such buildings is done by various software based on intelligent algorithms as well as the Internet of things (IoT) infrastructure, sensors, and equipment related to automation. Overall, these buildings have three important factors that distinguish them from other buildings (Waters, 2018):

1. Automated control by various types of equipment
2. Concatenation of tenant priority and feedback
3. Adaptability and learning ability (based on intelligent algorithms behind the automation mechanism, it can adapt to the environment and lifestyle of people in order to reduce cost and increase social welfare)

Energy management in ABs can be considered and evaluated from different viewpoints. However, in this chapter, the focus is on the most important issues which these kinds of buildings are confronted with. These contents will be explained in the following sections. Several objective functions like cost minimization, peak shaving, flexibility services, and different combinations of them in form of multi-objective functions will be presented in Sect. 10.2. Constraints that ABs must consider, like appliance constraints, thermal comforts, and grid power exchange limitation, have been presented in Sect. 10.3. Different AB's equations and solving approach for optimization problems will be explained in Sect. 10.4. Finally, in Sect. 10.5, the conclusion and summary of the whole chapter will be stated.

10.2 Energy Management Objectives of ABs

In the energy management programs in grid-connected ABs, various objective functions are considered. In most cases, this prioritization of objective functions is done according to the preferences of the occupants of the building. However, in some cases, these objective functions might be performed based on the needs of upstream network operators in order to achieve specific goals. In the following, we will discuss some of the most important objective functions related to energy management programs in grid-connected ABs. The objective functions that have been considered in these systems can be very diverse and extensive, for example, cost, peak shaving, and flexibility services (Beaudin & Zareipour, 2017). In Fig. 10.2, an overview on the EMS objectives and constraints could be found.

10.2.1 Cost Minimization

The energy management for grid-connected ABs can be expressed as both single-objective and multi-objective functions; this goal or goals can be considered from various aspects such as economic, technical, environmental, social, or a combination of two or more. Although a lot of research has been done on energy management in this type of buildings, it can be said that cost is one of the most important objective functions mentioned in most of the articles (of course in the presence of technical, economical, etc. constraints) (Costanzo et al., 2012). The total cost in such buildings

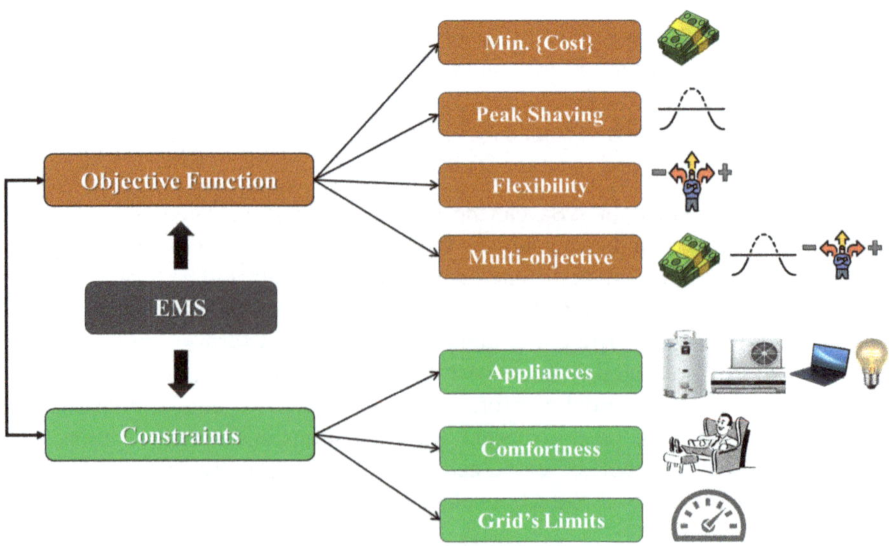

Fig. 10.2 Overview on the EMS objectives and constraints

can be in the form of or the start-up cost of different equipment such as the cost of energy storages failure or battery deterioration, penalties, or rewards for users due to following a specific consumption pattern (Agnetis et al., 2013; Clastres et al., 2010), the total cost of building's operation, maximizing profits of aggregators or residents, total cost and tax minimization caused by the emission of pollution, cost minimization of initial investment for construction such this buildings, and also annual operating cost and life cycle cost (Honarmand et al., 2014; Papari et al., 2018). For instance, in (Eq. 10.1), a single-objective function has been depicted; the aim of this equation is to minimize the total bill's cost of the residence and users, which is obtained by multiplying the energy price by the difference between the load profile and the energy sold by the AB to the upstream network (Badar & Anvari-Moghaddam, 2020).

$$\text{Min } \left(C_u \right) = \sum_{t=1}^{T} \left(P_u^n(t) - P_u^s(t) \right) . \lambda \left(t, P_n(t) \right) \tag{10.1}$$

C_u: Daily bill's cost
${P_u}^n$: Load profile of the building at time (t)
${P_u}^s$: Amount of power sold to the upstream network at time (t)
$P_n(t)$: Energy price

As shown above, residents can sell their extra power saved in energy storages in low demand hours and at peak load hours released the energy; the smaller the difference, the lower the cost of subscribers' bills. It should be noted that for this purposes constraints must also be met, so this difference cannot be less than one limit by considering different constraints, because the demand should be supplied in the first stage and then the surplus power might be sold to the network by user's decision.

10.2.2 Peak Shaving

The amount of load is not constant during a day and is constantly changing. Providing this load with high reliability is one of the concerns of operators. This is important especially during peak times, because in these cases there is double pressure on the network, and also increasing the generation capacity of resources is not cost-effective. Most of the time to overcome such problems, gas power plants or, in islanded regions or buildings, diesel generators can be used (they generated the surplus power for the response to the load increment of the network or building during peak time). Although the capital cost of these systems may be seen as low, the operation and maintenance (O&M) cost by the past imposes a lot of cost on the network and building management programs. For this reason, peak shaving programs seem to be important along with reducing the overall cost of the network and buildings, as well as increasing the overall reliability and robustness of the system simultaneously (Chua et al., 2013; Shirazi & Jadid, 2017).

Peak shaving is a strategy in which the load curve is flattened so that the load is transferred to the parts with less amount of load (valleys in load profile) during peak hours (peaks). This change is done by considering various factors such as users' urgent need for equipment at the moment, schedulable and un-schedulable equipment, peak hours on different days of different seasons, applying reward and penalty programs, etc.

There have been many articles that have discussed this issue recently, and the strategies that these papers have used to achieve this goal include the following:

1. Utilizing energy storage system (ESS) integration
2. Using electric vehicle (EV) for power exchange (V2G & G2V modes)
3. Demand-side management (DSM)

As one example, by using storage devices, energy can be stored at a low load zone (low price), and the required power can be discharged back to the grid or building during peak hours. Moreover, EVs can be agreed with their owners based on the contract, to provide their power to the network and the building during the peak load in order to compensate the required power in this way. On the other hand, demand-side management programs as the other strategies can be divided into two categories, as follows:

1. Reliability-based programs
2. Market-based programs

Reliability-based programs operate by utilizing economic incentives, which include offering low electricity prices or granting special credits to the responsive demands. This program is divided into three more detailed subcategories as follows:

1. Direct load control programs (DLCP)
2. Interruptible programs (IP)
3. Curtailable load programs (CLP)

In market-based programs, consumers voluntarily adjust their demand according to the economic offers given by the network operator, and in return, they can benefit from discounts on electricity prices or other incentive programs (Arens et al., 2008; York & Kushler, 2005).

It is important to note that considering peak shaving as an objective function indirectly has other benefits as well, such as improving power quality, increasing system efficiency, reduction in total cost, etc. (Uddin et al., 2018).

Grid-connected ABs, as their name suggests, have the ability to communicate with the upstream network and can participate in peak shaving programs in continuous communication with the network operator, in order to correct the load curve by using renewable energy resources and storage devices located in the buildings. Furthermore, by using automated systems and the IoT, they can receive signals from the operator, following demand response programs (DRP), and, thus in addition to modifying the load curve and helping to increase network reliability, also have benefits such as reducing their overall cost. In the following, for instance, an objective function that mainly focuses on peak shaving by using the EV charging station has been expressed. In this equation, peak shaving is achieved by minimizing the system's demand using the following objective function (Masoum et al., 2011).

$$\text{Min} \sum_{hs}^{he} \text{Demand}^h = \sum_{hs}^{he} \sum_{k=1}^{n} P_{k,\text{load}}^h \tag{10.2}$$

where

h_s: plug-in time
$h_{e:}$ plug-out time
$P_{k,\text{load}}^h$: Load demand of node k at hour h

According to the above function, by considering and limiting the charging and discharging time for EVs, it is possible to have proper control over the load regulator. For example, peaks can be flattened at a particular node in the network, so that if the load from the charging stations is limited and minimized (start and end times for EVs charging based on the operator's decision), demand's minimization and load curve flattening in that particular node of the system can be reached. By applying this program for nodes that face peak load obstacles, proper peak shaving can be done at the entire power system.

10.2.3 Flexibility Services

With the development of electrical networks from traditional and centralized to modern and decentralized networks, new challenges and problems were faced by planners and operators. These challenges have been due to the dense presence of renewable energy and its intermittency and uncertainties, power exchange in different directions (from upstream to downstream and reverse), more energy consumption due to the advent of electrical appliances more than before (e.g., electrical vehicles, new home appliances), etc. Since the power generation is no longer limited to the upstream network, and generation equipment (e.g., energy storages, renewable energy resources such as PV systems, etc.) have also reached the level of the distribution network (Laaksonen et al., 2020), the general nature of the power network has undergone extensive changes both on the generation and on the consumer (demand) side. Therefore, with careful and detailed studies, the dense presence of this equipment in the distribution network can be used in a direction that will help increase the overall efficiency of the power network (Amicarelli et al., 2017).

Flexibility services refer to those actions, which can be defined as a stable power regulator at the power grid level at a specific moment and for a specific period of time from a particular part of the network.

It can be said that flexible service means the adaptability of demand with the generated power, especially in buildings or grids with the presence of renewable energy sources (Khajeh et al., 2019a). In other words, with the increasing growth of demand and consequently the increase in costs due to the supply of this power by the upstream network, it is possible to use the generation resources and other devices available at the demand-side level (i.e., grid-connected ABs) or by other methods like DRPs so as to not only maintain system reliability but also provide continuous power and reduce overall costs for all parties (Geelen et al., 2013).

These services help to keep the power balance in the distribution network at an acceptable level and also aim to keep the load curve relatively smooth at different times of the day. Flexibility services have five general features: (1) the direction of the exchanged power, (2) their electrical structure in the network, (3) temporary features according to the start-up time, (4) duration of its presence, (5) and finally its location in the power system (Eid et al., 2016).

Some distributed energy resources (DERs) can be unidirectional like typical household loads like water heaters, dishwashers, etc. In contrast, some of them can exchange power in a bidirectional way like energy storage, EVs, etc.

It is important to pay attention to the network in which DER is used, to get the most out of these services. Some of the equipment is able to provide a lot of instantaneous power for a short period of time and cannot maintain their power level for a long time. In contrast, some of the other DERs can provide the power for longer periods, meaning that they can maintain their power at an acceptable level for a while.

Considering the time required to prepare and use this equipment is also important for the network, most DERs such as storage devices, photovoltaic, EVs, DRPs, etc. that can quickly be ready for use. However, some equipment such as small-scale combined heat and power (CHP) needs more time to prepare and cannot deliver the required power to the network in order to help it for better performance at the moment.

The duration of the presence of such resources is another point. This means that for how long each unit can stay in touch and cooperate with the power system for flexibility service works, for example, electrical vehicles cannot participate in these programs at a specific period of time (i.e., their owners need them).

Finally, the location of the DERs can be important since the operator decides to use this type in congestion programs or to provide power for part of the network or maybe to improve the power quality and flexibility and to decrease power losses. In other words, the location of the equipment in the power system affects how to use that particular equipment in a better and efficient way.

Moreover, as mentioned earlier, consumers can play an active role in flexibility services by using DRP, according to what was mentioned in Sect. 10.2.2. In the following, an equation that has shown this type of objective function has presented (Amicarelli et al., 2017), which is:

$$\text{Min} \sum_{AB,t} F\left(C, P^f\right) \tag{10.3}$$

In (10.3), the AB aims to minimize its total cost that resulted from providing flexibility services. The flexibility services can be provided to the system operators or aggregators. With this regards, the devices/appliances that are being utilized to provide flexibility will be prioritized based on their operational costs. Therefore, the objective would be a function of their power and their corresponding operational costs.

10.2.4 Multi-objective Function

As mentioned before, in the EMS of grid-connected ABs, power system operators usually seek to optimize and achieve more than one objective. Accordingly, several objective functions can be considered to achieve positive results also, to save time and money. In this case, the focus is on optimizing two or more objective functions, so that relative satisfaction is achieved. This creates a more desirable and complete plan for energy management in such buildings. There are some methods that can solve a multi-objective problem by converting them into single-objective functions, such as the weighting factor and lexicographic method, while others provide a set of optimal answers to the user according to priority or need, to choose among them, e.g., epsilon constraint and augmented epsilon constraint method. Some objective functions such as cost, environmental pollution, social welfare, power loss, etc. can be considered as well (Vergini & Groumpos, 2016; Jirdehi & Shaterabadi, 2020). In below, a bi-objective function has been expressed, which is the home's total energy cost and load profile deviation that should be minimized simultaneously (Sattarpour et al., 2018).

$$\begin{aligned} &\text{Min EC} = \sum_{t\in T} \left(E_t^{\text{total}} \times \Omega_t\right) \\ &\text{Min LPD} = \sum_{t\in T} \left|E_t^{\text{total}} - E^{\text{mean,total}}\right| \quad , \quad E^{\text{mean,total}} = \frac{\sum_{t\in T} E_t^{\text{total}}}{T} \end{aligned} \tag{10.4}$$

t: Index for time
T: Set of time intervals
Ω_t: Electricity tariff
E_t^{total}: Energy consumption time (t)

Concerning the functions which have been shown above, in this multi-objective problem, first, the total energy cost for homes should be minimized, and then to obtain the flattened load curve, the deviation in demand must be minimized too. It should be noted that in this multi-objective optimization problem, the energy cost for reaching the flattened load curve should not exceed the minimum operating cost of the first objective function; also according to the relationship of the second objective function, the amount of load should be shifted to the average daily consumption (for reaching to latter aim).

10.3 Constraints of Grid-Connected AB's Operation

In the previous sections, the number of important objective functions in the energy management of grid-connected ABs was discussed. In this section, some of the most important constraints are presented that those objective functions should be optimized by considering these constraints. In other words, optimization is meaningful under defined constraints. These constraints can be economical, technical, social, welfare, etc. The objective functions are optimized under a set of constraints, depending on the type of building, as well as its facilities, generation resources, geographical condition, appliances, and limitations of them; these constraints may vary from one building to another (Qayyum et al., 2015).

10.3.1 Controllable Load Appliances' Constraints

There are many different electrical appliances in the buildings like refrigerator, oven, washing machine, hair-dryer, dishwasher, EV, television, lights, computers, air conditioner, etc. These appliances, depending on the application of each and the users' needs, can be used at different hours of the day (Firoozi & Khajeh, 2017). The main constraints related to this type of device are related to the time period and the priority that each appliance is required to be used. These appliances, from the load scheduling point of view, can be categorized based on two groups:

1. Controllable appliances (operation of these can be the scheduled duration of their work, e.g., washing machine, dishwasher, heating system, etc.)
2. Non-controllable appliances (operation of these cannot be scheduled, e.g., hair-dryer, refrigerator, etc. (Leitao et al., 2020))

Previous studies have categorized devices into different categories and they did not reach a consensus on these categories. In other words, researchers have not yet agreed on the exact grouping that can categorize the appliances and just survey them from different perspectives.

Various constraints on the performance of building appliances have been explored in the articles, including ensuring the continuous operation of appliances and coordination between appliances, so that when one appliance starts operating, another appliance must be worked at the same time or should be turned off. For instance, if the TV is turned off, then the DVD player should also be turned off. Turn on or off at the specified time is another constraint that has been considered in many types of research. Other constraints include the power consumption limit of each device and also the time limitation for use of any device (Setlhaolo & Xia, 2015).

According to the above statements, another category can be obtained. In other words, appliances can be divided into six group, and their constraints may be written and considered based on this grouping (Jang et al., 2018). These types of groupings are as follows:

1. Discrete type (devices that fall into this category can be turned on and off without any restrictions like hair-dryer)
2. One-stop type (devices in this group can only be turned on and off once in a specified time period. In other words, they must operate continuously for a specified period of time like dish-washer)
3. Multiple-stop type (devices in this group can be turned on and off, M times in a specified period of time like water pump)
4. Stepped type (appliances that fall into this category have a variety of consumption patterns over the course of their operation and their energy consumption changes like washing machine)
5. Cool-down type (devices that fall into this category are not allowed to turn on immediately after being turned off and must be passed for a while like some electric motors)
6. Sequential type (devices in this group must start working after the end of the work of another device; in other words, to start working, they must wait for the end of the operation of another device like a dryer machine should work after washing machine)

In the following, an example of constraint related to the building's appliances is exhibited. This constraint shows the hot water temperature bounds specified by users (Du & Lu, 2011).

$$\begin{gathered} \theta_n^{low} \leq \theta_n \leq \theta_n^{up} \qquad n = 1 \ldots\ldots N \\ \text{For instance :} \\ \begin{Bmatrix} 132 \leq \theta_n \leq 150 & 13 \leq t_n \leq 17 \\ 142 \leq \theta_n \leq 160 & \text{otherwise} \end{Bmatrix} \end{gathered} \tag{10.5}$$

In this constraint, the temperature of the hot water can be controlled by a minimum and maximum limitation of boundaries, according to the user's decision based on various factors.

10.3.2 Thermal Comfortness

Aside from all that has been said about energy management and energy consumption at the grid-connected AB level, another important factor and constraint to consider is thermal comfortness. This means that all the buildings, in which people live or work, should provide comfortable conditions in terms of temperature and heat. Thermal comfortness is a mental and individual condition that people feel satisfied with the environment's temperature, and this is assessed by the person themselves. In other words, this feeling of satisfaction varies from person to person, and an environment that seems ideal to one person may not be appropriate to another. For this reason, sufficient attention should be paid to this point in the operation of ABs.

Factors that affect this feeling of satisfaction are divided into two categories: (1) environmental and (2) individual. Environmental factors include ambient temperature, relative humidity, level of radiation, air movement in the environment, etc. Individual factors include things like each person's body metabolism, sickness, clothing style in the environment, and generally sense of personal satisfaction (Vergini & Groumpos, 2016; Missaoui et al., 2014).

There are various standards for determining the average comfort and convenience in buildings that can be referred and included in the planning to greatly satisfy residents such as ANSI/ASHRAE Standard 55, ISO 7730 (ISO-International Organization for Standardization, 1994). In the following, a sample constraint related to thermal comfortness has been expressed (Missaoui et al., 2014):

$$D(i,k) = \left\{ \begin{array}{ll} \frac{T_{\text{opt}}(i,k) - T_{\text{in}}(i,k)}{T_{\text{opt}}(i,k) - T_{\text{min}}(i,k)} & \text{if } T_{\text{in}}(i,k) \leq T_{\text{opt}}(i,k) \\ \frac{T_{\text{in}}(i,k) - T_{\text{opt}}(i,k)}{T_{\text{max}}(i,k) - T_{\text{opt}}(i,k)} & \text{if } T_{\text{in}}(i,k) > T_{\text{opt}}(i,k) \end{array} \right\} \tag{10.6}$$

T_{opt} *(i, k)*: Optimal temperature at each point of time
T_{in} *(i, k)*: Actual temperature
T_{min} *(i, k)*: Minimum acceptable temperatures
T_{max} *(i, k)*: Maximum acceptable temperatures

Based on what is said and showed above, this constraint can control the environmental temperature concerning particular permanent service *i*, corresponding to a thermal zone, at each period *k*. Therefore, when the residents are in their buildings, satisfaction can be obtained by the difference between the ideal temperature and the actual temperature. This ensures that, in the energy management planning of this type of buildings, the ambient temperature remains acceptable and satisfactory for the residents.

10.3.3 Grid Exchange Limitation

Grid-connected ABs as the name implies have the ability to exchange power with the upstream network, which means that it can deliver power to the upstream network at times when it has surplus power excess of demand; as opposed to, at times when it is not able to supply the demands for some reason, it takes the required power from the upstream network.

These buildings connect to the upstream network by point of common coupling (PCC). This type of connection (i.e., indirect connection by common and detachable buses) helps and improves the protection and the reliability of the overall system in case of failure in the operation of ABs or vice versa (upstream network failure); failure isolation can be realized between the AB and the main grid.

As mentioned before, the presence of the grid in this type of buildings increases the reliability of buildings so that it is always possible to ensure continuous power supply without interruption. Moreover, by selling power to the network, it can make a profit for users and thus reduce its overall cost. The limitation of this exchangeable power depends on many factors like the structure of the network, the place of building in the electrical network, total surplus power that can be generated by different equipment in the building, minimum power required for vital equipment in the building, the amount of budget for construction of line (capacity of transmission line between AB and main grid), and the maximum acceptable cost for the operator and users (to build transmission line) (Monyei et al., 2018). A grid exchange limitation constraint can be written as follows:

$$\begin{cases} P^{\text{AB-to-grid}} \leq P^{\text{Line, max}} \\ P^{\text{AB-to-grid}} \geq P^{\text{Line, min}} \end{cases} \tag{10.7}$$

By considering the above constraint, it can be understood that there is a limitation for the amount of power that can be sell or buy to/from the upstream network. It is a quite crucial point that operators should pay good attention to it, since ignoring this constraint can lead to the collapse of the intended planning. This maximum limitation as we pointed to depends on many factors; sometimes the exchanged power must be more than the specified amount, which does not apply to all buildings.

10.4 Modeling Energy Management of Grid-Connected AB

The responsibilities of an EMS of the building include a range of tasks such as scheduling controllable appliances and devices, monitoring and forecasting the consumption of uncontrollable devices, forecasting the production of the building (e.g., solar production), and making bidding strategies on behalf of the building who wants to participate in local markets as an individual player. However, it is noticeable that the mentioned tasks are closely correlated with each other.

For example, the optimal scheduling of controllable and flexible devices of the building is highly dependent on the occupant's preferences and comfort, the real-time and forecasted consumption power of other appliances, the production of the building, and the grid's real-time need. In addition, the scheduling process performed by the EMS should be in line with the grid service that the building had promised to provide. However, the EMS requires to take into account the main objective of the building as the main factor.

If the household promised to provide grid services and, for example, contribute to the peak-reduction services, the working time of controllable devices should be shifted to other off-peak timeslots. On the other hand, the occupant might also define a time range as a permissible timeslot for appliances' operation. If the predefined time range overlaps with the peak timeslots, the EMS may send a message to the occupant. The message should warn the occupant about the consequences of using the appliances and remind them of their promises and the possible penalty costs of not adhering to them. Accordingly, the building's occupant can decide whether using the device is worth the extra cost or not.

There might be some local markets, where the building can participate as a seller. Regarding this structure, the building can individually sell energy and flexibility to the grid and/or other participants (Khajeh et al., 2020). In this way, the EMS should make bidding strategies for the building. However, finding the optimal bidding strategies is a complicated process for the EMS since it needs to analyze huge data and information on historical data, the behavior of other competitors, market prices, and the market-clearing process.

In general, an EMS of the building can utilize various scheduling approaches and methods so as to better control the flexible devices. According to recent research, the scheduling algorithms can be categorized into rule-based, artificial intelligence-based, and optimization-based approaches. The following subsections introduce the mentioned methods. Moreover, the most important features of these methods are briefly mentioned in Table 10.1.

10.4.1 Optimization-Based Approaches

Optimization-based approaches aim to mathematically model the system and find the optimal schedule of the controllable devices in the building. In these approaches, the EMS follows an objective that was defined by the occupant. According to this objective, the decision variables of the optimization problem are the optimal consumption and production power of controllable devices at each timeslot. In fact, the EMS should determine the optimal working time and working power of flexible appliances. However, if the power of an appliance is not controllable, the decision variable regarding this device is a binary variable (ON/OFF signal) for each time slot.

Table 10.1 Overview on energy management methods

Methodology	Objectives	Requirements	Examples
Optimization based	• Min. {Cost} • Max. {Profit}	• High user cooperation • Accurate predicted values • Consistent price following	DP, MILP, MINLP, QP, etc.
Rule based	• Min. {Cost}	• Users' preference • High user responsiveness	Rete, etc.
AI based	• Min. {Cost} • Max. {Profit} • Min. {User interaction}	• Low-latency ICT • Accurate AB's model • Highly accurate predicted values	ANN, ANFIS, FLC, etc.

In addition, the optimization problem has some constraints restricting the decision variables. The constraints are associated with the operation of each controllable appliance.

Some of them consider the inherent characteristics of the devices (e.g., the capacity, maximum and minimum working power, etc.), while others can be those related to the occupants' comfort and preferences (e.g., the time span related to the operation of the devices specified by the customers).

The first step to define an optimization problem is to model controllable appliances mathematically. Based on these mathematical models, different optimization problems can be built. Here are some examples (Beaudin & Zareipour, 2017):

1. The first model is to define a linear programming (LP) problem. LP problems are always convex and easy to solve in polynomial time. In these types of problems, the objective function and related constraints are defined to be linear affine functions. If the power consumed or produced by appliances is controllable and the other related functions are linear, it obtains an LP problem. In this situation, the variables should not be correlated with each other. However, as previously mentioned, most of the appliances consume constant power and are not controllable in terms of power. As a result, binary variables are required to implement the ON/OFF signals sent by the EMS. As a result, the problem is not LP anymore as it turns into a mixed-integer linear programming problem (MILP). However, one may relax integer variables to create an LP problem.
2. The EMS may define a quadratic programming problem. Solving this type of problem is relatively simple. A problem that has a quadratic objective function and some linear constraints is an example. In this case, if the objective function is a positive definite function, the defined optimization problem can be simply solved in polynomial time. Otherwise, solving this problem would be harder. In addition, the model may have a convex objective function as well as some linear

equality constraints and concave inequality constraints. This type of optimization problem is assured to be converged, leading to a unique solution.
3. The EMS may model the dynamic of the appliances and devices, creating a dynamic programming problem. A dynamic programming problem is capable of solving large complex problems by splitting the main problem into several smaller sub-problems. The sub-problems are then recursively solved.
4. The scheduling problem may define as a MILP problem in which the problem has a linear objective function along with some linear constraints. However, this problem consists of a mixture of continuous and integer decision variables (Khajeh et al., 2019b).
5. There exist some practical algorithms that help to solve the problem. One of the effective methods is the branch-and-bound algorithm. This method relaxes the problem and creates some linear sub-problems. Since most of the devices of a smart building operate with constant power, it would be more convenient for the EMS to define mixed-integer programming problems. In this way, the EMS can easily send the controllable appliances ON/OFF signals. These signals are modeled by binary decision variables. However, it is worth mentioning that integer variables may also make the problem complex which cannot be solved easily.
6. Defining a mixed-integer nonlinear programming (MINLP) problem is not an appropriate form of scheduling since it is relatively difficult to solve. In other words, unlike a convex optimization, obtaining a unique solution is not guaranteed in this problem. However, in some studies, they transfer MINLP problems to MILP ones through the use of linearization techniques (Firoozi et al., 2020).

10.4.2 Rule-Based Approaches

Rule-based algorithms define rules for different conditions of behavioral systems. One of the popular rule-based algorithms which can be utilized in EMS is a Rete algorithm. By using the Rete algorithm, the EMS can control the consumption and production of the building via smart taps (Kawakami et al., 2013). In this way, the flexible appliances of the building are distributed to the smart taps and go through the rule processing. Afterward, based on the priority defined by the occupant, the rule-based process defines some if/then rules (Yoshihisa et al., 2012).

For instance, the algorithm can define a rule to shift the operating time of high-priority flexible appliances from peak timeslots to off-peak timeslots providing that the building needs to provide peak-reduction services. The algorithm can also consider uncertainties associated with the dynamic prices and the consumption of uncontrollable devices of the building (Contreras-Aguilar et al., 2012). One of the disadvantages of the rule-based algorithm is that it fails to be extended to some advanced applications. In other words, the algorithm is unable to analyze large data when the EMS confronts several tasks and restrictions for the operation of appliances.

10.4.3 Artificial Intelligence and Machine Learning Approaches

An EMS may utilize AI techniques to optimally schedule the appliances of the smart building. In this way, controllers of appliances may use artificial neural network (ANN) algorithms and machine learning, the algorithms based on fuzzy logic control (FLC), and adaptive neural fuzzy inference system (ANFIS) so as to find the optimal schedule (Shareef et al., 2018). In these algorithms, the controller seeks to play the role of the human brain, imitate human thinking, and make optimal decisions.

Integrating ANN algorithms with EMS can empower the scheduling process of these systems. Non-linear decision making is the main application of these algorithms. In the process of decision making, the inputs data are processed, and the algorithm tries to simulate human brains to obtain the decision variables (Ahmed et al., 2017). The EMS can also be equipped with machine learning algorithms. In this way, according to the relation between historical inputs and decision variables, the rules and behaviors are learned.

If FLC algorithms are utilized in the EMS of a building, the input data should go through four steps, including fuzzification, defuzzification, rule base, as well as an inference engine. It is noticeable that FLC algorithm does not need any mathematical model to solve the scheduling problem, but it is still capable of obtaining a solution for non-linear and linear models (Wu et al., 2011; Suganthi et al., 2015). In addition, in this model the behavior of customers, its preferences, and comfort can be modeled using fuzzy parameters. In comparison, the ANFIS algorithm is a more complicated algorithm presenting several layers (Premkumar & Manikandan, 2015). The algorithm can define a set of fuzzy if/then rules, and its learning capability is able to approximate nonlinear functions.

10.5 Summary and Conclusions

Grid-connected ABs are those that not only utilize several renewable and non-renewable generation resources within themselves but also perform energy management levels in conjunction with a variety of automation systems. Thus, the EMS is the core of an AB that should schedule the controllable devices and resources of the building while following a certain objective. The objective of the EMS can be decided by the occupant (e.g., cost minimization) or be determined by the grid (e.g., peak-shaving and providing flexibility service), or it can be a hybrid of the mentioned objectives. In addition, there exist some key constraints which should be considered in the process of scheduling. The constraints may be related to the characteristics of the devices or appliances, those imposed by occupants in order to maintain their comfort level, and the restrictions associated with the external grid.

Thus, the EMS should take into account the mentioned limitation when scheduling controllable appliances.

The responsibilities of an EMS of the building include a range of tasks such as scheduling controllable appliances and devices, monitoring and forecasting the consumption of uncontrollable devices, forecasting the production of the building (e.g., solar production), and making bidding strategies on behalf of the building who wants to participate in local markets as an individual player. In order to fulfill these responsibilities, the EMS should model the internal system in the building. In fact, it should schedule appliances optimally using scheduling algorithms. In general, the scheduling algorithms can be categorized into rule-based, artificial intelligence-based, and optimization-based approaches. Regarding rule-based algorithms, the EMS defines several rules for different conditions, while the artificial intelligence-based approaches aim to imitate the human brain to find the optimal decision variables. In optimization-based methods, the system defines a mathematical model of the building in order to find the optimum control variables. These control variables can be ON/OFF signals sent to the appliances or the optimal working power of the devices. However, the defined model needs to be convex so as to get the best results.

References

Agnetis, A., De Pascale, G., Detti, P., & Vicino, A. (2013). Load scheduling for household energy consumption optimization. *IEEE Transactions on Smart Grid, 4*(4), 2364–2373. https://doi.org/10.1109/TSG.2013.2254506

Ahmed, M. S., Mohamed, A., Shareef, H., Homod, R. Z., & Ali, J. A. (2017). Artificial neural network based controller for home energy management considering demand response events. In *2016 international conference on advance electrical and electronic system engineering*. ICAEES 2016, pp. 506–509. https://doi.org/10.1109/ICAEES.2016.7888097

Al Dakheel, J., Del Pero, C., Aste, N., & Leonforte, F. (2020). Smart buildings features and key performance indicators: A review. *Sustainable Cities Society, 61*. https://doi.org/10.1016/j.scs.2020.102328

Amicarelli, E., Tran, T. Q., & Bacha, S. (2017). Flexibility service market for active congestion management of distribution networks using flexible energy resources of microgrids. In *2017 IEEE PES innovation smart grid technology conference European ISGT-Europe 2017 – Proc.*, vol. 2018-January, pp. 1–6. https://doi.org/10.1109/ISGTEurope.2017.8260198

Arens, E., Auslander, D., & Huizenga, C. (2008). *Demand response enabling technology development*.

Badar, A. Q. H., & Anvari-Moghaddam, A. (2020). *Smart home energy management system–A review*. *Advances in Building Energy Research*. https://doi.org/10.1080/17512549.2020.1806925

Beaudin, M., & Zareipour, H. (2017). Home energy management systems: A review of modelling and complexity. *Lectures Notes Energy, 33*, 753–793. https://doi.org/10.1007/978-3-319-26950-4_35

Chua, K. H., Lim, Y. S., & Morris, S. (2013). Battery energy storage system for peak shaving and voltage unbalance mitigation. *International Journal of Smart Grid Clean Energy*, 357–363. https://doi.org/10.12720/sgce.2.3.357-363

Clastres, C., Ha Pham, T. T., Wurtz, F., & Bacha, S. (2010). Ancillary services and optimal household energy management with photovoltaic production. *Energy, 35*(1), 55–64. https://doi.org/10.1016/j.energy.2009.08.025

Contreras-Aguilar, L., Garcia, N., Islas-Martinez, M. A., & Adame-Ortiz, R. (2012). Implementation of a VFT model in PSS/E suitable for power flow and transient stability simulations. In *IEEE power energy society general meeting*. https://doi.org/10.1109/PESGM.2012.6345395

Costanzo, G. T., Zhu, G., Anjos, M. F., & Savard, G. (2012). A system architecture for autonomous demand side load management in smart buildings. *IEEE Transactions on Smart Grid, 3*(4), 2157–2165. https://doi.org/10.1109/TSG.2012.2217358

Du, P., & Lu, N. (2011). Appliance commitment for household load scheduling. *IEEE Transactions on Smart Grid, 2*(2), 411–419. https://doi.org/10.1109/TSG.2011.2140344

Eid, C., Codani, P., Perez, Y., Reneses, J., & Hakvoort, R. (2016). Managing electric flexibility from distributed energy resources: A review of incentives for market design. *Renewable and Sustainable Energy Reviews, 64*, 237–247. https://doi.org/10.1016/j.rser.2016.06.008

Firoozi, H., & Khajeh, H. (2017). Optimal day-ahead scheduling of distributed generations and controllable appliances in microgrid. In *2016 smart grids conference, SGC 2016*, pp. 59–64. https://doi.org/10.1109/SGC.2016.7883453

Firoozi, H., Khajeh, H., & Laaksonen, H. (2020). Optimized operation of local energy community providing frequency restoration reserve. *IEEE Access, 8*, 180558–180575. https://doi.org/10.1109/ACCESS.2020.3027710

Geelen, D., Reinders, A., & Keyson, D. (2013). Empowering the end-user in smart grids: Recommendations for the design of products and services. *Energy Policy, 61*, 151–161. https://doi.org/10.1016/j.enpol.2013.05.107

Gholamzadehmir, M., Del Pero, C., Buffa, S., Fedrizzi, R., & Aste, N. (2020). Adaptive-predictive control strategy for HVAC systems in smart buildings – A review. *Sustainable Cities Society, 63*. https://doi.org/10.1016/j.scs.2020.102480

Honarmand, M., Zakariazadeh, A., & Jadid, S. (2014). Optimal scheduling of electric vehicles in an intelligent parking lot considering vehicle-to-grid concept and battery condition. *Energy, 65*, 572–579. https://doi.org/10.1016/j.energy.2013.11.045

ISO-International Organization for Standardization. (1994). EN ISO 7730. Moderate thermal environments-Determination of the PMV and PPD indices and specification of the thermal comfort. *Report*.

Jamil, M., & Mittal, S. (2018). Building energy management system: A review. In *2017 14th IEEE India Council for International Conference INDICON 2017*. https://doi.org/10.1109/INDICON.2017.8488004

Jang, Y., Park, L., Na, W., Lee, C., & Cho, S. (2018). Appliance type constraint design for demand response smart grid systems. In *International conference on ubiquitous future networks*, ICUFN, vol. 2018-July, pp. 407–409. https://doi.org/10.1109/ICUFN.2018.8436746

Jirdehi, M. A., & Shaterabadi, M. (2020). Incentive programs caused by the carbon capture utilization and storage technology profit's effect: Optimal configuration and energy planning of hybrid microgrid involving INVELOX turbine. *Energy Technology, 8*(10). https://doi.org/10.1002/ente.202000398

Kawakami, T., Yoshihisa, T., Fujita, N., & Tsukamoto, M. (2013). A rule-based home energy management system using the RETE algorithm. In *2013 IEEE 2nd global conference consumer electronics*. GCCE 2013, pp. 162–163. https://doi.org/10.1109/GCCE.2013.6664785

Khajeh, H., Laaksonen, H., Gazafroud, A. S., & Shafie-Khah, M. (2019a). Towards flexibility trading at TSO-DSO-customer levels: A review. *Energies, 13*(1), 165. https://doi.org/10.3390/en13010165

Khajeh, H., Foroud, A. A., & Firoozi, H. (2019b). Robust bidding strategies and scheduling of a price-maker microgrid aggregator participating in a pool-based electricity market. *IET Generation Transmission and Distribution, 13*(4), 468–477. https://doi.org/10.1049/iet-gtd.2018.5061

Khajeh, H., Firoozi, H., Laaksonen, H., & Shafie-Khah, M. (2020). A new local market structure for meeting customer-level flexibility needs. In *SEST 2020 – 3rd international conference on smart energy systems and technologies*, pp. 1–6. https://doi.org/10.1109/SEST48500.2020.9203499

Laaksonen, H., Parthasarathy, C., Hafezi, H., Shafie-khah, M., Khajeh, H., & Hatziargyriou, N. (2020). Solutions to increase PV hosting capacity and provision of services from flexible energy resources. *Applied Science, 10*(15), 5146. https://doi.org/10.3390/app10155146

Leitao, J., Gil, P., Ribeiro, B., & Cardoso, A. (2020). A survey on home energy management. *IEEE Access, 8*, 5699–5722. https://doi.org/10.1109/ACCESS.2019.2963502

Masoum, A. S., Deilami, S., Moses, P. S., Masoum, M. A. S., & Abu-Siada, A. (2011). Smart load management of plug-in electric vehicles in distribution and residential networks with charging stations for peak shaving and loss minimisation considering voltage regulation. *IET Generation Transmission and Distribution, 5*(8), 877–888. https://doi.org/10.1049/iet-gtd.2010.0574

Missaoui, R., Joumaa, H., Ploix, S., & Bacha, S. (2014). Managing energy smart homes according to energy prices: Analysis of a building energy management system. *Energy and Buildings, 71*, 155–167. https://doi.org/10.1016/j.enbuild.2013.12.018

Monyei, C., Viriri, S., Adewumi, A., Davidson, I., & Akinyele, D. (2018). A smart grid framework for optimally integrating supply-side, demand-side and transmission line management systems. *Energies, 11*(5). https://doi.org/10.3390/en11051038

Papari, B., Edrington, C. S., & Vu, T. (2018). Stochastic operation of interconnected microgrids. In *IEEE power energy society general meeting*, vol. 2018-January, pp. 1–5. https://doi.org/10.1109/PESGM.2017.8273898

Premkumar, K., & Manikandan, B. V. (2015). Fuzzy PID supervised online ANFIS based speed controller for brushless dc motor. *Neurocomputing, 157*, 76–90. https://doi.org/10.1016/j.neucom.2015.01.032

Qayyum, F. A., Naeem, M., Khwaja, A. S., Anpalagan, A., Guan, L., & Venkatesh, B. (2015). Appliance scheduling optimization in smart home networks. *IEEE Access, 3*, 2176–2190. https://doi.org/10.1109/ACCESS.2015.2496117

Rathor, S. K., & Saxena, D. (2020). Energy management system for smart grid: An overview and key issues. *International Journal of Energy Research, 44*(6), 4067–4109. https://doi.org/10.1002/er.4883

Saad Al-Sumaiti, A., Ahmed, M. H., & Salama, M. M. A. (2014). Smart home activities: A literature review. *Electric Power Components & Systems, 42*(3–4), 294–305. https://doi.org/10.1080/15325008.2013.832439

Sattarpour, T., Nazarpour, D., & Golshannavaz, S. (2018). A multi-objective HEM strategy for smart home energy scheduling: A collaborative approach to support microgrid operation. *Sustainable Cities and Society, 37*, 26–33. https://doi.org/10.1016/j.scs.2017.09.037

Setlhaolo, D., & Xia, X. (2015). Optimal scheduling of household appliances with a battery storage system and coordination. *Energy and Buildings, 94*, 61–70. https://doi.org/10.1016/j.enbuild.2015.02.051

Shareef, H., Ahmed, M. S., Mohamed, A., & Al Hassan, E. (2018). Review on home energy management system considering demand responses, smart technologies, and intelligent controllers. *IEEE Access, 6*, 24498–24509. https://doi.org/10.1109/ACCESS.2018.2831917

Shaterabadi, M., & Jirdehi, M. A. (2020). Multi-objective stochastic programming energy management for integrated INVELOX turbines in microgrids: A new type of turbines. *Renewable Energy, 145*, 2754–2769. https://doi.org/10.1016/j.renene.2019.08.002

Shirazi, E., & Jadid, S. (2017). Cost reduction and peak shaving through domestic load shifting and DERs. *Energy, 124*, 146–159. https://doi.org/10.1016/j.energy.2017.01.148

Suganthi, L., Iniyan, S., & Samuel, A. A. (2015). Applications of fuzzy logic in renewable energy systems – A review. *Renewable and Sustainable Energy Reviews, 48*, 585–607. https://doi.org/10.1016/j.rser.2015.04.037

Sun, B., Luh, P. B., Jia, Q. S., Jiang, Z., Wang, F., & Song, C. (2013). *Building energy management: Integrated control of active and passive heating, cooling, lighting, shading, and ventilation systems*.

Tabar, V. S., Jirdehi, M. A., & Hemmati, R. (2017). Energy management in microgrid based on the multi objective stochastic programming incorporating portable renewable energy resource as demand response option. *Energy, 118*, 827–839. https://doi.org/10.1016/j.energy.2016.10.113

Uddin, M., Romlie, M. F., Abdullah, M. F., Abd Halim, S., Abu Bakar, A. H., & Chia Kwang, T. (2018). A review on peak load shaving strategies. *Renewable and Sustainable Energy Reviews, 82*, 3323–3332. https://doi.org/10.1016/j.rser.2017.10.056

Vergini, E. S., & Groumpos, P. P. (2016). *A review on zero energy buildings and intelligent systems.*

Waters, K. (2018). Energy management for intelligent manufacturing. In *World Energy Engineering Congress*, 2018.

Wu, Y., Zhang, B., Lu, J., & Du, K.-L. (2011). Fuzzy logic and neuro-fuzzy systems: A systematic introduction. *International Journal of Artificial Intelligence and Expert Systems, 2*(2), 47–80.

York, D., & Kushler, M. (2005). *Exploring the relationship between demand response and energy efficiency: A review of experience and discussion of key issues.*

Yoshihisa, T., Fujita, N., & Tsukamoto, M. (2012). A rule generation method for electrical appliances management systems with home EoD. In *1st IEEE global conference consumer electronics 2012*, GCCE 2012, pp. 248–250. https://doi.org/10.1109/GCCE.2012.6379592

Zhou, B., et al. (2016). Smart home energy management systems: Concept, configurations, and scheduling strategies. *Renewable and Sustainable Energy Reviews, 61*, 30–40. https://doi.org/10.1016/j.rser.2016.03.047

Chapter 11
Active Building as an Electricity Network Service Provider

Hooman Firoozi, Hosna Khajeh, Hannu Laaksonen, and Mohammad Shaterabadi

Status Quo, Challenges, and Outlook
The future electricity networks will be changing from a centralized model to a more decentralized and distributed structure. Meanwhile, the growth of renewables in the system is making the electricity networks more unstable and intermittent. Therefore, the challenge of future electricity networks would be maintaining the normal condition of the system for any unpredicted condition. In this way, along with the generation-side resources, the potential demand-side resources, e.g., active buildings, could be deployed to achieve this target. This chapter aims to answer the question that how can demand-side resources such as active buildings contribute to electricity networks' needs? what are the characteristics and resources of an active building? and what services could active buildings provide to electricity networks?

11.1 Overview

Increasing the penetration of intermittent and uncertain distributed energy resources (DER) challenges the stability and security of power systems. As a result, network operators including distribution system operator (DSO) and transmission system operator (TSO) seek to deploy more flexibility services in order to better operate their networks as well as keeping the stability of the system between specific thresholds. In other words, system operators need to use the flexibility of different

H. Firoozi (✉) · H. Khajeh · H. Laaksonen
Flexible Energy Resources, School of Technology and Innovations, University of Vaasa, Vaasa, Finland
e-mail: hooman.firoozi@uwasa.fi

M. Shaterabadi
Department of Electrical Engineering, Kermanshah University of Technology, Kermanshah, Iran

V. Vahidinasab, B. Mohammadi-Ivatloo (eds.), *Active Building Energy Systems*, Green Energy and Technology, https://doi.org/10.1007/978-3-030-79742-3_11

flexible energy resources in order to maintain stable and reliable operation of the power system.

The flexibility of the power system is defined as the consistent adjustability of the operating point of the network to accommodate the variations and fluctuations of generation/demand in the system (Dalhues et al., 2019). Flexibility services are provided by flexible energy resources (Khajeh et al., 2019a). Flexible energy resources can be located at different levels of the power system including distribution levels (DSO level) or transmission levels (TSO level). TSOs can procure their required flexibility services from DSO level as well as TSO level flexible energy resources. In this way, the resources can provide the TSO with various types of reserves, helping it to operate its network more efficiently. However, traditionally, conventional generators have been the main flexible resources providing reserves for the TSO. It means that the TSO has not been able to exploit the flexibility potential of those resources connected to the distribution networks (Gerard et al., 2020).

Similarly, DSOs can utilize the flexibility potential of distribution network connected resources in order to operate their networks. These resources can provide DSOs with services related to congestion management like voltage control of the distribution network. However, currently, DSOs utilize traditional equipment such as transformer on-load tap changers (OLTCs) and re-dispatching in order to operate their networks. With the growing number of DG units in distribution networks as well as the high penetration of intermittent renewable power, these methods may not be applicable in the near future (Mahmud & Zahedi, 2016), and they should increasingly be coordinated with DER control principles (Laaksonen et al., 2020).

ABs, as demand-side resources located in distribution network, can have several flexible resources such as storage-based resources (e.g., electric vehicles (EVs) and batteries) and controllable appliances, enabling it to actively participate in the flexibility provision (see Fig. 11.1). These flexible resources enable the AB to respond to the grid flexibility requirements and provide the system operators with their required flexibility. In this regard, this chapter aims to analyze the potential of ABs for the provision of flexibility services.

This chapter firstly categorizes the controllable appliances of an AB into high-flexible, medium-flexible, and low-flexible appliances. Afterwards, the existing and potent flexibility services are assessed. These services will be split up into TSO level (system-wide flexibility services) including those procured for normal operations and services obtained for disturbances and special circumstances. The DSO level (local flexibility) services are also discussed in the chapter. These services need to be adopted by the DSO in order to preserve the reliability and power quality of the distribution network. Moreover, the AB's capability to provide each introduced service is analyzed. Finally, the chapter assesses the role of AB's energy management system (EMS) in the provision of flexibility services.

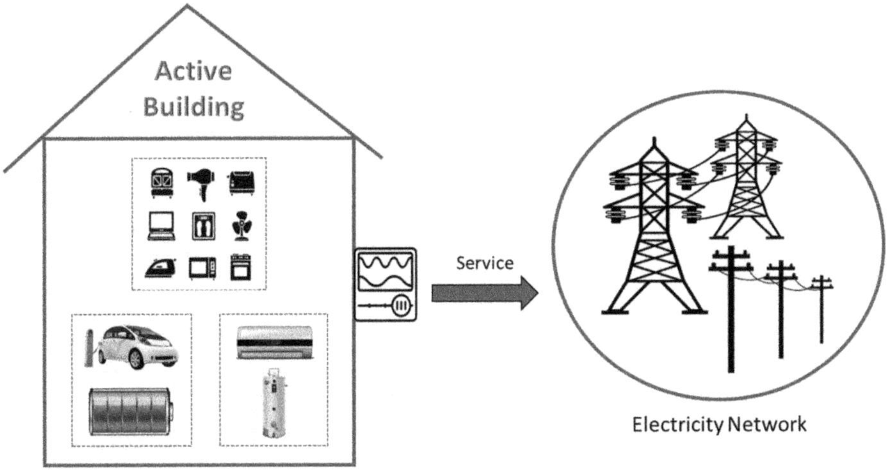

Fig. 11.1 Overview on service provision by active building (AB)

11.2 Flexible Appliances and Devices in ABs

An AB that is equipped with advanced information and communication technologies (ICT) has several flexible appliances that could be utilized for providing flexibility services to the system operators. Generally, the appliances and devices of an AB can be categorized into flexible and non-flexible appliances. Non-flexible appliances need to be only scheduled by the resident of the AB. In other words, non-flexible appliances are not capable to provide flexibility services for the DSOs and TSOs.

In comparison, the working time and/or the power of flexible appliances can be controlled based on external signals sent by the system operators. However, the control and management of these appliances should not disturb the comfort of the customer. Regarding thermostatically controllable loads (TCL), for instance, the desired temperature of the customer should be considered when the building offers flexibility services to the system operators.

Each flexibility service requires control of specific flexible appliances in ABs. For example, some services require high-flexible appliances that are able to react to the changes very fast. In comparison, some services can be activated by less-flexible resources that may be cheaper. In this regard, we categorize appliances into high-flexible, medium-flexible, and low-flexible appliances based on their degree of flexibility. Each type of flexible appliance is briefly introduced in the following subsections. In addition, Fig. 11.2 overviews the introduced flexible appliances of an AB. As can be seen in the figure, the EMS of the AB can control its flexible appliances by sending control signals to these devices.

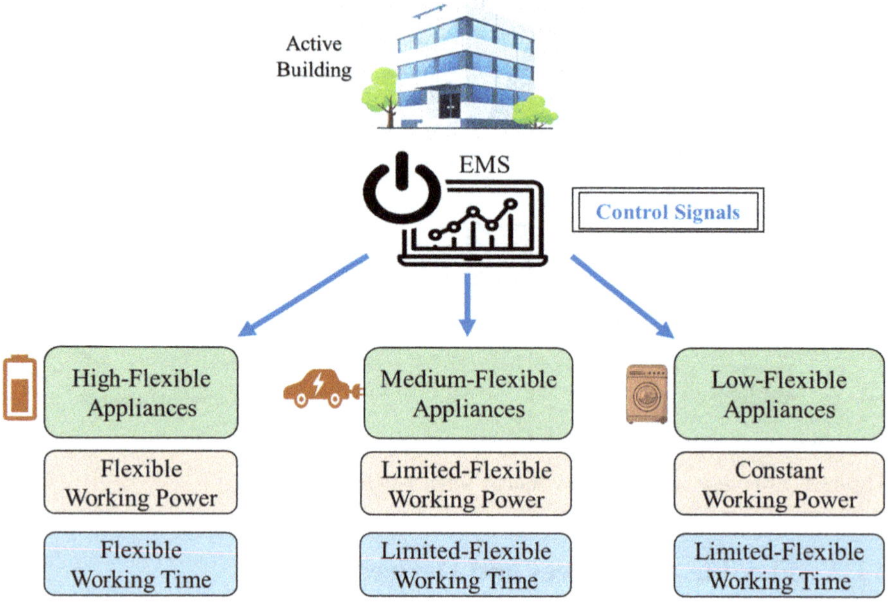

Fig. 11.2 Different categories for appliances/devices of an AB

11.2.1 High-Flexible Appliances and Devices

An AB may have some appliances and devices, which are highly flexible. These devices should have two features explained below:

- They should have high-flexible working time. In other words, the working time of devices can be adjusted according to the need of the grid. Moreover, the owner of these devices should not interrupt the predefined operation of these appliances.
- They should have high-flexible working power. It means that the operating power of these devices can be controlled in a specific range. The only constraints restricting the operating power are those associated with the inherent characteristics of the device. In addition, the injection of the working power of high-flexible devices may be bidirectional. In other words, the device may provide the grid with upward flexibility services as well as downward ones. It should be noted that the occupant must not have any impacts on the operating power of these devices.

Battery energy storage is an example of high-flexible devices. The owner of an AB may have a battery in order to sell flexibility services to the grid. In this way, the owner can enjoy the monetary benefits of this device. The working time and power of the battery can be totally controlled based, e.g., on the system operators need. The only constraints are those related to the stored energy and capacity of the battery.

However, the owner does not interfere in the operation of the battery. In addition, the battery can provide the grid with both upward and downward flexibility. In case of upward flexibility, the battery injects power into the grid (i.e., discharging mode), while for downward flexibility the battery consumes power (i.e., charging mode).

11.2.2 Medium-Flexible Appliances and Devices

Medium-flexible appliances are defined to have the following features:

- The working time of medium-flexible appliances is also flexible. However, it is affected by the owners' comfortness and preference settings.
- The working power of these appliances can be managed according to the need of the grid. However, the flexibility obtained from these appliances is restricted to the limits specified by the owner. The power injection of these devices may be bidirectional as well.

Inverter-based and thermostatically controllable loads whose power is controllable can be included in this category. The operating power and working time of these appliances are dependent on the desired range of temperature that is determined by the occupant of the AB. In order to model the power of TCL appliances, the heat transfer process of these devices needs to be modeled based on the variation of ambient temperatures. Considering an HVAC as a TCL, the temperature of the AB must vary within a dead band, denoted by T^{DB}. The room temperature falls and rises based on the operating power and on/off status of the HVAC. Given that, the occupant sets the upper and lower limit to be equal to $T^{+}$and T^{-}, and thus the dead band would be $T^{DB} = T^{+} - T^{-}$. In this way, the working time and operating power should be set considering that the room temperature should not violate this range.

EVs that can be charged with flexible power can be another example. In this regard, the owner sets the desired temporal period during which the EV can be charged as well as its preferred status (e.g., the final state of charge (SOC) of the battery of the EV). The grid operators can deploy the flexibility of the charging power of EVs by controlling the charging power within the predetermined period. However, the EV should reach its desired status when the charging period is up. In addition, the EV may also work in the vehicle-to-grid (V2G) mode. In this regard, it can also inject power into the grid when needed. In conclusion, the management of medium-flexible resources is of vital necessity since they highly affect occupant comfort.

11.2.3 Low-Flexible Appliances and Devices

Other flexible devices that are not included in the previously introduced ones are categorized as low-flexible devices. These devices are mainly controllable in terms of their working time. However, the owners may apply some constraints in order to restrict the operation of these appliances as well.

EVs with constant charging power are considered low-flexible devices. This is due to the fact that the charging power cannot be controlled over charging period. On the other hand, the EV owner determines a range of time slots during which the vehicle is allowed to be charged meaning that the EV should be managed to be charged in the period specified by the owners. Moreover, the owner may apply another constraint related to the final SOC of the EV's battery. In this situation, charging the EV is highly restricted by the owner's preferences.

A number of household appliances such as washing machines and lighting devices can be included in the low-flexible category as well. Regarding these appliances, their operating power is not controllable. However, they can be managed in terms of their working time meaning that the owner may apply some constraints restricting the appliances' operating time. Regarding lighting devices, the management system should take into account the occupancy status, the daylight, and the personal preference of the building owner (Kruisselbrink et al., 2020). Thus, the full control of the lighting devices of the building requires smart equipment and photo-sensors that can manage the restrictions regarding the users' comfortness.

11.3 Flexibility Services Provided by ABs

ABs can provide various types of services to the grid in order to help system operators to operate their networks effectively. As a result, ABs can sell flexibility services to the DSO and/or TSO. If the AB sells services to the DSO, it provides DSO level or local services. The flexibility services sold to the TSO are mainly procured to control the frequency of the whole system. Hence, TSO-level services are also called system-wide services. A brief overview on the different services that ABs can provide to the grid is illustrated in Table 11.1. In the following subsections, we introduce different types of services that can be procured from ABs.

11.3.1 TSO-Level Flexibility Services

The TSO needs to ensure the secure operation of the power system by maintaining the balance between generation and load closely, moment by moment. Any type of imbalances results in frequency deviation from the nominal value. Hence, to control the frequency of the power system in different conditions of the system,

Table 11.1 Comparison of different service provision by ABs

Service	Application	Activation speed	Bid size	Network level
FFR (new)	Very low-inertia situations	1 s	Not defined yet	TSO
FCR-D	Huge frequency deviations	< 1 min	> 1 MW	TSO
FCR-N	Always in use	1–5 min	> 0.1 MW	TSO
aFRR	Certain hours	5 min	> 5 MW	TSO
mFRR	Incidents/imbalances	15 min	> 5 MW	TSO
Voltage Control	Weak grid/components' failure/etc.	Not defined yet	Not defined yet	DSO
Congestion Management	Overload/excessive generation	Not defined yet	Not defined yet	DSO

TSO-level or system-wide flexibility services are deployed. In this regard, the TSO aims to procure a suitable type of reserves that suits the existing condition of the system. Accordingly, various flexibility services and reserves have been designed for normal, disturbance, and low-inertia situations. If ABs want to contribute to the provision of TSO-level flexibility, they should be aggregated via an aggregator who plays the role of a broker between ABs and the TSO. Without the aggregator, ABs are not able to participate in providing system-wide services since these services require a minimum flexible capacity that is more than the capacity of the building. In the following, TSO-level flexibility services associated with Nordic countries (especially Finland) are introduced, and we assess the capability of ABs to provide these services. Figure 11.3 reviews different services that can be provided for the grid.

11.3.1.1 Services for Normal Operation

Frequency containment reserve in normal condition (FCR-N) and manual frequency restoration reserve (mFRR) are two reserves used in normal operation of the system. In the following, the characteristics of these services are discussed.

FCR-N

Regarding normal conditions, the TSO requires to constantly control the frequency of the system by activating fast reserves such as spinning reserves (in the USA (Xu et al., 2016)) or frequency containment reserves (FCR) (in Europe (Thien et al., 2017, Gomez-Gonzalez et al., 2020)). FCRs for normal conditions (i.e., FCR-N) are utilized as primary frequency control. Activating FCR-N is the first action taken by the TSO in a decentralized fashion (Oureilidis et al., 2020). It is worth mentioning

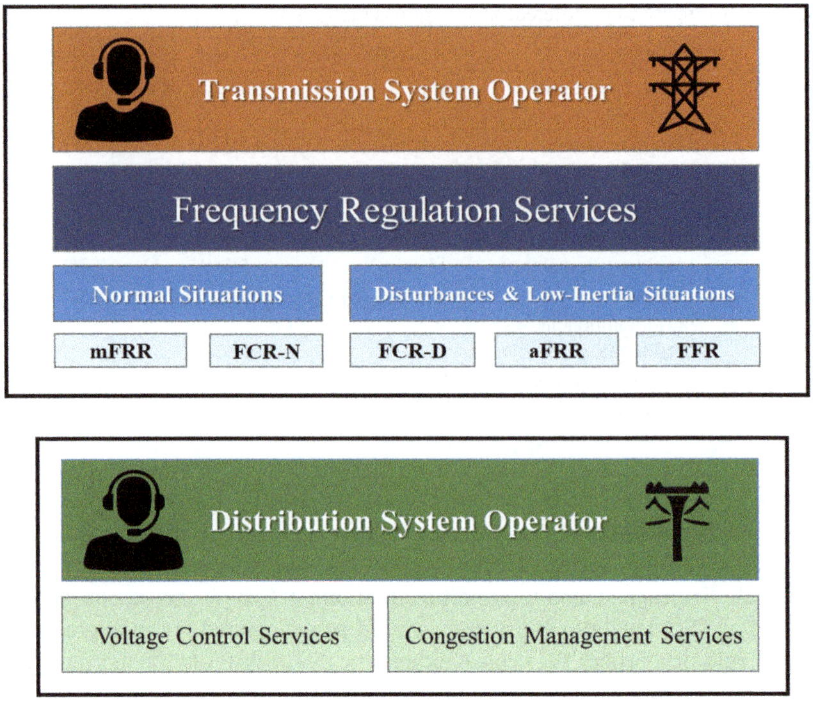

Fig. 11.3 Overview of the existing gird services in Finland (as an example of Nordic countries)

that the purpose of FCR-N is not to mitigate the consequences of a disturbance since it is used only in normal situations. FCR-N is an automatic control based on the local frequency measurements. Moreover, the reserve resource (e.g., an AB) deployed for the provision of FCR-N should pass the technical tests and be prepared for providing this service.

Regarding Finland, as an example of Nordic countries, the reserve resources providing FCR-N should be a symmetrical product (Divshali & Evens, 2020a, 2020b). It means that the resources should be able to inject power as well as consuming it. In addition, the full amount of the upward reserve capacity needs to be activated in situations that the frequency is 49.9 Hz or less. Similarly, the full amount of the downward reserve capacity should be activated when the frequency is 50.1 Hz or more. When the frequency range is between 49.9 and 50.1 Hz, the amount of the activated capacity is proportional to the magnitude of the frequency deviation (Divshali & Evens, 2020a), as illustrated in Fig. 11.4.

A flexible resource may be capable of controlling its power continuously such as batteries or relay-based ones that can be managed in a piecewise manner. EV, which can be charged with several different power, can be an example of piecewise-based resources. Considering FCR-N, the flexibility of the resource should be activated in 3 min. In addition, if the capacities of upward and downward flexibility are of

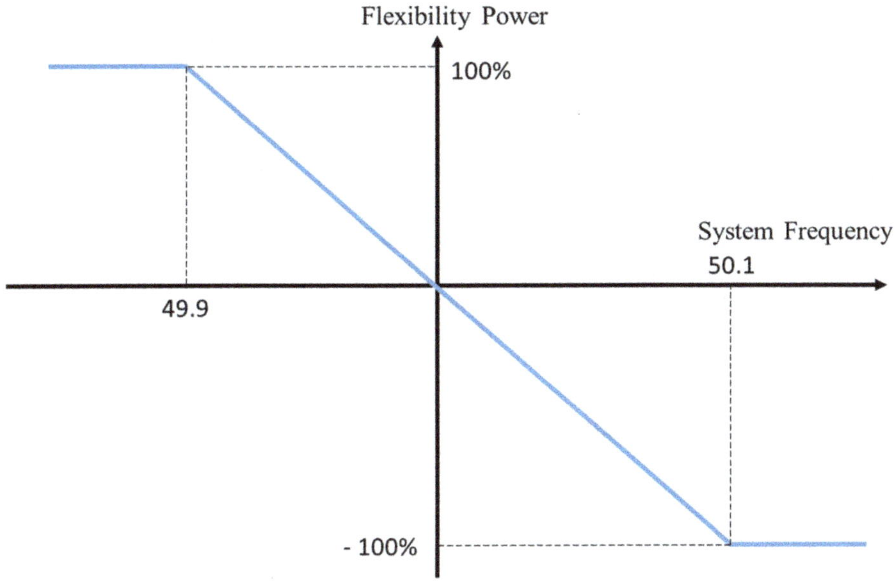

Fig. 11.4 The activation of FCR-N for each flexibility resource

a different magnitude, the smaller volume is taken for the provision of FCR-N (Fingrid, 2019).

Based on the technical considerations of the FCR-N, ABs with both upward and downward flexible resources can provide this service to the TSO. In this way, the minimum flexible capacity (minimum of upward and downward capacities) can be offered to the TSO. However, the corresponding device should be fast and flexible. Hence, the hinder actions taken by the owner are avoided so that the reserve capacity can be freely activated based on the instant need of the system. Nevertheless, the AB should take into account the constraints associated with the operation of the device. For example, if a battery is utilized for the provision of FCR-N, the capacity of the battery, its current SOC, and the minimum and maximum SOC should be considered when the AB is calculating its available flexible capacities.

mFRR

The mFRR is used to control the flow of the grid and helps to restore the faster reserves such as FCRs. In addition to this, manual FRR can be also utilized in the case of expected and deterministic frequency deviations (Fingrid, 2020). The mFRR product should be adopted in normal operation (Fingrid, 2020). Regarding Nordic reserve services, mFRR is localized to the extent that the synchronous system can be balanced at any moment (Spodniak et al., 2019). Manual FRR is procured by the TSO according to its assessment of the local flexibility requirements. In this regard, the TSO should consider bottlenecks of its network and the dimensioning faults and procure mFRR accordingly. The reserve products associated with mFRR can be manually activated in 15 min (Fingrid, 2020). Thus, it can be suitable for slower flexible resources since they have more time to respond to the TSO's requests.

ABs can contribute to the provision of mFRR. In this way, they need to be aggregated by an aggregator in order to reach the permissible capacity. In addition, they should inform the aggregator about their available flexible capacities for providing mFRR and adhere to their schedules in real time. The activated flexibility power is determined by the TSO in real time. Firoozi et al. (2020), as an example, introduces the participation of a local energy community which consists of several residential households in providing mFRR services.

11.3.1.2 Services for Disturbances and Low-Inertia Situations

There might be several unexpected occurrences that cause frequency deviations and instability of the power system. In these situations, the TSO needs to be prepared by deploying different types of reserves. In Finland and Nordic countries, frequency containment reserve in disturbance (i.e., FCR-D) is being deployed for disturbances, fast frequency reserve (FFR) is procured for low-inertia situations, and automatic frequency restoration reserve (aFRR) is one of the main measures taken to avoid the weakening of the frequency quality of FCR-D (Fingrid, 2020). In the following, these services are introduced according to the Finnish reserve products and markets. An overview of the reserves for non-normal operation of distribution networks can be found in Fig. 11.5.

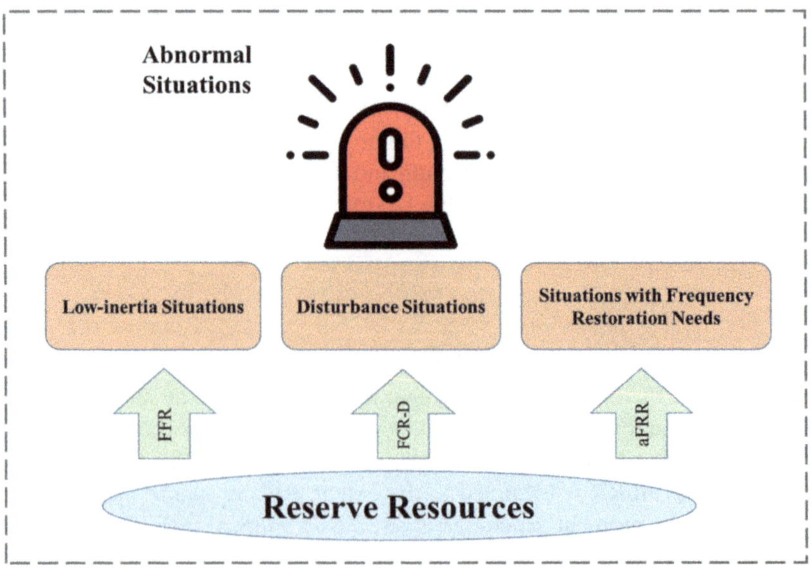

Fig. 11.5 Overview of reserves for abnormal situations

FCR-D

Currently, the needed FCR-D capacity should be equal to the largest possible imbalance resulted from the loss of individual components such as generation units, lines, transformers, and bus bars. Thus, this type of reserve is required for low-frequency situations in the range between 49.5 and 49.9 Hz. As a result, upward flexible capacities are the only products required for providing FCR-D. If the frequency drops within the range of 49.5–49.9 Hz, the amount of the activated upward capacity should be proportional to the magnitude of the frequency deviation, similar to the FCR-N activation. Both linear-activated and relay-based devices can contribute to the provision of FCR-D. However, currently, the Finnish TSO restricts the total volume of relay-connected reserves procured for FCR-D (Fingrid, 2019). In addition, the relay-connected reserve resources are not allowed to be aggregated with those activating linearly (Fingrid, 2019). Regarding FCR-D, at least half of the reserve capacity needs to be activated in 5 s, and the full volume should be activated in 30 s in situations that the frequency drops by 0.5 Hz (Fingrid, 2019).

Considering the mentioned technical considerations, ABs can participate in providing FCR-D using their upward flexible capacities. However, again the devices need to be flexible and so fast in order to be activated in less than 30 s without the intervention of the owner. Moreover, the device should be automatically controlled, and the AB should be equipped with decent ICT technology so as to respond as fast as possible. It should be noted that for some medium flexible resources, the request of the system should be within the predetermined range of the utilization of the devices. For example, EVs as a flexible resource of ABs can also provide upward flexibility for FCR-D if the need of the system is compliant with the charging need determined by the owner in terms of time. As an example, (Divshali & Evens, 2020b) proposed charging stations that can provide FCR-D and FCR-N services for the grid.

aFRR

Unlike mFRR, which is a locally adopted reserve, aFRR is a centrally controlled reserve in Nordic countries. In other words, the activation of aFRR is based on the frequency deviation within the synchronized Nordic area (Modig et al., 2019). This product is considered as an automatic complement to aFRR services in the process of restoring frequency (Modig et al., 2019). In fact, aFRR products need to restore the frequency to the nominal value while releasing FCRs which have been activated beforehand (Modig et al., 2019). All Nordic TSOs specify the hours at which aFRR services should be dimensioned. In these hours, the frequency variations should be very challenging.

A reserve resource that desires to provide aFRR should activate its entire flexible capacity within 5 min after receiving the activation signal. In addition, it should start to respond no later than 30 s from the moment it receives the signal. Accordingly, the flexible devices of ABs are also able to provide aFRR if they are smart and equipped with decent communication technology. In this way, when the TSO requests for the activation of the flexibility, the flexible device should immediately react to the request regardless of the owner's preference. It should be noted that most of the TSO-level services have capacity markets in which the reserve resource can submit its available flexible capacities. According to its offers, the reserve resource should adhere to its schedule that had been submitted to the capacity market and provide the TSO with its required flexibility.

FFR

FRR should handle disturbances occurring in low-inertia situations (Modig et al., 2019). These flexibility services are defined exclusively for under frequency circumstances since these low-inertia situations are considered much more critical than those with over frequency (Modig et al., 2019). Therefore, upward flexible capacities are required for the provision of FFR. There exist two different FFRs defined in terms of activation time. One is the long support duration FFR that the reserve resource should support the system for a duration of at least 30 s. In another type, which is named short support duration FFR, the resource supports the system for a duration of at least 5 s. In addition, there exist three different combinations of activation levels and their corresponding time for the full activation. Thus, the providers of FFR are able to freely select the most appropriate combination based on the characteristics and features of their flexibility resources. These combinations are as follows (Modig et al., 2019):

- 0.7 s for the activation of the full capacity at the activation level of 49.5 Hz
- 1.0 s for the activation of the full capacity at the activation level of 49.6 Hz
- 1.3 s for the activation of the full capacity at the activation level of 49.7 Hz

An AB should also submit its available upward capacity to the corresponding upstream aggregator. The type of the FFR service (in terms of the duration and the combination of the reserve) should be selected based on the flexible devices used for this purpose. However, as stated, the resource need to be deactivated based on the type of service. Hence, the AB cannot deploy those low-flexible appliances that are not interruptible, e.g., washing machines.

11.3.2 DSO-Level Flexibility Services

DSO-level flexibility services are mainly referred to those services helping DSOs operate their networks more actively. The main responsibilities of DSOs include congestion management as well as voltage control of the distribution networks (Khajeh et al., 2019a). ABs are able to provide DSOs with these two services. In the following subsections, the possibility of the provision of DSO-level services is discussed.

11.3.2.1 Services Related to Congestion Management

Modern power systems are heading toward smart grids along with the high penetration of intermittent power produced by renewable energy resources and distributed generation (DG) units (Khajeh et al., 2019b). Additionally, the number of flexible loads such as storage-based resources is increasing in distribution systems, which can be automatically controlled with recently emerging advanced ICT technologies. The mentioned transition in electrical networks creates serious challenges for the operation of distribution systems (Ghazvini et al., 2019). On the one hand, the DG units may alleviate congestion in the existing transmission grids as they decrease the required power transmitted to the end-users (Bayod-Rújula, 2009). On the other hand, the excessive power produced by DGs along with the bidirectional flow of power in distribution networks can cause congestion in these networks (Linna et al., 2017). For instance, high demand resulted from the simultaneous charging of EV causes overloading of the network lines (Hu et al., 2013). In contrast, the high power production of DGs and other small-scale resources located in distribution networks could reach the maximum capacity of lines and components. This could consequently result in a phenomenon called congestion within the distribution system. Hence, the uncoordinated operation of distributed network-located resources can increase unexpected congestions in the distribution network. The appropriate actions and plans for avoiding the congestion in distribution grids are known as congestion management.

In order to mitigate the congestion-related issues, the DSO can reinforce the network based on its identified and forecasted needs in the future (Liu et al., 2016). This solution is a long-term solution and requires costly actions and equipment. Consequently, the DSO may enhance the grid's hosting capacity for the high number of DGs and intermittent renewable resources by increasing investments in the infrastructure of the grid (Dalhues et al., 2019).

In short term, the DSO performs some actions to operate the network efficiently. One conventional approach is to change the setting and reconfigure the set points of elements of the network. The other method for the DSO is re-dispatching the DGs and generation resources of the network and requests for curtailment, if needed. The mentioned approaches are not economically efficient for the DSO. In addition, some actions discourage the growth of DGs in distribution networks. However, in order to realize the green power system, investments in renewable-based DGs should increase.

One of the effective methods of managing congestion in the distribution network is using dynamic prices and tariffs in the network (Gu et al., 2017). In this way, the DSO needs to predict congestion based on the forecasted load and generation of the network and determine dynamic prices accordingly. Thus, for example, by specifying greater electricity prices for the nodes associated with the congested feeders, the users located at these particular nodes are encouraged to decrease their demand. In addition, the DSO may design several local markets with the target to minimize its operational costs while causing the least discomfort to the participants. Another approach is to purchase flexibility from the end-users

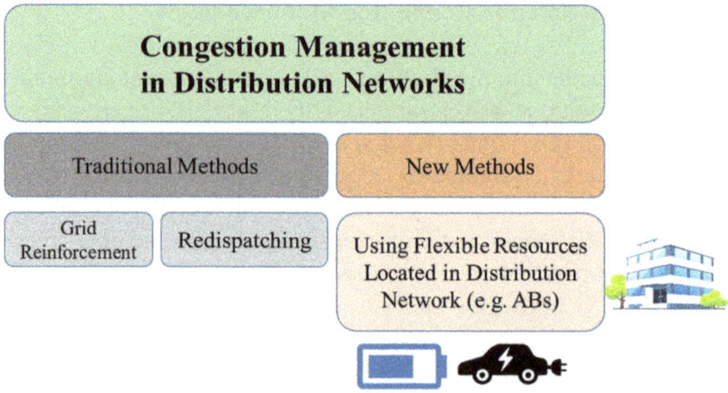

Fig. 11.6 An overview of congestion management methods in distribution networks

directly. In fact, the DSO may negotiate a contract with the end-user to control its flexible energy resources in some situations. It also can design flexibility markets asking participants to offer their flexible resources. The methods of congestion management in distribution networks are reviewed in Fig. 11.6.

Therefore, ABs can be regarded as potential congestion management-related service providers since they are able to react to the local market prices actively. It should be mentioned that the AB requires being equipped with an EMS so to manage the building's flexible resources based on the local market prices or tariffs that were previously determined. Thus, according to the price signals and the comfort of the occupant, the EMS scheduled flexible energy resources of the building aiming to minimize the total costs of the AB. Every kind of flexible resources of the AB can participate in the provision of congestion-management services. This is due to the fact that compared to the fact flexibility services, such as FCRs, these services are not highly fast and is more predictable.

11.3.2.2 Services Related to Voltage Control

The growing number of DGs in distribution networks creates some regulation issues for the DSO that can affect voltage profiles as well as the power flow of this network. Moreover, the existing voltage regulation devices are mostly designed for the traditional distribution system without the high penetration of DGs (Mahmud & Zahedi, 2016). Therefore, the voltage magnitude, which was reduced along the distribution feeder, increases at the nodes with DGs (Ibrahim, 2018). Accordingly, DSO needs to deploy more effective actions to regulate the voltage in the distribution network.

Traditionally, DSOs deploy OLTC transformers, switched capacitors (SC), and step voltage regulator (SVR) to regulate voltages of their networks (Long & Ochoa, 2016). These devices should automatically adjust the voltage using the estimation of voltage drops along with the network feeders. However, the power produced by renewable-based DGs is uncertain and volatile, adding uncertainties to the estimation of voltages in the network. In addition, the traditional voltage-regulation devices fail to rapidly track the variation of voltage resulting from the intermittent power of renewable DGs. If they change their status rapidly, for example, for tap changing devices, it decreases their lifetime and increases the maintenance costs (Mahmud & Zahedi, 2016).

Regarding LV networks, active power can play a more significant role in regulating the voltages of the network (Laaksonen et al., 2005). Hence, the DSO can actively control and regulate the voltage of the distribution network utilizing the active power of flexible energy resources (Khajeh et al., 2020). ABs with flexible appliances can be deployed as one of the resources to manage and control voltages in the distribution network. In this regard, the DSO should predict the flexibility need of the system associated with the voltage regulation. Afterward, the DSO should send signals to ABs at different nodes to react according to the signals. The ABs who respond to these requests should be compensated. Accordingly, the DSO needs to determine decent incentives for the ABs so as to adopt the maximum flexibility potential of these resources.

The control of voltage and flexibility management of ABs can be in a centralized or decentralized manner (Mahmud & Zahedi, 2016). In a centralized method, the DSO receives the information on the online status of the network and determines the amount of flexibility required for each node. However, a comprehensive supervisory control and data acquisition (SCADA) system is required in order to manage the distribution network and deal with a huge amount of data receiving from sensors and meters. These data should be analyzed as well to find the set point for each flexible device with regard to the constraints imposed by each flexible appliance and devices.

In a decentralized approach, the DSO may shift the responsibility of the voltage control to local controllers. However, the controllers should be actively in touch with each other and act coordinately. For example, the management system of ABs who act as local controllers should communicate with each other in order to recognize the states of the network. In this way, they exchange information regarding their individual states and the control actions and try to reach a global solution to regulate the voltage of their local network. Multiagent systems (MAS) are considered a potent technology that can implement the decentralized voltage regulation in distribution networks (Yorino et al., 2014). There also might be a central controller for each region with the responsibility of coordinating controllers and assisting the DSO with operating the network. In this way, ABs can exchange information with the central controller. Hence, the central controller aims to find the best operating point for each flexible device of the ABs. It is noticeable that the ABs should be located in the central controller's territory.

11.4 The Role of Energy Management System in ABs

As previously stated, ABs are potent flexible resources that can provide system operators (both DSOs and TSOs) with flexibility services. Not only ABs can assist network operators, but also the smart technologies used in ABs can enhance the comfort level of the occupants by the automatic control of appliances and devices. Moreover, the ABs can bring considerable financial profits obtained from saving electricity costs as well as selling their flexible capacities. In addition to monetary benefits, ABs can pave the way for the transition of the power system to a sustainable, smart, and environmentally friendly electrical system.

The recently advanced ICT technologies enable ABs to follow a certain target and help the system operators to obtain their required flexibility from the ABs. Generally, the main target of the ABs is to minimize their electricity costs or maximize their profits by selling their productions. These productions can be in terms of energy or flexibility. If the ABs are integrated with production resources such as PV panels and small-scale storage-based resources, they are able to sell energy as well as flexibility. Otherwise, the ABs can schedule their flexible energy, aiming to sell flexibility. In this regard, the EMS of the AB is in charge of scheduling its production and flexible devices. The owner of the AB may impose some constraints regarding their medium- and low-flexible energy resources. In this situation, the EMS should consider these constraints as well as the operational limits related to each device. In other words, the EMS of the AB should control and manage the appliances and flexible energy resources of the AB according to the types of appliances, the operating constraints of devices, the user's need and preferences, the objective function defined for the system, and the external signals associated with the particular service.

In general, the EMS of ABs comprises several tasks including monitoring, scheduling, and controlling the flexible devices. Figure 11.7 introduces some of these tasks. The EMS may also be responsible for making bidding strategies for the AB or finding the most profitable and suitable services that the AB can provide for the system operators. This system may also able to change the consumption pattern of the occupant while considering his/her comfort and preference (Silva et al., 2020). Each AB may have its own EMS working autonomously. In another case, neighboring ABs can form a local energy community whose their EMSs are working cooperatively. In addition to the mentioned architecture, the flexible resources of ABs may be controlled centrally by the manager of the community (Sousa et al., 2019). In this case, there exists one EMS for the whole community's members (i.e., ABs).

Generally, the EMS of the AB should consider several factors when scheduling the flexible resources of the building. In the following, we list these factors:

1. *The objective of the scheduling*: The EMS requires taking into account the main objective of the household predefined by the occupant. The objective could be minimizing the total cost or maximizing the monetary profits of the building from providing flexibility services.

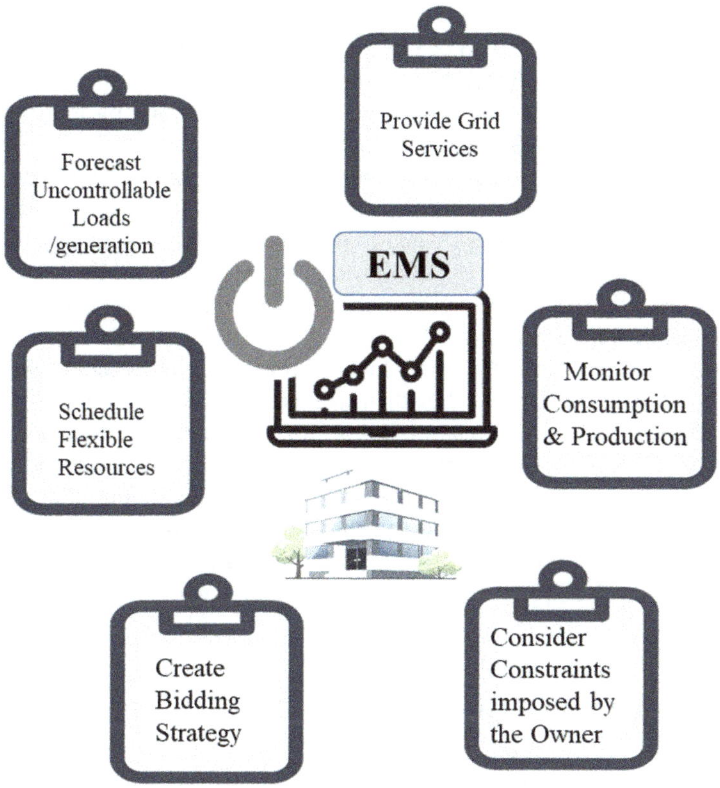

Fig. 11.7 Tasks related to EMS of an AB

2. *The types of appliances and the related constraints*: The EMS should consider the type of appliances in every process of management. The degree of flexibility (low, medium, high) is of importance. In addition, the preference of the occupant in utilizing the appliance, the working time, and the operational constraints are some factors that can vary based on the type of the devices. Moreover, constraints associated with the comfort level of occupants need to be regarded as well. It is worth mentioning that non-flexible appliances and devices of the AB should not be controlled by the EMS. However, the consumption power of these devices should be forecasted, measured, and monitored by the system in order to schedule flexible appliances and make bidding strategy in a more effective way.
3. *The type of flexibility service(s)*: The management system needs to select the services that are suitable for the AB. This decision should be made based on the available flexibility and the type of flexible resources of the building. Furthermore, the selected services should bring profits for the building. Thus, the EMS may choose the most profitable services. It may select one or several services. However, regarding TSO-level services, there should be an aggregator to aggregate several small-scale flexible resources. Thus, the AB can submit

its available flexible capacity to the aggregator. On the other hand, in terms of DSO-level flexibility, the AB can be directly in touch with the DSO or through a DSO-level aggregator, depending on the structure defined in the network.

4. *The promised services*: The scheduling process performed by the EMS should be in line with the service that the AB had promised to provide. For instance, if the AB promised to reduce its consumption at peak hours, the working time of flexible appliances need to be shifted to other time slots. However, the occupant may apply some constraints related to the working hours of the appliances that are not in line with the promised services. In this situation, the system should warn the occupant against the huge penalty costs that he/she would be incurred if it does not adhere to the schedule.
5. *Bidding strategy:* There may exist a local market for energy and/or flexibility in which ABs participate and offer their availability (Khajeh et al., 2020). In this way, the EMS of the AB creates bidding strategies based on the objective of the building. The process of making a bidding strategy can be simultaneously done with the scheduling process. However, it is worth mentioning that the process of making an optimized bidding strategy can be extremely complicated since it requires analyzing information on the behaviors of the competitors and the need of buyers as well as predicting prices of the local market.

11.5 Summary and Conclusion

ABs as demand-side agents located in distribution networks have several flexible resources such as storage-based resources (e.g., EVs and batteries) and controllable appliances, enabling active participation in flexibility provision. These flexible resources assist the AB to respond to the grid's flexibility requirements and provide the system operators with their required flexibility.

The flexible resources owned by ABs can be categorized based on their degree of flexibility. This categorization can help the AB's EMS to manage and operate these devices more effectively. In addition, it can select potent grid services according to the category of the available appliances and devices. In this regard, the devices can be divided into high-flexible, medium-flexible, and low-flexible ones. The high-flexible devices are controllable in terms of their working power and operating time. The only constraints restricting their operation are those related to the technical characteristics of the device. However, the working power and operating time of medium-flexible devices can be limited by the constraints imposed by the building's occupants as well. On the other hand, not only the working power of low-flexible appliances is inflexible, but also they are also highly restricted by their owners and the inherent operational limits.

Grid services that can be provided by the ABs can also be categorized according to their buyers. DSOs and TSOs are the main buyers of these services. TSO-level services that aim to help TSOs can be deployed for both normal operation, disturbances, and low-inertia situations. In contrast, DSO-level services are procured by DSOs to control voltage and manage congestion in the distribution networks. ABs can provide both TSO-level and DSO-level services based on their available resources and devices.

In order to provide grid services, firstly, ABs need to be equipped with advanced ICT technology in order to react to the external signals sent by the operators. Moreover, the EMS of the AB requires considering the main objective of the household predefined by the customer. Besides, the scheduling process performed by the EMS should be in line with the service that the AB had promised to provide. Otherwise, if the AB fails to adhere to the schedule, it incurs huge penalty costs. In some situations, the DSO may form local markets in order to procure flexibility from customers in which the EMS of ABs is also responsible for creating bidding strategies based on the status of the local market. Finally, in order to expand this chapter, the related future directions could be as follows:

- One can consider different types of grid services for ABs and analyses the possibility of participation in a marketplace in which it is financially more profitable for the AB.
- Another possibility would be considering different types of ABs beside residential units such as commercial ABs, office ABs, etc. in service provision to upstream network and its challenges.
- Moreover, the contribution of ABs to grid service provision could be problematic in near future due to high penetration of distributed small-scale energy resources. Therefore, the required grid codes as well as regulations for this purpose could be further investigated.

References

Bayod-Rújula, A. A. (2009). Future development of the electricity systems with distributed generation. *Energy, 34*(3), 377–383.

Dalhues, S., et al. (2019). Research and practice of flexibility in distribution systems: A review. *CSEE Journal of Power Energy Systems, 5*(3), 285–294.

Divshali, P. H., & Evens, C. (2020a). Optimum operation of battery storage system in frequency containment reserves markets. *IEEE Transactions on Smart Grid*.

Divshali, P. H., & Evens, C. (2020b). Stochastic bidding strategy for electrical vehicle charging stations to participate in frequency containment reserves markets. *IET Generation Transmission and Distribution, 14*(13), 2566–2572.

Fingrid. (2019). *The technical requirements and the prequalification process of Frequency Containment Reserves (FCR)*. January, p. 17.

Fingrid. (2020). *TSO report on balancing in accordance with Article 60 of Commission Regulation (EU) 2017/2195 of 23 November 2017 establishing a guideline on electricity balancing*. November 2017.

Firoozi, H., Khajeh, H., & Laaksonen, H. (2020). Optimized operation of local energy community providing frequency restoration reserve. *IEEE Access*. https://doi.org/10.1109/access.2020.3027710

Gerard, H., Rivero, E., & Vanschoenwinkel, J. (2020). TSO-DSO interaction and acquisition of ancillary services from distribution. In *TSO-DSO Interactions and ancillary services in electricity transmission and distribution networks*. Springer, pp. 7–23.

Ghazvini, M. A. F., et al. (2019). Congestion management in active distribution networks through demand response implementation. *Sustainable Energy, Grids Networks, 17*, 100185.

Gomez-Gonzalez, M., Hernandez, J. C., Vera, D., & Jurado, F. (2020). Optimal sizing and power schedule in PV household-prosumers for improving PV self-consumption and providing frequency containment reserve. *Energy, 191*, 116554.

Gu, C., Yan, X., Yan, Z., & Li, F. (2017). Dynamic pricing for responsive demand to increase distribution network efficiency. *Applied Energy, 205*, 236–243.

Hu, J., You, S., Lind, M., & Østergaard, J. (2013). Coordinated charging of electric vehicles for congestion prevention in the distribution grid. *IEEE Transactions on Smart Grid, 5*(2), 703–711.

Ibrahim, S. K. (2018). *Distribution system optimization with integrated distributed generation.*

Khajeh, H., Laaksonen, H., Gazafroud, A. S., & Shafie-Khah, M. (2019a). Towards flexibility trading at TSO-DSO-customer levels: A review. *Energies, 13*(1), 1–19. https://doi.org/10.3390/en13010165

Khajeh, H., Foroud, A. A., & Firoozi, H. (2019b). Robust bidding strategies and scheduling of a price-maker microgrid aggregator participating in a pool-based electricity market. *IET Generation Transmission Distribution.* https://doi.org/10.1049/iet-gtd.2018.5061

Khajeh, H., Firoozi, H., Laaksonen, H., & Shafie-khah, M. (2020).*A new local market structure for meeting customer-level flexibility needs.* https://doi.org/10.1109/sest48500.2020.9203499

Kruisselbrink, T. W., Dangol, R., & van Loenen, E. J. (2020). A comparative study between two algorithms for luminance-based lighting control. *Energy and Buildings, 228*, 110429.

Laaksonen, H., Saari, P., & Komulainen, R. (2005). Voltage and frequency control of inverter based weak LV network microgrid. In *2005 international conference on future power systems*, p. 6.

Laaksonen, H., Parthasarathy, C., Hafezi, H., Shafie-khah, M., Khajeh, H., & Hatziargyriou, N. (2020). Solutions to increase PV hosting capacity and provision of services from flexible energy resources. *Applied Sciences, 10*(15), 5146.

Linna, N. I., Fushuan, W. E. N., Weijia, L. I. U., Jinling, M., Guoying, L. I. N., & Sanlei, D. (2017). Congestion management with demand response considering uncertainties of distributed generation outputs and market prices. *Journal of Modern Power Systems and Clean Energy, 5*(1), 66–78.

Liu, Z., Wu, Q., Oren, S. S., Huang, S., Li, R., & Cheng, L. (2016). Distribution locational marginal pricing for optimal electric vehicle charging through chance constrained mixed-integer programming. *IEEE Transactions on Smart Grid, 9*(2), 644–654.

Long, C., & Ochoa, L. F. (2016). Voltage control of PV-rich LV networks: OLTC-fitted transformer and capacitor banks. *IEEE Transactions on Power Apparatus and Systems, 31*(5), 4016–4025. https://doi.org/10.1109/TPWRS.2015.2494627

Mahmud, N., & Zahedi, A. (2016). Review of control strategies for voltage regulation of the smart distribution network with high penetration of renewable distributed generation. *Renewable and Sustainable Energy Reviews, 64*, 582–595.

Modig, N. et al. (2019). *Technical requirements for fast frequency reserve provision in the Nordic synchronous area*, May, pp. 1–20.

Oureilidis, K., et al. (2020). Ancillary services market design in distribution networks: Review and identification of barriers. *Energies, 13*(4), 917.

Silva, B. N., Khan, M., & Han, K. (2020). Futuristic sustainable energy management in smart environments: A review of peak load shaving and demand response strategies, challenges, and opportunities. *Sustainability, 12*(14), 5561.

Sousa, T., Soares, T., Pinson, P., Moret, F., Baroche, T., & Sorin, E. (2019). Peer-to-peer and community-based markets: A comprehensive review. *Renewable and Sustainable Energy Reviews, 104*, 367–378.

Spodniak, P., Ollikka, K., & Honkapuro, S. (2019). *The relevance of wholesale electricity market places: The Nordic case.*

Thien, T., Schweer, D., vom Stein, D., Moser, A., & Sauer, D. U. (2017). Real-world operating strategy and sensitivity analysis of frequency containment reserve provision with battery energy storage systems in the German market. *Journal of Energy Storage, 13*, 143–163.

Xu, B., Dvorkin, Y., Kirschen, D. S., Silva-Monroy, C. A., & Watson, J.-P. (2016). A comparison of policies on the participation of storage in US frequency regulation markets. In *2016 IEEE Power and Energy Society General Meeting (PESGM)*, pp. 1–5.
Yorino, N., Zoka, Y., Watanabe, M., & Kurushima, T. (2014). An optimal autonomous decentralized control method for voltage control devices by using a multi-agent system. *IEEE Transactions on Power Apparatus and Systems, 30*(5), 2225–2233.

Chapter 12
Participation of Active Buildings in Peer-to-Peer and Local Transactive Energy Markets

Vahid Vahidinasab, Mohammad Nasimifar, and Behnam Mohammadi-Ivatloo

12.1 Introduction

A *collaborative economy* which is sometimes called *sharing economy* is an economic model that is highly affecting the way of trading products and services between corporations, start-ups, and people. This results in a more efficient marketplace which brings new products, services, and business growth (Owyang, 2013). According to this concept, all members of a community might be able to have access to certain goods or services and can share them through a proper infrastructure (Sousa et al., 2019). *Airbnb* and *Uber* are two examples of many platforms that have been formed upon this paradigm. The collaborative economy is the opposite of traditional concepts in which a limited number of players are in charge of providing services and goods.

The energy systems are also affected by the collaborative economy paradigm. Especially with the evolving development of the distributed energy resources (DERs), especially Renewable Energy Sources (RESs), and their deployment on the demand-side of the energy sector, the issue of potential energy sharing and

---

V. Vahidinasab (✉)
Department of Engineering, School of Science and Technology, Nottingham Trent University, Nottingham, UK
e-mail: vahid.vahidinasab@ntu.ac.uk

M. Nasimifar
Faculty of Electrical Engineering, Shahid Beheshti University, Tehran, Iran
e-mail: m.nasimifar@mail.sbu.ac.ir

B. Mohammadi-Ivatloo
Department of Electrical and Electronics Engineering, Muğla Sıtkı Koçman University, Muğla, Turkey
e-mail: mohammadi@ieee.org

V. Vahidinasab, B. Mohammadi-Ivatloo (eds.), *Active Building Energy Systems*, Green Energy and Technology, https://doi.org/10.1007/978-3-030-79742-3_12

small-scale transactions between the peers are highlighted (Nezamabadi & Vahidinasab, 2019a). While the prosumers (i.e. agents who are able to produce, consume, or store the energy) were able to trade their energy and ancillary services with the main grid, the paradigm shift achieved by the collaborative economy concept paved the way for energy trading according to their preferences not only with the main grid but also with the peers in the neighbourhood area. Also, consumers should be able to choose the provider and the type of energy to buy.

All these points along with advances in Information and Communication Technologies (ICT) have led to the creation of new concepts of Transactive Energy (TE) (GridWise Architecture Council (GWAC), 2021) and Peer-to-Peer (P2P) energy market (Rahimi et al., 2016a). In such markets, prosumers and consumers can make use of a distributed, secure, and transparent platform to trade energy and service with all of the players in the network. A variety of players can be considered, namely, small-scale users, community/microgrid managers, system operators, and large-scale producers. Hence, players at all levels of the network have the possibility of participating in the provision of energy and ancillary services, while they seek to reach their own goals. These goals can be individuals, such as reducing the total cost of energy, or social goals such as using green energy, or even philanthropic purposes, such as helping people who have energy poverty.

In such an environment, active buildings are able to perform a pivotal role in future P2P energy transactions. By employing the different generation technologies and smart appliances, a community of active buildings has the potential to not only procure internal demand at the minimum cost but also provide flexibility and other services for grid operators (Nezamabadi & Vahidinasab, 2019b; Sharifi, 2017). Hence, in this chapter, the role of active buildings and their potential to provide internal and external services through P2P energy transactions are investigated. Figure 12.1 shows the graphical overview of the presented topics in this chapter.

12.2 Transactive Energy and P2P Markets

In this section, an overview of the definition, concept, and challenges of the TE and P2P markets will be introduced, and their potential applications at active building level are discussed.

12.2.1 Transactive Energy: Overview, Concept, and Challenges

Among the various definitions proposed for TE, the following definition which is provided by the GridWise Architecture Council (GWAC) (2021) (Abrishambaf et al., 2019) is the most popular definition used in the literature: *"a set of economic and control mechanisms that allows the dynamic balance of supply and demand across the entire electrical infrastructure using value as a key operational parameter"*. According to this definition, TE exploits the whole power system

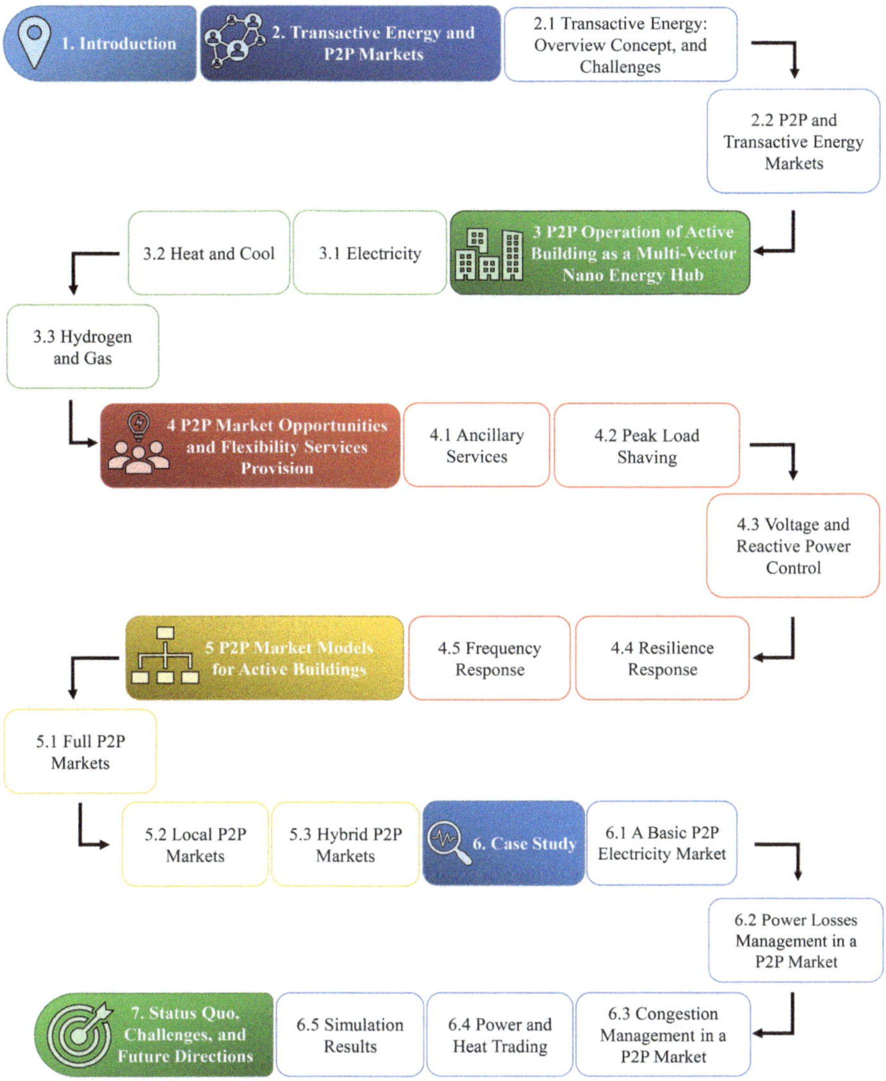

Fig. 12.1 A graphical overview of the presented topics

potentials, including demand-side response (DSRs) and DERs to achieve a more sustainable, affordable, and secure network. While TE and DSR share the same objective of seeking to balance demand and supply across the network, they are acting differently in approach. As its name reveals, DSR focuses on the demand part of the network. Hence, by providing economic incentives, they lead demand-side potentials to match with the available energy generation (Sharifi et al., 2019a, 2017a). Although this approach has been around for many years, it seems that DSR

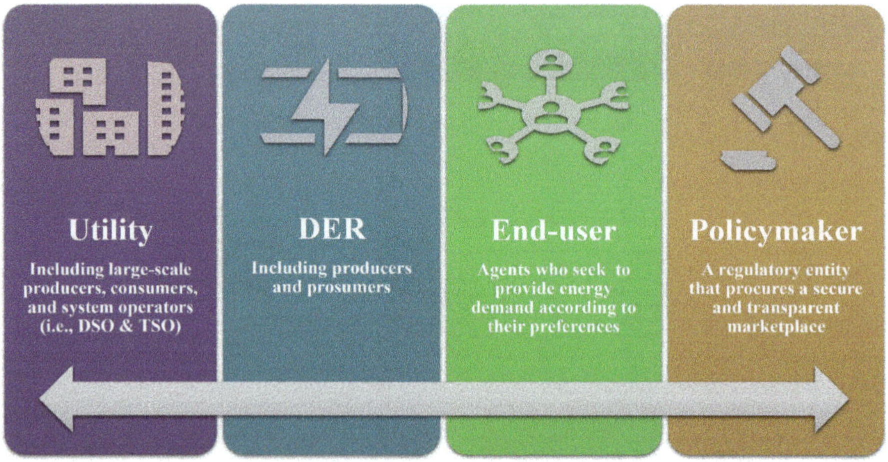

Fig. 12.2 Main participants of a TE marketplace

programs are not able to respond to the current requirements of the movement towards a smarter network (Sharifi et al., 2017b). Thus, by utilizing the generation potential of the grid along with the demand-side resources, TE defines its decisive role in the current energy sector evolutions.

It should be mentioned that other variations of electrical energy, such as ancillary services, might be also considered as a transactive product. Hence, each agent at different levels of the grid is able to perform a transaction with other agents. Hence, transactions may occur on the lower levels of the grid between two prosumers, or between a prosumer and the distribution system operator (DSO), or even on the higher levels between distribution utilities and wholesale energy market (Rahimi et al., 2016b). In this regard, participants in TE markets might be categorized as follows (Fig. 12.2) (Abrishambaf et al., 2019):

- **Utilities:** Including large-scale producers and consumers along with system operators, such as DSO and transmission system operator (TSO);
- **DERs:** Including producers and prosumers with the ability to procure energy;
- **End-Users:** Who seeks to provide energy demand according to their preferences;
- **Policymaker:** A regulatory entity that procures a secure and transparent marketplace.

According to GWAC and as illustrated in Fig. 12.3, the entire power system can be segmented into four layers of *residential*, *microgrid*, *local* and *regional* as explained in the following.

In the residential layer, consumers and prosumers might participate in DSR programs to decrease their energy costs (Sharifi et al., 2016) or even make a profit through direct energy trading with other users. Small-scale energy transactions can unlock the opportunity for energy to be seen as a heterogeneous commodity.

Fig. 12.3 TE-based classification of the power systems from the GWAC point of view

Through this feature, user preferences are considered, and customers can choose desired technologies and trade parties for energy provision.

In the microgrid layer, advanced control and management of resources, such as DERs and smart appliances, lead to a more flexible and resilient network (Sharifi et al., 2020). By exploiting different DER technologies together with intelligent devices, active buildings can play a pivotal role in this layer. In the local layer, enhanced data exchange would result in new services for end-users to participate as active agents in the energy market. In this layer, DSO operates as an intermediary agent between the retail and wholesale market participants while matching the demand and supply at the distribution level.

Finally, in the regional layer, by increasing interoperability among local and regional markets, the efficiency and reliability of the system would be improved. In this layer, TSO plays a key role in enhancing interoperability between these two layers.

TE brings significant benefits to electric energy systems. Due to the smart control of resources, TE would suggest new solutions to address the intermittency of RESs, which results in a more reliable network. Additionally, thanks to price signals and the available price information, end-users have the opportunity to not only minimize energy costs but also consider their preferences in energy provision. Furthermore, by providing the right framework for integrating more DERs into the grid, TE contributes to developing more sustainable grids. Besides, economic signals for different generation technologies ensure efficient production and delivery of energy.

Although TE would benefit the grid in several ways, several challenges must be addressed. Due to the necessity of more involvement of end-users, increasing their knowledge about this new framework would affect their engagement with the TE architecture. In other words, they must be convinced why it is important to put aside their passive role and act as an active agent in the system. Furthermore, since all participants can make transactions at any time and the number of users would increase over time, managing all communications and transactions result in scalability issues, which must be tackled. Moreover, since successful TE deployment highly depends on ICT infrastructures, investment, and maintenance, costs of these devices would be another issue to consider (Daneshvar et al., 2019).

12.2.2 P2P and Transactive Energy Markets

While TE concept comprises a variety of applications at each level of the electric energy system, it can be categorized into the following three main areas (Abrishambaf et al., 2019).

- **Transactive network management:** Managing the electricity supply chain through the centralized, decentralized, or distributed operation of microgrids and aggregators;
- **Transactive control:** Exploiting DSR approaches, such as Time-of-Use (ToU) and direct load control, along with price signals for operators to keep the balance of supply and demand;
- **P2P markets:** Enabling direct energy trading among small-scale participants.

Among the aforementioned areas, this chapter focuses on the P2P energy markets and discusses its potentials at active building level.

Among the different DER technologies, RESs are among the promising energy sources at the distribution level that enable the end-users such as buildings and houses to produce energy and share the redundant energy with their neighbours (Nezamabadi & Vahidinasab, 2020). As a result, they are no longer passive consumers and are able to take a more proactive role as a prosumer. While RESs address environmental concerns and are beneficial from different aspects, they bring some challenges to the operation of the electric power networks (Sharifi et al., 2018).

Among all TE solutions, the concept of P2P market is the one with the promising future for facilitating the contribution of the RESs into the energy networks.

P2P markets can be beneficial for prosumers as well. In conventional electricity markets, prosumers with energy surplus have only three options of curtailing production, storing energy, or selling the surplus to the grid. However, by employing P2P markets, prosumers would have another option of trading the surplus with neighbours and other network participants. As a result, instead of being a challenge for the network, DERs become a new opportunity for both grid operators and end-users. Accordingly, following modes can be considered for P2P markets operation in a transactive structure (Rahimi et al., 2016b):

- Autonomous operation based on user preferences;
- Based on available bids and offers of other participants;
- Based on incentives and price signals from network operators;
- Based on instructions from the network operators.

In the first operation mode, the only factor that needs to be considered in a trade is the user or community preferences. For instance, for the sake of environmental concerns, a particular consumer may only prefer to buy green energy generated from Photovoltaic (PV) panels. Furthermore, a philanthropic organization/community might decide to perform P2P transactions to provide low-cost energy for those who are struggling with energy poverty.

The second operation mode suggests a more economic-based structure in which demand and supply and bid-ask price strategies are the only deciding factor for participants (Sharifi et al., 2019b). Consequently, a P2P energy market would be similar to stock exchanges where each person can submit buy-sell orders for a certain amount of shares at the desired price. Accordingly, an order book is constructed in which relevant transactions would be matched with each other.

In the third operation mode, DSO is able to affect the P2P market and lead transactions by providing price signals. For example, by increasing the energy price at peak hours or providing economic incentives, DSO can encourage P2P trading at those periods to lower down peak load of the network. Likewise, in the occurrence of congestion on grid lines, by assigning a penalty value to transactions on those lines, DSO can perform congestion management in the P2P market (Nasimifar et al., 2019). Furthermore, aggregators can participate in DSR programs by allowing P2P trading among their clients (Abrishambaf et al., 2019).

Finally, in the last operation mode, DSO has full control and management over the P2P market. In this mode, all transactions must be validated and accepted by DSO before taking place. Furthermore, by occurring grid problems, such as contingency or voltage regulation, local P2P markets may be employed as service providers for the main grid.

It can be noticed that among these operation modes, the first two gives agents the maximum privilege for performing a transaction, even if it may not be the suitable choice for the grid according to no supervision from DSO. On the other hand, the last two modes would be more secure and reliable for the grid while agents are more constrained. Consequently, in a TE framework, agents must be able to change between these four modes according to the market and grid circumstances (Rahimi et al., 2016b).

In the rest of this chapter, various potentials inside active buildings and methods of using them for different P2P operation modes are discussed.

12.3 P2P Operation of Active Building as a Multi-vector Nano-energy Hub

The building is the intersection of different energy sectors such as electricity, gas, heat and cool, and water (see Fig. 12.4). In this section, we focus on the P2P operation of such an energy hub.

12.3.1 Electricity

In this subsection, different technologies for producing and storing electricity in building area is introduced, which enable buildings to participate in P2P electricity trading.

12.3.1.1 Photovoltaic and Battery Storage Systems

Since solar PV provides clean and low-cost energy, it is one of the most popular technologies for energy production. Because of the intermittent nature of renewable energies, most of the case buildings exploit battery storage systems (BSS) to save energy for other times of the day. Through the traditional structures, these buildings are only allowed to trade energy with the main grid. For instance, each active building makes use of DERs to supply its demand. If the internal generation does not cover the demand, active building must buy energy at fid-in tariffs from the grid. On the other hand, by covering the demand and having a surplus of energy, each

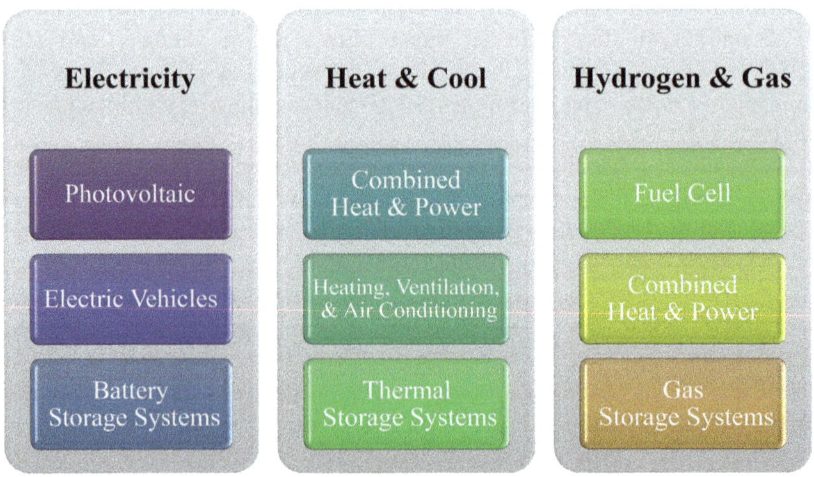

Fig. 12.4 Different technologies inside a community of active buildings

active building is only allowed to sell energy at export tariffs to the grid (Long et al., 2018).

On the other hand, by integrating a P2P market, each active building has the opportunity to store its excess energy to use it later that day or sell it to the other buildings at a price higher than export tariffs but also lower than fid-in tariffs. Similarly, other buildings can provide their demand at a lower price compared with fid-in tariffs. As a result, a revenue flow would be shaped in the community in which benefits all participants.

In summary, PV storage systems in active buildings can result in the following opportunities for the grid operation:

- P2P energy trading during peak hours would result in eliminating the stress from the grid and reducing the need for spinning reserves that mostly work with fossil fuels.
- Active buildings with only solar PV can sell the surplus to the building, which has BSS. Accordingly, the building with BSS either can use the energy to cover its load or participate in the P2P market.
- A higher-level actor, such as a community manager or even DSO, can use BSS to buy energy from buildings with solar PV and sell it later to the local consumers. The resulted profit may be used in line with community interests or enhancing its infrastructures.

12.3.1.2 Electric Vehicles

Due to concerns related to greenhouse gas production, by providing different incentives, governments encourage electric vehicles (EVs) deployment. Despite environmental advantages, EVs would cause several challenges for electricity grids, e.g. frequency regulation, voltage stability, and operational costs (Alvaro-Hermana et al., 2016). Moreover, because of the random behaviour of drivers, EVs would create an unpredictable load for the grid, which increases the need for spinning reserves and new grid investments.

In conventional methods under Vehicle-to-Grid (V2G) concept, an optimized schedule for charging/discharging of each EV is determined to enhance load profile and grid stability. Through this concept, each EV would be controlled by an aggregator, which uses the potential of EV batteries to alleviate peak demand.

Although the V2G concept is a suitable method to manage large-scale EV integration, P2P energy sharing would provide new opportunities for EVs to not only provide grid services but also make a profit out of local energy trading without an intermediary agent.

According to Alvaro-Hermana et al. (2016) in a P2P energy scheme, EVs can be divided into two groups. The first group comprises EVs that have excess energy after performing their scheduled trips. On the other hand, other EVs need to be charged at different intervals to be able to reach their destinations. In a community of active buildings and having a shared parking spot for residents of each building, all EVs would be able to trade energy with one another. Moreover, each parking spot can be connected to other spots in the community to form a larger-scale local market. Hence, each EV is able to trade a particular amount of energy at an agreed price. It should be mentioned that the energy price in the P2P market should be lower than grid tariffs at that specific time to incentivize local trading.

12.3.2 Heat and Cool

Heating and cooling are among the most essential needs of each building. In this subsection, it is described how different units can be employed to enable buildings to trade energy not only as electricity but also as heat power.

12.3.2.1 Heating, Ventilation, and Air-Conditioning System

According to a report, buildings are responsible for about 40% of the end-use energy, which half of that is consumed by heating, ventilation, and air-conditioning system (HVAC) systems (Cui & Xiao, 2020). Hence, by controlling the indoor temperature through the HVAC system in each building, a flexible community of buildings would be shaped, which can participate in different DSR programs. Besides, allowing P2P energy sharing among buildings with HVAC systems would provide several opportunities as follows:

- By participating in DSR programs or intentionally deviating from comfortable temperature, energy usage would be decreased at particular times. Hence, buildings with DERs, such as solar PV, have more energy surplus to offer in the P2P market, and consequently, they can make more profit.
- By reducing the consumption of the HVAC system according to the desired temperature range, each building can sell its demand right at particular times to other buildings in a P2P manner.
- According to Cui and Xiao (2020), integrating a P2P energy market into a community of buildings would result in more efficient use of HVAC systems, and indoor temperature would get closer to a comfortable temperature.

12.3.2.2 Combined Heat and Power Systems

Along with electricity, thermal energy is also an important demand for buildings that need to be supplied. By using the extracted heat from electricity generation, combined heat and power (CHP) systems can provide heat and electricity simultaneously. The resulted heat in conventional generators gets wasted which causes a reduction in energy efficiency, while by using a CHP system, it can be converted to hot water or steam to supply thermal demand.

By using CHP systems through a P2P market, each building is able to not only sell electricity but also exchange thermal energy to other buildings. Through interconnecting buildings using pipelines, the hot water can flow to other buildings. Hence, along with electricity, the surplus of thermal energy can also be traded locally, which results in both power and heat optimization.

12.3.3 Hydrogen and Gas

In this subsection, the role of gas and hydrogen units in P2P trading of the buildings is discussed.

12.3.3.1 Fuel Cell-Combined Heat and Power System

Despite the high system costs, fuel cell-combined heat and power (FC-CHP) systems provide several benefits for the power grid, e.g. producing power and heat efficiently, reducing GHG, and lowering energy costs. Hence, FC-CHP has been one of the attractive DERs for supplying local loads, which reduces the dependency of buildings on the main grid. Along with power production, the exhausted heat from FC would be used to heat the input water of the building. By storing the hot water in a storage tank, it can be used for supplying hot water needs of the building, e.g. conditioner, bath, and toilet. As a result, in many buildings and dwellings, the FC-CHP system is the main source of energy production.

Integrating a P2P energy market can also leverage the benefits of the FC-CHP system. In a community of buildings with FC-CHP, each fuel cell working under its rated power output can participate in local energy trading and sell its surplus to other building that reached their rated power but still not fulfilling their demand. Hence, the need for BSS and consequently the investment costs would be reduced. Furthermore, the opportunity of local trading would result in increasing FC output, which results in more efficient power and heat production.

According to the results of Nguyen and Ishihara (2021), without employing a P2P market, most of the buildings are forced to provide energy from the main grid, while fuel cells of other buildings might work in low power output. On the other hand, by integrating a local energy market, the import of energy would be decreased, and most of the demand is provided with the FC-CHP system of the building and the P2P market.

12.3.3.2 Gas Storage Systems

Power-to-gas (P2G) technology has attracted a lot of attention in recent years. P2G units can convert energy into hydrogen gas with an approximate efficiency of 80%. On the other hand, hydrogen fuel cell is able to convert gas to electricity with an efficiency of around 40%–60% (Basnet & Zhong, 2020). By comprising a P2G unit along with a gas storage unit and hydrogen fuel cell, gas storage systems (GSS) can store electricity in gaseous form and convert it back to electricity at particular times.

Exploiting GSS would result in the proper integration of RES and addressing the intermittent nature of renewable energies. For instance, when export tariffs get much lower than import tariffs, the energy surplus from PV panels can be converted to hydrogen and be transformed back to electricity in times with lower PV generation. Hence, the export and import of the main grid would be reduced considerably.

Moreover, by employing a P2P market, each building with GSS can participate in the local market as a prosumer. At certain times with available PV generation, it can act as a consumer and buy the excess energy. While at other times with zero or lower PV generation, it can be a producer by selling energy back to other buildings.

According to results from Basnet and Zhong (2020), without employing GSS and a P2P market, PV prosumers would sell their surplus to the grid and in other times a high amount of energy would be imported from the utility. However, by using GSS through a local market, a portion of the excess generation of PV panels would be sold to building with GSS, and at other times, GSS would decrease energy import from the grid by supplying the community through the P2P market.

12.4 P2P Market Opportunities and Flexibility Services Provision

In the previous section, different technologies in active buildings and their possible interactions through P2P markets have been proposed. P2P transactions in a community can also bring new opportunities for grid services provision. According to the potentials in the community and the diversity of technologies, several services can be realized to be provided for the main utility, as shown in Fig. 12.5. Hence, in the following subsections, the contribution of P2P markets in providing different services is discussed.

12.4.1 Ancillary Services

Ancillary services can be considered as a variety of actions that help the grid operator to maintain the secure operation of the power system while balancing demand and supply. the number of ancillary services and the purpose of them might be different in each part of the world. Several common services include energy imbalance service, operating reserves, frequency response, etc.

Ancillary services provision can be realized on both the demand and supply sides of the power system. On the supply side, most of the time, these services would be provided by large-scale power plants which result in a considerable increase in

Fig. 12.5 Different flexibility services through P2P transactions inside a community of active buildings

operational costs. On the other hand, ancillary services can also be provided through DSR programs on the demand side of the system.

The diversity of DERs and appliances in a community of buildings would bring a new option of providing ancillary services through P2P energy transactions. This concept is called "*federated power plants*", which can unlock new opportunities for the P2P energy market while addressing several challenges faced by top-down hierarchical structures (Morstyn et al., 2018).

In a P2P market, prosumer buildings can sell their surplus to other consumers and prosumers. However, for reaching demand and supply balance at each time-step, some buildings might need to trade energy with the main grid. Accordingly, all participants can also be allowed to trade ancillary services with the utility. For instance, to provide peak shaving/valley filling services, each consumer/prosumer can adjust its consumption/production to support the operational needs of the grid. Consequently, it not only contributes to a more secure operation and management of the utility but also results in more profits for the small-scale participants.

The operation of such a structure can be performed through three mechanisms, namely, P2P, P2G, and ancillary services. At first, each active building would use DERs and flexible loads to procure the required energy at the minimum cost. Then, after supplying its internal demand, the excess energy would be traded locally with other buildings through a P2P mechanism.

However, due to different preferences, prices, or conditions for energy transactions, likely, some asks and bids do not get matched with each other, forcing buildings to trade energy with the main grid. In this situation, active buildings would participate in a P2G mechanism to sell their excess energy or buy more energy to supply internal loads. Therefore, the power utility would supply loads at retail prices and buy energy at export tariffs which are relatively lower than retail prices. It makes the community of buildings a price taker which can be considered as a typical customer from the perspective of the grid (Zhou et al., 2020).

Then, according to the other two mechanisms, the utility would evaluate its operational challenges (e.g. congestion, frequency deviation, voltage deviation) to provide price signals and incentives for ancillary services provision. Accordingly, the grid operator would determine the actions that contribute to reaching a secure operation, namely, peak shaving, and request these actions from P2P market participants through ancillary services. This mechanism would be performed in three steps (Zhou et al., 2020):

- **Submission of asks for ancillary services:** According to the current status of the system and future assessments, the grid operator would specify the type, the amount, and the price of required ancillary services and announce them to P2P market participants.
- **Submission of bids from P2P market participants:** After determining services and their prices, all buildings in the P2P market aiming to maximize their economic benefits would respond to broadcasted services and submit their bids for ancillary services provision.
- **The final decision on ancillary services provision:** After receiving all bids from the P2P market, the grid operator would decide to accept/reject orders and purchase services according to the grid conditions and offers submitted by buildings.

Hence, three different markets for P2P transactions, peer-to-grid trading and ancillary services can be realized. These markets might be operated sequentially or simultaneously. In the case of sequential operation, some levels of sub-optimality may appear in the system. On the other hand, a simultaneous operation would lead to a more complex structure. Moreover, in some countries, such as Great Britain, ancillary services may be procured by different entities, such as DSO or electricity provider, which makes the simultaneous operation of energy and ancillary services more difficult. In such situations, designing and operating separate mechanisms seems to be a more feasible and flexible option (Zhou et al., 2020).

It should also be noted that according to the type of ancillary services, some buildings might be forced to deny their transactions in the P2P market. In such circumstances, there should be a procedure to assign an extra payment to compensate for the revenue losses of the service provider due to the violated transactions.

12.4.2 Peak Load Shaving

Although the challenge of peak load periods has been around for many years, integrating DERs, such as EVs, would make this challenge more crucial to address. In conventional approaches, grid reinforcements would be exploited to solve the peak load problem. However, it is not a cost-effective approach and due to the high dependency on fossil fuels, it increases environmental concerns.

On the other hand, unlocking the potential flexibility of demand-side resources can be a relatively low-cost solution for peak load shaving. The demand-side flexibility is conventionally provided by large-scale customers. However, using local P2P markets can be a suitable enabler to provide market access for small-scale customers willing to offer flexible services. Hence, not only small-scale users would minimize their energy costs, but also DSO would use this capacity to maintain the secure operation of the grid. As a result, in recent years, exploiting transactive approaches and local market designs to provide system flexibility has attracted a lot of attention from policymakers.

The local market can be operated in two modes of with or without an intermediary agent. By the lack of a third party, the P2P market would incentivize buildings to trade locally at peak hours due to the high energy prices at the wholesale market. Therefore, in peak periods, buildings would sell their surplus at a lower rate compared to the main grid and at the same time at a higher rate compared to export tariffs. Furthermore, by enabling prosumers to perform direct P2P transactions with the main grid, they can sell energy at higher prices in peak periods. As a result, prosumers would make more profits, while the main grid would acquire more generation at peak periods.

On the other hand, with the existence of intermediaries, several aggregators can control and manage the DERs of the customers and representing them in P2P transactions with DSO to reach peak load shaving. Moreover, an entity, called TE operator, can be considered to provide a trading platform between aggregators and DSO. Accordingly, three main roles can be categorized as follows (Masood et al., 2020):

- **Aggregators:** At first, aggregators would evaluate the energy needs of their customers by scheduling and optimizing DERs and appliances of each building. Then the energy profile would be broadcasted to DSO for further analysis. Next, if DSO announces flexibility requirements, aggregators would assess the flexibility potentials of their customers and accordingly will submit bids to DSO through the TE platform for flexibility provision.
- **DSO:** This entity is responsible to maintain the secure and reliable operation of the grid at each given time or condition. At first, DSO would evaluate the announced energy profiles from aggregators to assure no network violations. By foreseeing any network issues, DSO would request flexibility services for peak periods through orders in the TE platform with specific price and quantity of energy to be decreased.

- **TE operator:** An agent that can be a physical entity or even a virtual platform through ICT, such as blockchain. The main responsibility of the TE operator is to provide a trading platform for aggregators and DSO to submit ask/bid orders. TE operator would match orders and clear the market. Then, the final results and information would be broadcasted to both parties.

12.4.3 Voltage and Reactive Power Control

Voltage deviation is one of the most challenging issues in power systems which invites the use of reactive power compensation strategies. Nowadays, by the proliferation of unexpected loads, such as EVs, and due to the intermittent generation of RES, Volt/VAr control strategies are getting more important. On the other hand, since most DERs are connected to the grid through inverters, they can be suitable resources to compensate for reactive power.

Volt/VAr feedback control strategies, such as Volt/VAr droop control, are promising choices to address reactive power compensation (Ortmann et al., 2020). As a result, the reactive power injection of each inverter can be controlled according to the voltage at the point of common coupling.

Two types of centralized and decentralized can be considered for feedback control. In both types, the only parameter to measure is the voltage at the point of common coupling, and the difference is through the way these measurements are processed. Exploiting P2P markets to provide reactive power is a decentralized way to perform feedback control for compensating reactive power.

In a community of active buildings, due to the diversity of loads and generation, there are lots of inverters that can be employed as a feedback Volt/VAr controller for contributing to reactive power compensation when it is needed. The architecture for such a mechanism might be similar to the one explained in Sect. 12.4.2. Inverters of active buildings can be managed by an intermediary agent, called aggregator or community manager. This agent is responsible to provide reactive power services when there is a call from DSO. Hence, a transactive platform can be realized that DSO can submit asks orders and aggregators, according to the inverters of their customers, can bid for participating in reactive power compensation. Hence, voltage control would be performed in a decentralized way, while providing benefits for both the community of building and DSO.

12.4.4 Resilience Response

Natural disasters and storms, such as hurricanes, can occur anywhere in the world, and humans do not have the control to stop them from happening. Since these events are responsible for many blackouts around the world, power systems should evolve to a more resilient network. For instance, hurricanes Irma and Maria caused massive damage to the power grid of Puerto Rico which resulted in a blackout for around 80% of end-users for 1 month (Mehri Arsoon & Moghaddas-Tafreshi, 2020). Resilience in a power system is defined by the ability of the network to withstand high-impact and low-frequency (HILF) events and recover rapidly to a stable operation. Although HILF events rarely occur, they can result in immense damage to a power system which makes them an important challenge to solve.

The enhancement of resilience response might be categorized into two types of planning-based and operation-based. In terms of planning-based resilience, although activities like the optimal placing of grid equipment would enhance resilience response, it leads to much higher investment costs.

On the other hand, operation-based resilience, such as reconfiguration and rescheduling, might be a more cost-effective option. However, to reach a proper amount of resiliency, a variety of back-up generators and reserves should be considered which would increase investment costs as well (Mehri Arsoon & Moghaddas-Tafreshi, 2020). Coupling microgrids together is another operation-based approach to address resilience response which has proven to be a low-cost solution. Thus, in occurring HILF events and switching to islanding mode, each microgrid can share its surplus with other microgrids.

The concept of P2P markets would be best applied to a structure of networked microgrids. In our case, each community of buildings can be considered as a microgrid that can trade energy not only internally but also with other communities. Hence, by occurring any natural disasters and losing connectivity from the main grid, the networked microgrids can support each other to stop blackouts or at least minimize the load curtailment. The diversity of DERs in the community of buildings can facilitate this process. Accordingly, to modes of P2P operation can be realized during HILF events:

- A full P2P market can be realized in which all participants of all communities can trade energy directly with one another. Hence, during an unexpected event and disconnecting from the main grid, each prosumer would optimize its demand and generation and then share its surplus through the P2P market with other prosumers and consumers in all communities. This operation mode results in considering energy as a heterogeneous product and each participant can trade according to its preferences. However, scalability issues are the main characteristic of this design which should be addressed.
- A community-based market can be realized in which each community has a manager or aggregator. These agents are responsible to assess the internal needs and potentials of their communities to not only provide the energy of buildings but also represent them in P2P transactions with the aggregators of other communities. Thus, by occurring HILF events and disconnecting from the utility, all communities can be connected through a common coupling point while they are able to trade energy with each other. As a result, each community aims to maximize its export to reach minimize load curtailment in the whole system which results in the enhancement of resilience response (Mehri Arsoon & Moghaddas-Tafreshi, 2020).

12.4.5 *Frequency Response*

Maintaining system frequency at the desired set-point is a crucial task for grid operators. In some countries, such as the UK, frequency should be maintained around 60 Hz, while in some other countries, like the USA, it is 50 Hz. Hence, all generators in a national grid should spin at the same speed or otherwise it would cause serious damage to the system. Therefore, grid operators consider different approaches for frequency regulation to prevent frequency deviations and maintain the secure operation of the grid.

Frequency deviation is a result of mismatched demand and supply. If energy generation gets higher compared to the loads, it causes the frequency to increase. On the other hand, in the situation of not being enough generation to feed the loads, the frequency would start to drop. Due to the proliferation of unexpected loads, such as EVs, and also the intermittent generation of RES, nowadays, frequency regulation is getting more critical for operators.

In the case of higher frequency compared to the set-point, the operator can perform frequency regulation by decreasing the output of some generators. On the other hand, if frequency gets lower than the standard set-point, frequency regulation would be more challenging for the grid operator; Especially, when there is no more room to increase the output of generators. In such circumstances, besides ancillary services and regulation markets, considering DSR programs would be a proper option to use demand-side potentials for frequency regulation.

By exploiting different generation technologies and a variety of smart loads, a community of active buildings is capable of providing frequency-responsive services for the grid. Interestingly, in the existence of a P2P market, buildings are not limited to load shifting/shedding to provide DSR, but they can procure their loads from the local market in the community to remove the burden from the grid. Furthermore, the whole community of buildings can also participate in frequency services provision. For instance, if the frequency drops below the set-point, a community or microgrid can switch to a P2P operation mode in which all the loads would be supplied through P2P transactions inside/among the buildings. As a result, a considerable amount of loads would be mitigated from the grid, and it contributes to frequency regulation. It should be noted that considering such concepts for frequency response highly depends on the internal capabilities of each community along with price signals and incentives from DSO.

12.5 P2P Market Models for Active Buildings

In a community of active buildings, a P2P energy market is a network of buildings capable of exchanging energy with each other. According to the type of connectivity between buildings and the degree of decentralization, three structures can be

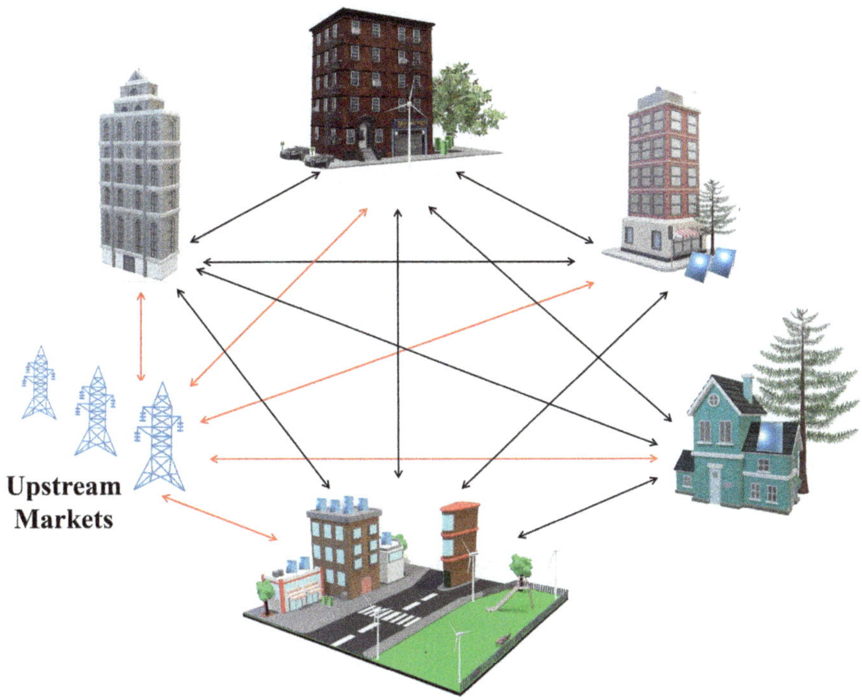

Fig. 12.6 A full P2P market (arrows represent cash flow and communication links; black arrows, among buildings; red arrows, between each building and other existing markets)

considered for a P2P market. In the following sub-sections, a brief introduction of these architectures is presented.

12.5.1 Full P2P Markets

A full P2P market is characterized by direct energy trading between peers. Through this mechanism, buildings are able to negotiate to buy/sell a specific amount of energy at an agreed price. This design is the most decentralized structure compared to other P2P architectures since it eliminates the need for an intermediary agent.

Figure 12.6 shows a full P2P market and the connectivity between the agents. It should be mentioned that these connections illustrate communication links between peers, while electricity connectivity can be different (Sorin et al., 2019). Due to the focus of this chapter, each peer can be considered as an active building with DERs and smart appliances that are capable of consuming, producing, and storing energy.

In a full P2P energy market, the exchangeable product is the excess energy. Thus, a building that has an available energy surplus from its DERs can sell it to other buildings. Accordingly, a transaction would be performed among buildings to exchange a certain amount of energy at an agreed price. Therefore, all buildings would compete on asking/bidding processes to reach their desired objective, which is buying energy at a low price or selling it at a high rate.

Full P2P markets have diverse advantages for a community of active buildings. First of all, it considers energy as a heterogeneous product. Accordingly, each building might decide to buy energy according to its preferences for the type of resources (e.g. renewable) and trading parties. Furthermore, due to its distributed design, a P2P market has a modular structure which makes it work even by collapsing any peers of the network. Moreover, new buildings can be added to the network at any time, without the need to change operation methods (Giotitsas et al., 2015). This modular structure not only results in a more secure operation of the network but also enables buildings to share updated data and information more conveniently. Furthermore, since energy would produce and consume locally, line usage and transmission losses would decrease considerably.

Due to the possibility for each peer to trade with any other peers in the network in semi-real-time, scalability is the main challenge of this design. High investment and maintenance costs for ICT infrastructures are other challenges that should be addressed. It is also necessary to consider that according to the lack of central control, grid operators, such as DSO, have the most limited supervision on the P2P market comparing to two other designs.

12.5.2 Local P2P Markets

In this design, instead of individual peers, groups of peers are able to trade energy with each other. Figure 12.7 illustrates this structure. In a local P2P market, each group/community has a manager which not only manages the entire community by evaluating their needs and potentials but also represents internal peers in trading with other groups/communities in the local area. Although this agent might be referred to with different phrases, such as community/district/microgrid manager, the responsibility of this intermediary agent stays the same.

According to its mechanism, local P2P markets can readily be applied to a community of buildings in which each community has a manager. As a result, several communities can be shaped in the local area which is capable of sharing energy and services not only with each other but also with the main grid. Moreover, peers of each community may have common interests and preferences through energy provision that would be considered in their transactions. These common interests might be environmental goals, such as using green-only production to contribute to the reduction of carbon emission, or social goals, such as supporting those who are in energy poverty by sharing energy at the lowest possible rates.

A local P2P market has several advantages over a full P2P design. First of all, by assigning a manager to each community/group, it has a semi-decentralized structure which can enhance scalability considerably. Furthermore, it increases the collaboration of community members and mobilizes them to reach certain goals.

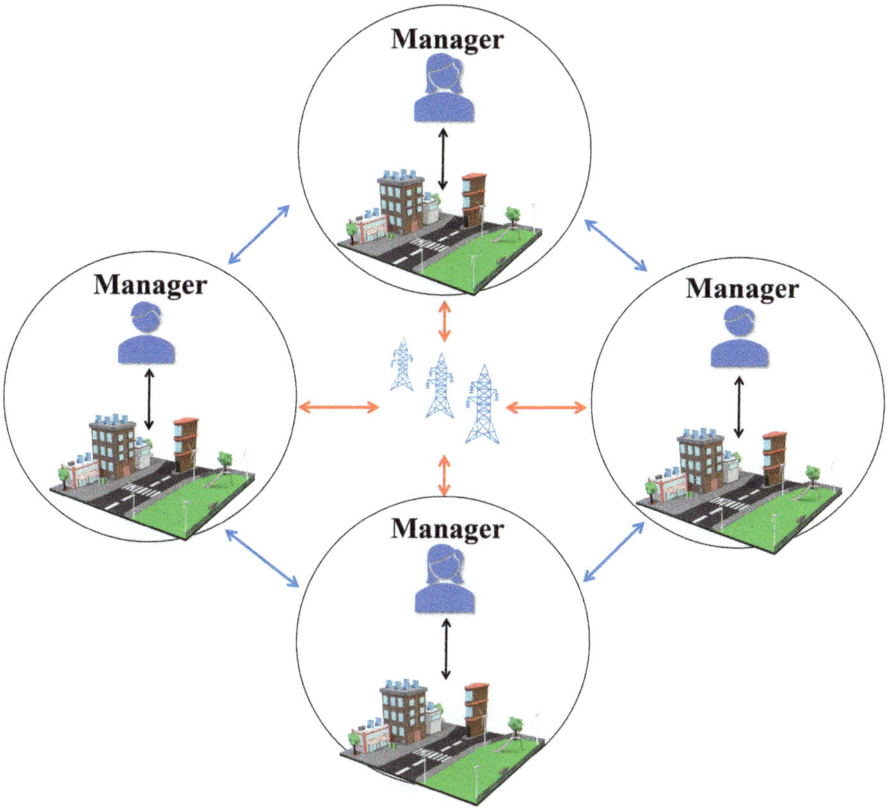

Fig. 12.7 A local P2P market (arrows represent cash flow and communication links: black arrows, among buildings and the manger; blue arrows, between local communities of buildings; red arrows, between each community and other existing markets)

Through their collaboration, different services can also be provided for the main grid, e.g. peak shaving, voltage regulation, and congestion management.

Despite the advantages, this design has also some challenges that should be addressed. Comparing to the full P2P architecture, the community manager makes the last decision for energy trading. Hence, this is difficult to consider the interest and preferences of all individual peers in the community simultaneously. Moreover, collecting the information of each peer and managing their needs and expectations would be challenging for the community manager. However, exploiting distributed ledgers, such as blockchain, seems to be a promising solution for such challenges.

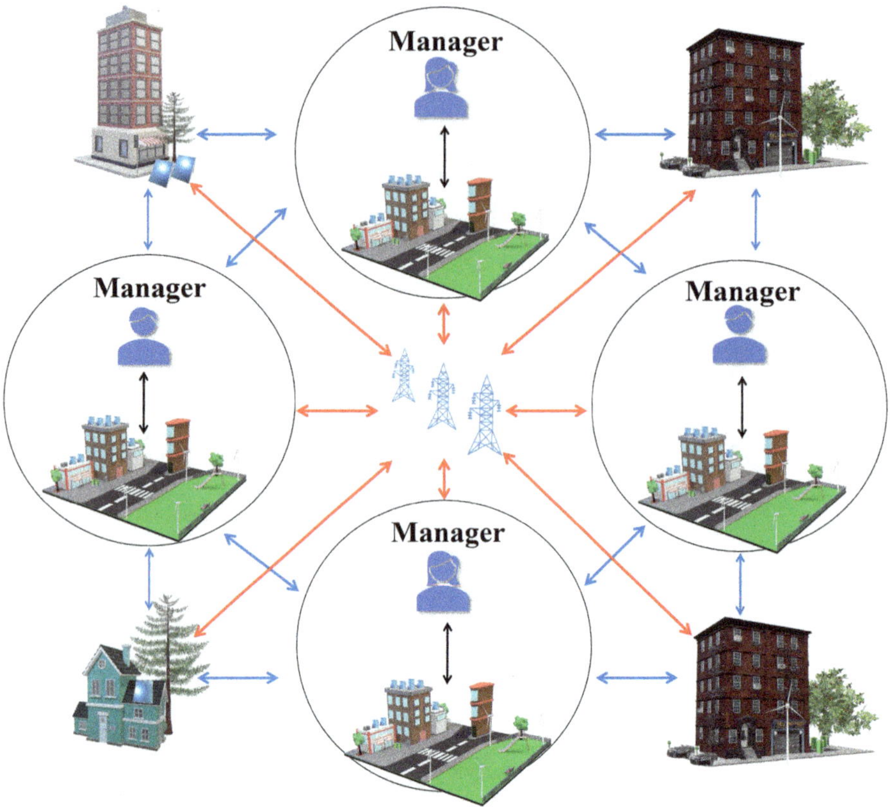

Fig. 12.8 A Hybrid P2P market (arrows represent cash flow and communication links: black arrows, between buildings and the manger; blue arrows, between all of the buildings and communities; red arrows, transactions of buildings and communities with the main grid)

12.5.3 Hybrid P2P Markets

A hybrid P2P market can be realized as a combination of two other designs. Hence, in the lower level of the market, peers would trade directly to one another while at the upper-level communities would interact with each other through their community manager (Sousa et al., 2019). Figure 12.8 shows the architecture of this design.

A hybrid design would be beneficial since it combines the advantages of two other structures. By using communities as the upper level, it would result in scalability enhancement while this structure can be more predictable for grid operators. Simultaneously, each peer can trade with other peers and even communities according to their preferences. However, implementing proper coordination between the two levels of this design is a challenging task that should be addressed.

12.6 Case Study

In previous sections, different structures for P2P markets and their advantages for not only a community of buildings but also the main grid has been discussed. Now, in this section, a P2P market for buildings area is modelled to show the basics of such markets and giving ideas for further research on the subject.

12.6.1 A Basic P2P Electricity Market

Consider a P2P market with a set of peers Γ, which includes two groups of producers Γ_p and consumers Γ_c, capable of trading electricity directly with each other. To model this market, the power injection of each peer P_i is assumed as cumulative trades with neighbouring peers:

$$P_i = \sum_{j \in \mu_i} P_{ij} \ , \ \forall (i, j) \in (\Gamma, \mu_i) \tag{12.1}$$

where $\left\{ P_{ij} | i \in \Gamma \ , j \in \mu_i \right\}$ is the set of decision variables that represent the amount of energy traded between peers i and j. Furthermore, μ_i is the set of participants that peer i can trade with. Hence, P_i can also be considered as the amount of all power transactions of peer i. Furthermore, the injected power of each peer is limited by the following constraint:

$$P_i^{min} \leq \sum_{j \in \mu_i} P_{ij} \leq P_i^{max} \ , \ \forall i \in \Gamma \tag{12.2}$$

where P_i^{min} and P_i^{min} are upper and lower bounds for the injected power of peer i. Furthermore, according to the role of each peer at any given time, the sign of P_{ij} would be determined as follows:

$$P_{ij} \begin{cases} Positive, & i \in \Gamma_p \\ Negative, & i \in \Gamma_c \end{cases} . \tag{12.3}$$

Thus, according to the role of each peer, $P_i^{min} P_i^{max} \geq 0$. Furthermore, the objective function of the problem would be formulated according to the production cost/willingness to pay function of each peer as follows:

$$c_i (P_i) = \sum_{i \in \Gamma} \frac{1}{2} a_i \left(\sum_{j \in \mu_i} P_{ij} \right)^2 + b_i \left(\sum_{j \in \mu_i} P_{ij} \right) + c_i \ , \ a_i, b_i, c_i > 0. \tag{12.4}$$

Due to the limited insight about the actual utility and cost functions of small-scale agents, these functions are modelled in a quadratic form. However, any other types of functions can be used as long as it is convex (Sorin et al., 2019). Finally, the balance of each transaction would be ensured through:

$$P_{ij} + P_{ji} = 0 \ , \ \forall (i, j) \in (\Gamma, \mu_i) \,. \tag{12.5}$$

Thus, the optimization problem for minimizing operation costs of a P2P market would be formulated as follows:

$$\min \ F = \sum_{i \in \Gamma} \frac{1}{2} a_i \left(\sum_{j \in \mu_i} P_{ij} \right)^2 + b_i \left(\sum_{j \in \mu_i} P_{ij} \right) + c_i \ , \ a_i, b_i, c_i > 0 \tag{12.6}$$

subject to:

$$P_i^{min} \le \sum_{j \in \mu_i} P_{ij} \le P_i^{max} \ , \ \forall i \in \Gamma \tag{12.7}$$

$$P_{ij} \ge 0 \ , \ \forall (i, j) \in \left(\Gamma_p, \mu_i\right) \tag{12.8}$$

$$P_{ji} \le 0 \ , \ \forall (i, j) \in (\Gamma_c, \mu_i) \tag{12.9}$$

$$P_{ij} + P_{ji} = 0 \ , \ \forall (i, j) \in (\Gamma, \mu_i) \,. \tag{12.10}$$

This is a cost allocation problem for the forward market mechanism and does not consider network constraints and grid services, such as reserves and ancillary services. Moreover, for the sake of simplicity, a single time-step formulation is simulated to focus on interactions between different buildings and further developing to include grid considerations. However, it can readily be employed for a multi time-step implementation for any operation horizon.

12.6.2 Power Losses Management in a P2P Market

In Eqs. 12.6–12.10, the only factor for performing a power transaction is production costs and the amount of willingness to pay. Hence, each peer aims to minimize its energy costs regardless of the effect on the grid operation and network constraints. Hence, in this subsection, a method based on penalty functions is proposed which results in considering network constraints and using the potentials of a P2P market to obtain different grid services.

In this approach, buildings not only consider their preferences but also contribute to power loss reduction by providing energy from transactions that cause minimum energy losses. As one of the most important reasons for energy losses is the resistance of lines, a penalty function based on the total resistance existing in each transaction would be added to the objective function as the cost of power losses (Nasimifar et al., 2019). Hence, total resistance between peers i and j can be calculated as:

$$R_{ij} = \sum_{l \in \Lambda_{ij}} r_l \ , \ \{\Lambda_{ij} | (i, j) \in (\Gamma, \mu_i)\} \subset L = \{1, \ \ldots, l\} \tag{12.11}$$

where L is the set of all lines of the network and Λ_{ij} is the set of lines between peers i and j. Hence, R_{ij} would be the total resistance between each pair of peers. It should be mentioned that the amount of R_{ij} would be derived according to the network topology. Moreover, a radial network is considered since P2P markets usually are implemented at the distribution level of the network. Hence, all lines are series and R_{ij} can be easily calculated. In the next step, a positive variable called the resistance-based penalty α_i would be defined which represents the amount of penalty that peer i should pay according to the amount of existing resistance in its transactions with other participants. Accordingly, resistance-based penalty function would be defined as:

$$C_i^R (P_i) = \sum_{j \in \mu_i} (R_{ij}\alpha_i) \times S_i^R \times P_{ij} \tag{12.12}$$

where S_i^R is the sign parameter which would be determined according to the role of each peer:

$$S_i^R \begin{cases} Positive, & i \in \Gamma_p \\ Negative, & i \in \Gamma_c \end{cases} . \tag{12.13}$$

It should be mentioned that the amount of α_i would be decided by the grid operator according to the operational conditions and network situation at any given time. Finally, by adding Eqs. 12.13 to 12.6, the objective function would turn into:

$$F^R = \sum_{i \in \Gamma} \left[\frac{1}{2} a_i \left(\sum_{j \in \mu_i} P_{ij} \right)^2 + b_i \left(\sum_{j \in \mu_i} P_{ij} \right) + c_i + \sum_{j \in \mu_i} (R_{ij}\alpha_i) \times S_i^R \times P_{ij} \right] . \tag{12.14}$$

According to this model, along with minimizing operational and energy costs, peers would also consider energy losses in their transactions. Furthermore, since these penalties show the usage of lines in each transaction, it can also be used as a cost recovery tool for grid operators to assign fees particularly to each peer instead of socializing it to the whole participants. Additionally, due to the reduction in line usage, this method would also result in some levels of congestion management. However, by occurring congestion on a particular line or a group of lines, there should be a more exact method to eliminate congestion, which is the subject of the next sub-section.

12.6.3 Congestion Management in a P2P Market

Congestion management is an essential part of the power systems operation. Due to the decentralized nature of P2P markets, it is hard to predict the behaviour of each peer and their transactions. Hence, grid operators need a tool for particular situations to ensure the secure operation of the grid at any given time. By using the approach proposed in Nasimifar et al. (2019), congestion management in a P2P market can be performed according to price signals and incentives broadcasted from the grid operator. At first, the set of congested lines σ should be specified according to the power flow of the network. Then, a binary variable θ can be defined as:

$$\theta_{ij} = \begin{cases} 0 & \sigma \not\subset \Lambda_{ij} \\ 1 & \sigma \subseteq \Lambda_{ij} \end{cases}, \quad (i, j) \in (\Gamma, \mu_i) \tag{12.15}$$

which determines the existence of congested lines between peers i and j. Then, the same as the previous sub-section, a congestion-based penalty function is considered as follows:

$$C_i^C = \sum_{j \in \mu_i} (\theta_{ij}\beta_i) \times S_i^C \times P_{ij} \tag{12.16}$$

where S_i^C is the sign parameter which would be determined according to the role of each peer:

$$S_i^C \begin{cases} Positive, & i \in \Gamma_p \\ Negative, & i \in \Gamma_c \end{cases}. \tag{12.17}$$

Now, according to different operational conditions, the grid operator can signal an amount of penalty β_i for transactions on congested lines. Therefore, the power flow on congested lines would be reduced. Finally, the objective function for the problem turns into:

$$F^C = \sum_{i \in \Gamma} \left[\frac{1}{2} a_i \left(\sum_{j \in \mu_i} P_{ij} \right)^2 + b_i \left(\sum_{j \in \mu_i} P_{ij} \right) + c_i + \sum_{j \in \mu_i} (\theta_{ij}\beta_i) \times S_i^C \times P_{ij} \right]. \tag{12.18}$$

It should be mentioned that both penalty functions F^C and F^R can exist at the same time to achieve simultaneous management of congestion and power losses

(Nasimifar et al., 2019). Moreover, since developed models are convex, they can be solved with a variety of centralized and decentralized optimization approaches.

12.6.4 Power and Heat Trading

A community of active buildings may employ different technologies for producing energy. In previous subsections, a basic model for electricity trading in the community has been developed. As mentioned in Sect. 12.3.2.2, P2P transactions are not limited to electricity. Hence, other forms of energy, such as heat, can be traded in the community of buildings. Hence, in this subsection, the proposed model in Eqs. 12.6–12.10 would be developed to also consider CHP technology and heat transactions between peers.

Consider a community with a set of peers Γ, which includes three groups of heat and electricity producers Γ_p, heat and electricity consumers Γ_c, and CHP units Γ_{chp}, which are able to trade energy directly with one another. The heat injection of each peer H_i would be:

$$H_i = \sum_{j \in \mu_i} H_{ij}, \ \forall (i, j) \in (\Gamma, \mu_i) \tag{12.19}$$

where $\{H_{ij} | i \in \Gamma \ , j \in \mu_i\}$ is the set of decision variables that represent the amount of energy traded between peers i and j. Moreover, the injected heat of each peer would be limited by:

$$H_i^{min} \leq \sum_{j \subset \mu_i} H_{ij} \leq H_i^{max} \ , \ \forall i \in \{\Gamma_p, \Gamma_c\} \tag{12.20}$$

where H_i^{min} and H_i^{min} are upper and lower bounds for the injected heat of peer i. Furthermore, the sign of H_{ij} would be determined according to the role of the peer at the given time:

$$H_{ij} \begin{cases} Positive, & i \in \Gamma_p \\ Negative, & i \in \Gamma_c \end{cases} . \tag{12.21}$$

Next, the objective function in Eq. 12.4 would turn into:

$$\min F = C^p(i) + C^h(i) = \sum_{i \in \Gamma} \left[a_i^p \left(\sum_{j \in \mu_i} P_{ij} \right)^2 + b_i^p \left(\sum_{j \in \mu_i} P_{ij} \right) + c_i^p \right] \tag{12.22}$$

$$+ \sum_{i \in \Gamma} \left[a_i^h \left(\sum_{j \in \mu_i} H_{ij} \right)^2 + b_i^h \left(\sum_{j \in \mu_i} H_{ij} \right) + c_i^h \right]$$

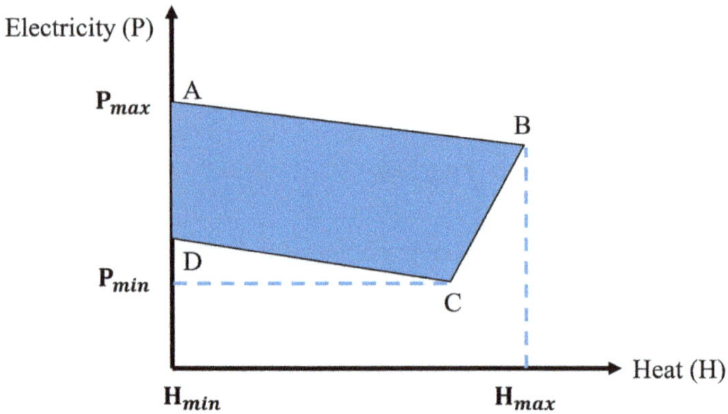

Fig. 12.9 An example of the feasible region for the operation of CHP units

where $C^p(i)$ and $C^h(i)$ are cost/willingness to pay functions of peer i for electricity and heat, respectively. Furthermore, the balance of each heat transaction would be ensured through:

$$H_{ij} + H_{ji} = 0\ ,\ \forall (i, j) \in (\Gamma, \mu_i) \tag{12.23}$$

Next, CHP units, with the ability to produce electricity and heat simultaneously, can be added to the model. The production of these units is limited to a feasible operation region, as shown in Fig. 12.9. Hence, P_i and H_i for CHP units are limited to this region and can be modelled as:

$$\sum_{j\in\mu_i} P_{ij} - D_p \geq \left(\sum_{j\in\mu_i} H_{ij} - D_h\right) \times \left(D_p - C_p\right) / \left(D_h - C_h\right)\ ,\ \forall i \in\ \Gamma_{chp} \tag{12.24}$$

$$\sum_{j\in\mu_i} P_{ij} - A_p \leq \left(\sum_{j\in\mu_i} H_{ij} - A_h\right) \times \left(A_p - B_p\right) / \left(A_h - B_h\right)\ ,\ \forall i \in\ \Gamma_{chp} \tag{12.25}$$

$$\sum_{j\in\mu_i} P_{ij} - B_p \geq \left(\sum_{j\in\mu_i} H_{ij} - B_h\right) \times \left(B_p - C_p\right) / \left(B_h - C_h\right)\ ,\ \forall i \in\ \Gamma_{chp}. \tag{12.26}$$

Therefore, the optimization problem for minimizing operation costs of a P2P market, with consideration of heat trading and CHP units, is formulated as follows:

$$\min F = C^p(i) + C^h(i) \tag{12.27}$$

subject to:

$$P_i^{min} \leq \sum_{j \in \mu_i} P_{ij} \leq P_i^{max}, \ \forall i \in \{\Gamma_p, \Gamma_c\} \tag{12.28}$$

$$H_i^{min} \leq \sum_{j \in \mu_i} H_{ij} \leq H_i^{max}, \ \forall i \in \{\Gamma_p, \Gamma_c\} \tag{12.29}$$

$$\sum_{j \in \mu_i} P_{ij} - D_p \geq \left(\sum_{j \in \mu_i} H_{ij} - D_h \right) \times (D_p - C_p) / (D_h - C_h), \ \forall i \in \Gamma_{chp} \tag{12.30}$$

$$\sum_{j \in \mu_i} P_{ij} - A_p \leq \left(\sum_{j \in \mu_i} H_{ij} - A_h \right) \times (A_p - B_p) / (A_h - B_h), \ \forall i \in \Gamma_{chp} \tag{12.31}$$

$$\sum_{j \in \mu_i} P_{ij} - B_p \geq \left(\sum_{j \in \mu_i} H_{ij} - B_h \right) \times (B_p - C_p) / (B_h - C_h), \ \forall i \in \Gamma_{chp} \tag{12.32}$$

$$P_{ij} \geq 0, \ \forall (i, j) \in (\Gamma_p, \mu_i) \tag{12.33}$$

$$P_{ji} \leq 0, \ \forall (i, j) \in (\Gamma_c, \mu_i) \tag{12.34}$$

$$H_{ji} \geq 0, \ \forall (i, j) \in (\Gamma_p, \mu_i) \tag{12.35}$$

$$H_{ji} \leq 0, \ \forall (i, j) \in (\Gamma_c, \mu_i) \tag{12.36}$$

$$P_{ij} + P_{ji} = 0, \ \forall (i, j) \in (\Gamma, \mu_i) \tag{12.37}$$

$$H_{ij} + H_{ji} = 0, \ \forall (i, j) \in (\Gamma, \mu_i). \tag{12.38}$$

This is a cost allocation problem that can be used for the forward market mechanism. However, despite the basic P2P model, it considers technical constraints of the generation technology, which is the CHP unit in this example. Following the same method, one may further develop this model by adding technical constraints of other generation technologies and different appliances that are available in active buildings.

12.6.5 Simulation Results

To simulate the proposed models, an 11-bus radial distribution network is considered as the case study (see Fig. 12.10). The simulation is performed at a particular

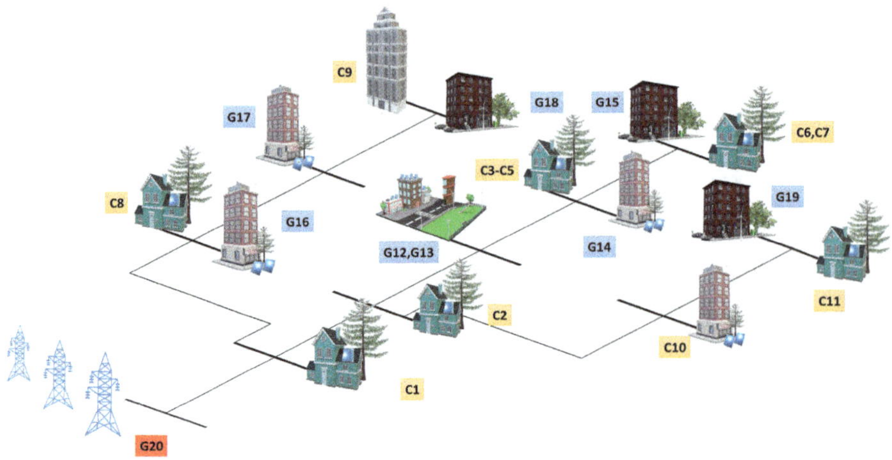

Fig. 12.10 The 11-bus radial distribution network test case

time-step in which the role of each building is specified as there are 11 consumers $\{C_1, C_2, \ldots, C_{11}\}$ and 9 producers $\{G_{12}, G_{13}, \ldots, G_{20}\}$, including the main grid. This would help to focus on the interactions among participants and their responses to signals from grid operators. However, as mentioned in Sect. 12.6.1, it can easily be developed to perform a multi time-step simulation which invites the consideration of technical constraints, such as ramp up/down of producers and the flexibility of consumers. It should be mentioned that each peer can be an entire building or even a small office or household inside the building. For further information about the case study and input data, the reader is referred to Nasimifar et al. (2019).

12.6.5.1 Transactions in the Basic Model

By using Eqs. 12.6–12.10, peers interact with each other to minimize their energy costs. Hence, each producer that sells energy at a lower price would be the first choice of consumers. Table 12.1 shows P2P transactions among different participants. It is clear from Table 12.1 that most of the buildings are willing to trade with the main grid since it offers energy at a relatively lower price than the other producers. However, this might be not desirable for the grid, since these transactions may violate several network constraints or cause power losses and congestion.

12.6.5.2 Power Losses Management

Now, by using Eq. 12.14 as the objective function and through different amounts of resistance-based penalty α_i, it is possible to achieve power loss reduction in the

Table 12.1 Transactions among buildings and the main grid in the basic model

	C1	C2	C3	C4	C5	C6	C7	C8	C9	C10	C11
G12	0	0	220	0	0	235	0	0	195	140	0
G13	0	0	220	0	0	235	0	0	285	53	18
G14	0	0	0	246	262	0	122	30	0	191	0
G15	0	166	0	46	49	0	29	29	0	357	74
G16	0	0	0	0	0	0	0	0	0	0	0
G17	0	0	0	0	0	0	0	0	0	0	0
G18	0	0	0	0	0	0	0	0	0	0	0
G19	0	0	0	0	0	0	0	0	0	0	0
G20	420	344	0	77	80	0	60	481	0	140	748

Table 12.2 The amount of power losses in different structures

α_i (\$/kWΩ)	Power losses (kW)
α_i=0[a]	110.76
α_i=0[b]	26.95
α_i=5	8.56
α_i=10	8.56
α_i=20	7.97
α_i=30	7.38
α_i=35	7.07

[a] Pool-based
[b] Basic P2P

P2P market. For a better understanding of the impact of this model, a pool-based structure is also considered as a base for comparison. By considering $G20$ as the only producer of the network, the basic P2P model would turn into an economic dispatch problem for a pool-based market. Hence, Table 12.2 shows the amount of power losses for each structure and different amount of α_i. It can be seen that the maximum reduction in power losses occur in $\alpha_i = 35\left(10^{-3}\$/\text{kW}\Omega\right)$ which is particular to this case study. However, according to the diversity of generation technologies, the number of peers and also the topology of the network, lower or higher power loss reduction can be achieved in other cases.

12.6.5.3 Congestion Management

One of the advantages of P2P markets is its decentralized approach to solving problems. By using the model in Eq. 12.18, with encountering congestion or natural disasters in the power system, the grid operator has a tool to control transactions on congested or damaged lines. Hence, according to the type of incident and the scale of the community, some level of flexibility can be achieved for the main grid. According to the line capacities, in both previous structures, lines 6 and 9 were congested. However, by using the proposed method and $\beta = 5$ (10^{-3}\$/kW), power transactions on these lines would be reduced which result in eliminating the

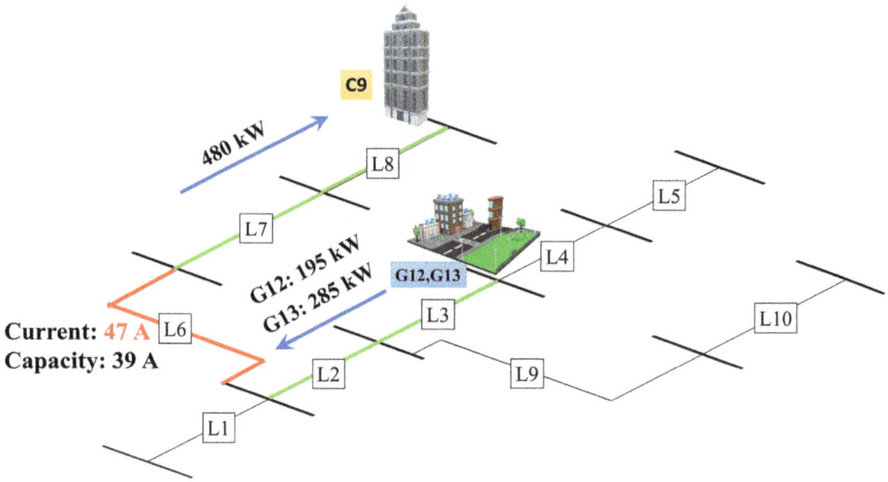

Fig. 12.11 The power flow between $C9$, $G12$, and $G13$ before performing congestion management

congestion. To investigate the reason for these results, Figs. 12.11 and 12.12 show the transactions of the building $C9$ before and after congestion management. It can be seen that in the basic P2P market, $C9$ prefers to procure its energy from $G12$ and $G13$ which involve one of the congested lines ($L6$). The reason is, in the basic structure, $C9$ only seeks to buy energy at the minimum cost and does not consider the grid situation in its transactions. On the other hand, by using the proposed method for congestion management, $C9$ buys all the energy from $G17$ which only involves one line ($L8$). The same analysis can be performed for $C11$ which results in congestion management on $L9$. Hence, by assigning penalties to congested lines and their transactions, the grid operator can achieve congestion management through the operation of the P2P market.

12.6.6 Heat and Electricity Trading

As mentioned before, electricity is not the only tradable product in a community of active buildings. By employing different technologies, such as CHP, buildings are able to trade heat energy with each other. Hence, in this subsection, we consider Eqs. 12.27–12.38 to model a P2P market with the existence of CHP units. To simulate this market, the 11-bus network, shown in Fig. 12.10, is considered as the test case. Despite the earlier subsections, along with electricity, $C1$–$C11$ have also heat demand, which should be procured through the P2P market. On the other hand, $G12$–$G15$ are able to produce electricity and heat simultaneously through CHP units. Furthermore, $G16$–$G20$ are considered as electricity-only producers in

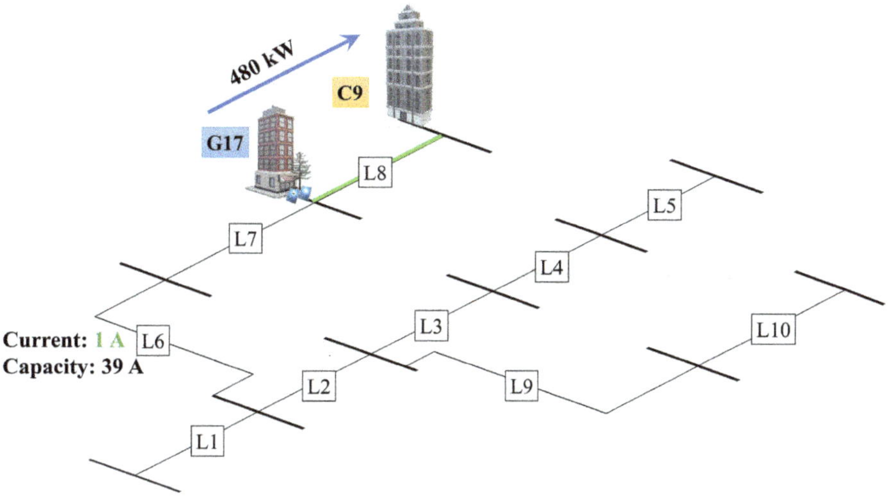

Fig. 12.12 The power flow between $C9$ and $G17$ after performing congestion management

Table 12.3 Electricity transactions among peers with the existence of CHP units

	C1	C2	C3	C4	C5	C6	C7	C8	C9	C10	C11
CHP12	0	0	0	0	0	0	0	0	338	0	0
CHP13	0	124	288	0	0	0	58	0	0	150	0
CHP14	0	386	152	0	0	0	152	0	0	0	0
CHP15	0	0	0	0	0	0	0	0	142	117	261
G16	0	0	0	59	284	408	0	0	0	0	0
G17	345	0	0	258	53	62	0	0	0	0	101
G18	0	0	0	0	0	0	0	540	0	0	260
G19	0	0	0	0	0	0	0	0	0	460	218
G20	175	0	0	53	53	0	0	0	0	152	0

the network. Input data and further information about the case study are available in Input Data (2021).

Finally, Tables 12.3 and 12.4 show the heat and electricity transactions between all peers of the network, respectively. This model can be considered as an example for integrating different technologies in the P2P market. It can also be combined with the models proposed in Sects. 12.6.2 and 12.6.3 to also achieve power losses reduction and congestion management.

Table 12.4 Heat transactions among peers with the existence of CHP units

	C1	C2	C3	C4	C5	C6	C7	C8	C9	C10	C11
CHP12	28	0	0	87	31	0	0	0	31	31	31
CHP13	54	0	170	23	31	39	110	240	31	31	0
CHP14	128	200	0	40	41	82	0	0	41	56	0
CHP15	0	0	0	0	56	69	0	0	46	181	279

12.7 Status Quo, Challenges, and Future Directions

Decarbonization of the energy systems is a priority all around the world and affected most of the parts of the energy systems. In parallel, the advances in renewable energy generation technologies and BSS and promising reduction in their production costs, along with the smart grid solutions, accelerate this evolution. These developments are on both the bulk energy systems (which is mostly managed by the system operators) and the demand-side (which is mainly a result of public awareness about the environmental issues). This evolution on the demand side has led to an increased rate of uptake of PVs, BSS, and EVs at the level of building and houses, which as a result has changed the role of these entities from a passive consumer to an active agent of the energy systems who is able to provide energy services to the network. The recent studies on the peer-to-peer transactions along with the TE provide a proper platform that enables trading of a small amount of energy without any needs to have a third party who guarantees these kinds of transactions. The following directions need to be more investigated to pave the way for the new generation of energy systems and cope with the current challenges:

- Design of an active energy management system (AEMS) that enables a wide range of transactions inside the buildings among the IoT-enabled appliances and residents and also among the buildings and houses in a district area is a promising direction in this field that need to be more investigated.
- Considering the dynamic and changing state of the demand-side energy sources and energy demands, artificial intelligence and machine learning techniques would be an important direction for making a distributed data-driven decision instead of the conventional optimization-based approaches.
- By defining a tradable energy packet (ePKT) instead of current interpretation for energy, the transactions would not be limited to the electricity and are able to be done on any types of energy.
- Another direction is to develop a special cryptocurrency for energy transactions (e.g. we can call them EGY-coin or EGC) or defining a trading mechanism based on the current cryptocurrencies.

References

Abrishambaf, O., Lezama, F., Faria, P., & Vale, Z. (2019). Towards transactive energy systems: An analysis on current trends. *Energy Strategy Reviews, 26*, 100418.

Alvaro-Hermana, R., Fraile-Ardanuy, J., Zufiria, P. J., Knapen, L., & Janssens, D. (2016). Peer to peer energy trading with electric vehicles. *IEEE Intelligent Transportation Systems Magazine, 8*(3), 33–44.

Basnet, A., & Zhong, J. (2020). Integrating gas energy storage system in a peer-to-peer community energy market for enhanced operation. *International Journal of Electrical Power & Energy Systems, 118*, 105789.

Cui, S., & Xiao, J.-W. (2020). Game-based peer-to-peer energy sharing management for a community of energy buildings. *International Journal of Electrical Power & Energy Systems, 123*, 106204.

Daneshvar, M., Pesaran, M., & Mohammadi-ivatloo, B. (2019). 7-transactive energy in future smart homes. In W. Su, & A. Q. Huang (Eds.) *The Energy Internet* (pp. 153–179). Woodhead Publishing.

Giotitsas, C., Pazaitis, A., & Kostakis, V. (2015). A peer-to-peer approach to energy production. *Technology in Society, 42*, 28–38.

GridWise Architecture Council (GWAC). (2021). Transactive energy-gridwise architecture council. [Online]. Available: https://www.gridwiseac.org/about/transactive_energy.aspx.

Input data. (2021). [Online]. Available: http://vahidinasab.com/data/P2P_AB.zip.

Long, C., Wu, J., Zhou, Y., & Jenkins, N. (2018). Peer-to-peer energy sharing through a two-stage aggregated battery control in a community microgrid. *Applied Energy, 226*, 261–276.

Masood, A., Hu, J., Xin, A., Sayed, A. R., & Yang, G. (2020). Transactive energy for aggregated electric vehicles to reduce system peak load considering network constraints. *IEEE Access, 8*, 31,519–31,529.

Mehri Arsoon, M., & Moghaddas-Tafreshi, S. M. (2020). Peer-to-peer energy bartering for the resilience response enhancement of networked microgrids. *Applied Energy, 261*, 114413.

Morstyn, T., Farrell, N., Darby, S., & Mcculloch, M. (2018). Using peer-to-peer energy-trading platforms to incentivize prosumers to form federated power plants. *Nature Energy, 3*, 02.

Nasimifar, M., Vahidinasab, V., & Ghazizadeh, M. S. (2019). A peer-to-peer electricity marketplace for simultaneous congestion management and power loss reduction. In *2019 Smart Grid Conference (SGC)* (pp. 1–6).

Nezamabadi, H., & Vahidinasab, V. (2019a). Micro-grids bidding strategy in a transactive energy market (invited paper). *Scientia Iranica, 26*, 3622–3634. Special Issue on machine learning, data analytics, and advanced optimization techniques in modern power systems [Transactions on Computer Science & Engineering and Electrical Engineering(D)], 3622–3634.

Nezamabadi, H., & Vahidinasab, V. (2019b). Market bidding strategy of the microgrids considering demand response and energy storage potential flexibilities. *IET Generation, Transmission & Distribution, 13*(8), 1346–1357.

Nezamabadi, H., & Vahidinasab, V. (2020). Arbitrage strategy of renewable-based microgrids via peer-to-peer energy trading. *IEEE Transactions on Sustainable Energy, 12*(2), 1372–1382.

Nguyen, D. H., & Ishihara, T. (2021). Distributed peer-to-peer energy trading for residential fuel cell combined heat and power systems. *International Journal of Electrical Power & Energy Systems, 125*, 106533.

Ortmann, L., Prostejovsky, A., Heussen, K., & Bolognani, S. (2020). Fully distributed peer-to-peer optimal voltage control with minimal model requirements. *Electric Power Systems Research, 189*, 106717.

Owyang, J. (2013). The collaborative economy: Products, services and market relationships have changed as sharing startups impact business models. To avoid disruption, companies must adopt the collaborative economy value chain. Altimeter. Retrieved august 8, 2014.

Rahimi, F., Ipakchi, A., & Fletcher, F. (2016a). The changing electrical landscape: End-to-end power system operation under the transactive energy paradigm. *IEEE Power and Energy Magazine, 14*(3), 52–62.

Rahimi, F., Ipakchi, A., & Fletcher, F. (2016b). The changing electrical landscape: End-to-end power system operation under the transactive energy paradigm. *IEEE Power and Energy Magazine, 14*(3), 52–62.

Sharifi, R., Fathi, S. H., & Vahidinasab, V. (2016). Customer baseline load models for residential sector in a smart-grid environment. *Energy Reports, 2*, 74–81.

Sharifi, R. (2017). Economic demand response model in liberalised electricity markets with respect to flexibility of consumers. *IET Generation, Transmission & Distribution, 11*, 4291–4298.

Sharifi, R., Fathi, S. H., & Vahidinasab, V. (2017a). A review on demand-side tools in electricity market. *Renewable and Sustainable Energy Reviews, 72*, 565–572.

Sharifi, R., Anvari-Moghaddam, A., Fathi, S., Guerrero, J., & Vahidinasab, V. (2017b). Dynamic pricing: An efficient solution for true demand response enabling. *Journal of Renewable and Sustainable Energy, 9*, 10.

Sharifi, R., Fathi, S. H., Anvari-Moghaddam, A., Guerrero, J. M., & Vahidinasab, V. (2018). An economic customer-oriented demand response model in electricity markets. In *2018 IEEE International Conference on Industrial Technology (ICIT)* (pp. 1149–1153).

Sharifi, R., Anvari-Moghaddam, A., Fathi, S. H., & Vahidinasab, V. (2019a). A flexible responsive load economic model for industrial demands. *Processes, 7*, 1–13.

Sharifi, R., Anvari-Moghaddam, A., Fathi, S. H., Guerrero, J., & Vahidinasab, V. (2019b). An optimal market-oriented demand response model for price-responsive residential consumers. *Energy Efficiency, 12*, 03.

Sharifi, R., Anvari-Moghaddam, A., Hamid Fathi, S., & Vahidinasab, V. (2020). A bi-level model for strategic bidding of a price-maker retailer with flexible demands in day-ahead electricity market. *International Journal of Electrical Power & Energy Systems, 121*, 106065.

Sorin, E., Bobo, L., & Pinson, P. (2019). Consensus-based approach to peer-to-peer electricity markets with product differentiation. *IEEE Transactions on Power Systems, 34*(2), 994–1004.

Sousa, T., Soares, T., Pinson, P., Moret, F., Baroche, T., & Sorin, E. (2019). Peer-to-peer and community-based markets: A comprehensive review. *Renewable and Sustainable Energy Reviews, 104*, 367–378.

Zhou, Y., Wu, J., Song, G., & Long, C. (2020). Framework design and optimal bidding strategy for ancillary service provision from a peer-to-peer energy trading community. *Applied Energy, 278*, 115671.

Chapter 13
Non-intrusive Load Monitoring and Its Application in Energy Flexibility Potential Extraction of Active Buildings

Elnaz Azizi, Mohammad T. H. Beheshti, and Sadegh Bolouki

13.1 Introduction

The ever-increasing energy consumption in the residential sector as shown in Fig. 13.1 is of great importance in terms of carbon emissions, global warming, and sustainability issues. Due to the fact that residential demand accounts for 30% of the total demand (Li & Dick, 2018), increasing the role of residential buildings in the power grid from passive consumers to active ones (Clarke, 2021) plays a crucial role in addressing the aforementioned issues. Numerous studies in the area of building energy confirm that reducing energy consumption in buildings requires effective and efficient residential energy management (REM) (Faustine et al., 2020). Therefore, the concepts of residential demand-side management (RDSM) and home energy management (HEM) appear and take an important place in the research and development efforts of science and industry.

RDMS aims to shape customers' consumption patterns to enable more efficient use of energy system resources, improve grid reliability, and reduce emissions by peak-shaving and valley-filling, which results in increasing the second-by-second balance between demand and generation. In this regard, one of the main goals of the active buildings is proposing a cost-efficient way of minimizing energy demand of buildings (Wang et al., 2017). The implementation and success of such programs depend on advanced knowledge of characteristics of loads and the usage pattern of consumers. Therefore, modelling and predicting these factors could be an important support for RDSM and HEM (Zoha et al., 2012).

In this trend, researchers and industry have worked over the last decades to obtain information on household consumption. This information is beneficial for both the

E. Azizi (✉) · M. T. H. Beheshti · S. Bolouki
Tarbiat Modares University, Tehran, Iran
e-mail: e.azizi@modares.ac.ir; mbehesht@modares.ac.ir; bolouki@modares.ac.ir

V. Vahidinasab, B. Mohammadi-Ivatloo (eds.), *Active Building Energy Systems*, Green Energy and Technology, https://doi.org/10.1007/978-3-030-79742-3_13

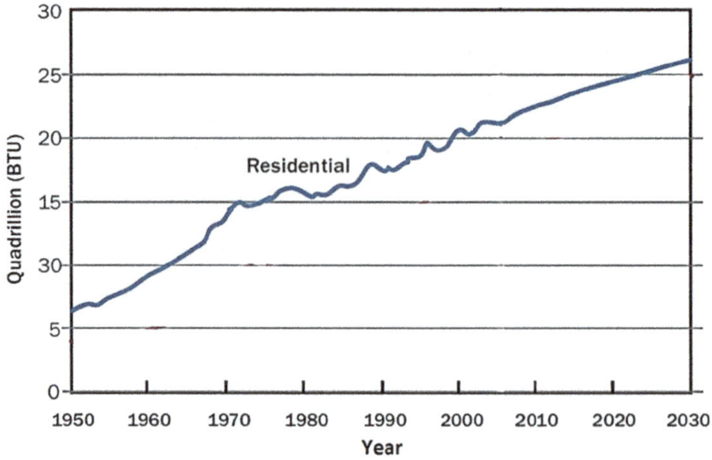

Fig. 13.1 Total primary energy consumption for building in the USA (Hafemeister, 2009)

supplier and the consumer (Devlin & Hayes, 2019). The utility company can use this information to propose different energy management strategies or for control purposes. Consumers can use this information to identify appliances with high electricity consumption and faulty or inefficient appliances. They can also control their consumption to reduce their costs while ensuring their comfort based on this information. Overall, this information can reduce consumption by more than 9% (Aydin et al., 2018; Darby et al., 2006).

Monitoring the consumption of each appliance which is entitled as load monitoring (LM) is one of the useful methods to extract this information (Zhou et al., 2020). Figure 13.2 illustrates the benefits of LM algorithms for consumers and the utility. As it is mentioned in Fig. 13.2, LM provides an opportunity to the utility to extract and characterize the energy flexibility potential of each consumer. Furthermore, studies show that the active building concept is associated more with energy flexibility than self-generation of electricity (Bulut et al., 2016). Therefore, extracting the flexibility via LM plays a crucial role in the efficiency of active buildings.

In the first phase of this chapter, fundamental concepts of load monitoring will be discussed. Then, its application in flexibility extraction of residential buildings will be shown. The schematic of this technique is illustrated in Fig. 13.3.

13.2 Different Types of Load Monitoring

Load monitoring algorithms based on the number of meters and appliances can be categorized into three main groups which will be discussed below.

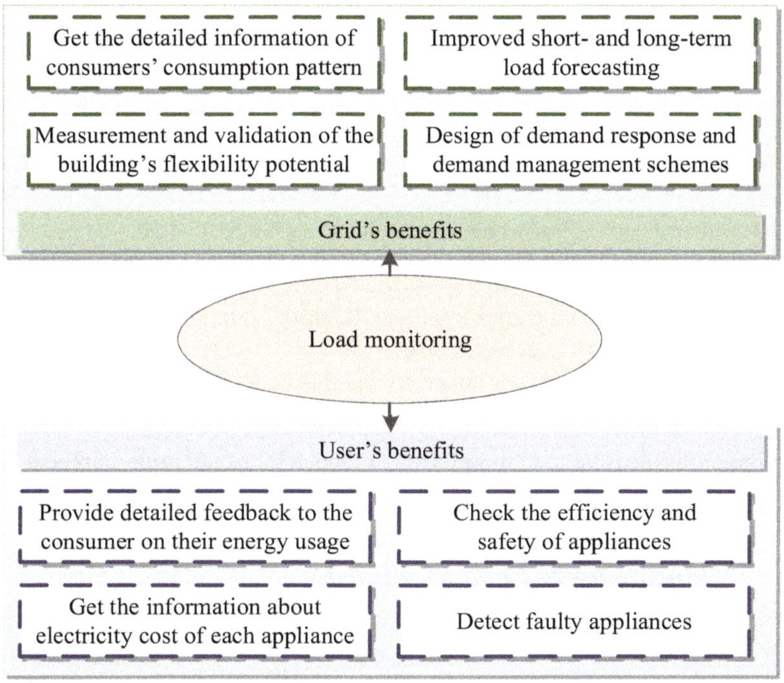

Fig. 13.2 LM benefits to consumers and grid

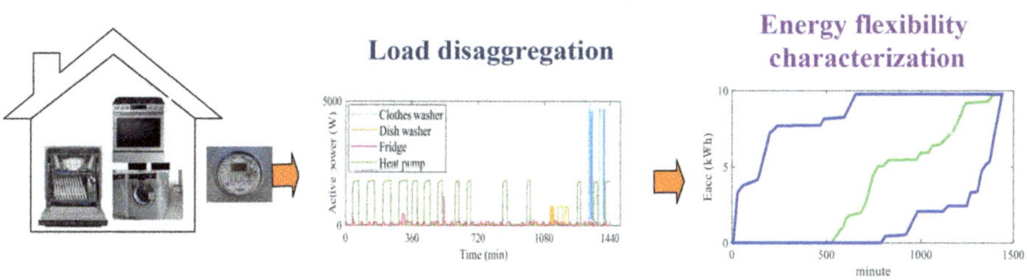

Fig. 13.3 The framework of this study

13.2.1 Intrusive Load Monitoring

The simplest and most accurate approach to obtain the consumption pattern of appliances is to install a meter on each of them to measure, record, and report their consumption which is called intrusive load monitoring (ILM) (Azizi et al., 2020). Figure 13.4c shows the schematic of ILM. However, this approach causes the following concerns:

1. Installing a meter on each appliance is time-consuming (Liu et al., 2019).

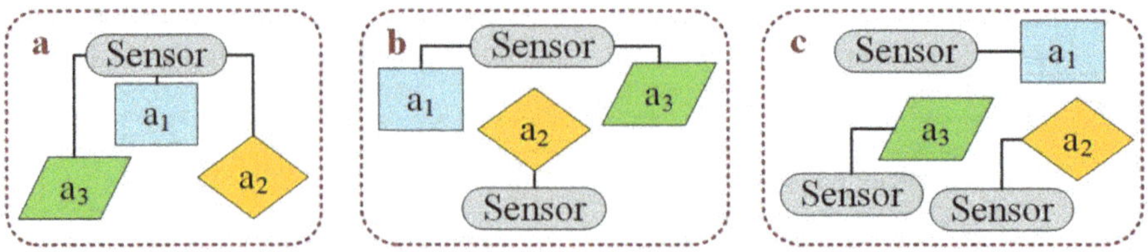

Fig. 13.4 Different types of load monitoring, (**a**) NILM, (**b**) SILM, (**c**) ILM

2. It incurs high costs to the consumer and the utility due to the need for additional meters, cables, etc. (Liu et al., 2019).
3. This approach causes privacy concerns and most consumers are unwilling to take the risk of this approach (Erkin et al., 2013).

Therefore, non-intrusive load monitoring attracts a lot of attention in recent years.

13.2.2 Non-intrusive Load Monitoring

Rather than installing a meter on each appliance individually and measuring its consumption intrusively, Hart proposed non-intrusive load monitoring (NILM) in 1992 (Hart, 1992). The NILM technique aims to obtain the consumption pattern of each appliance by analysing and disaggregating a given aggregate signal measured by existing meters using purely analytical approaches. The schematic of this technique is illustrated in Fig. 13.4a. However, its accuracy is lower than ILM methods. The concept of semi-intrusive load monitoring (SILM) emerged in this field which is the trade-off between cost and accuracy.

13.2.3 Semi-intrusive Load Monitoring

In the semi-intrusive load monitoring (SILM) technique, the set of appliances are divided into some subgroups, and a meter is installed in each subgroup to measure and record the aggregated consumption signal of appliances. Then, the disaggregation approach is applied to each subgroup, separately (Dash et al., 2019). The schematic of this approach is illustrated in Fig. 13.4b. In comparison with ILM methods, SILM reduces the cost and in comparison with NILM increases the accuracy of load disaggregation. The main challenge of these approaches is obtaining a proper number of subgroups based on the consumer's willingness to pay the cost of meters, the infrastructure of the building, and number of appliances.

Due to the fact that NILM does not interfere with the existing building infrastructure, omits the requirement of costly sub-metering, and saves time and protects consumer's privacy, it is more practical and is preferred from different viewpoints

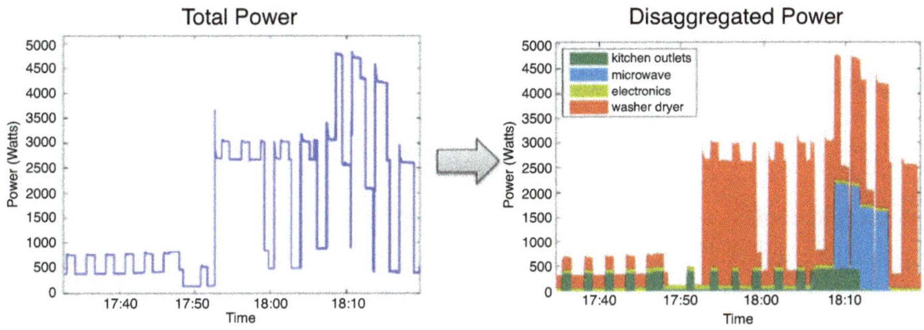

Fig. 13.5 Appliances' consumption extraction based on NILM (Esa et al., 2016)

(Zhang et al., 2019). Therefore, in this chapter, the basic concepts of this technique will be discussed.

13.3 NILM Problem Statement

Non-intrusive load monitoring is a well-known problem that involves disaggregating the total electrical energy consumption of a building into its individual electrical load components without the need for extensive metering equipment at each appliance. This problem can be classified as a blind identification problem, Abed-Meraim et al. (1997) where given the observed output of the entire system (i.e. the household total power consumption), the unobserved sub-states (i.e. the electricity consumption of individual appliances) are to be estimated. The schematic of NILM is shown in Fig. 13.5. Mathematically, NILM can be formulated as an optimization problem and a learning-based problem, which are described below.

13.3.1 Optimization-Based Formulation

NILM algorithms aim to extract the consumption signal of each appliance from the aggregated consumption signal of each consumer, herein denoted by $P(t)$. Considering $\{1, 2, \ldots, N\}$ as the dataset of appliances, and N the total number of appliances, $P(t)$ is sum of the consumption signal of each appliance, $P_i(t)$ as (13.1).

$$P(t) = \sum_{i=1}^{N} P_i(t). \tag{13.1}$$

However, each appliance i, $1 \leq i \leq N$, has m_i operating modes. Due to the presence of noise, fluctuations, overshoots, and spikes, the power value in these operating modes is not fixed in practice. However, by ignoring them and considering fixed values for mode j of appliance i, denoted as x_{ij}, the estimated piecewise constant consumption power of appliance i, $\tilde{P}_i(t)$ is obtained as (Piga et al., 2016)

$$P_i(t) \simeq \tilde{P}_i(t) = \Theta_i^{\mathrm{T}}(t) X_i, \tag{13.2}$$

where element j of the vector X_i shows the fixed consumption of mode j of appliance i. $\Theta_i = \left[\theta_{i1} \ldots \theta_{im_i}\right]^T$ is a binary-valued vector indicating the operating mode of appliance i at time t, defined as

$$\theta_{ij}(t) = \begin{cases} 1 & \text{if } i \text{ operates in mode } j \text{ at time } t \\ 0 & \text{otherwise.} \end{cases} \tag{13.3}$$

In other words, $\Theta_i(t)$ is a binary-valued vector indicating the mode of appliance i at time t that satisfies

$$\sum_{j=1}^{m_i} \theta_{ij}(t) = 1. \tag{13.4}$$

We note that the condition (13.4) guarantees that an appliance cannot operate in two modes simultaneously.

NILM method aims to obtain the consumption signal of each appliance, $\tilde{P}_i(t)$, by determining $\Theta_i(t)$ via minimizing the following error

$$e(t) = P(t) - \sum_{i=1}^{N} \tilde{P}_i(t) = P(t) - \sum_{i=1}^{N} \Theta_i^{\mathrm{T}}(t) X_i, \quad 0 \le t \le T \tag{13.5}$$

13.3.2 Learning-Based Problem Statement

In learning-based NILM problem definition, a training dataset of appliances' consumption exists. The training dataset consists of the consumption or the ON/OFF status of all appliances for a specific period of time. Based on this training dataset, the operation mode of each appliance in each event of the aggregated signal exists. Mathematically, let $A = \{1, \ldots, i, \ldots, N\}$ be the set of appliances, in which N stands for the total number of appliances. TD the training dataset consist of labelled events

$$TD = \{(e_i, l_i) : e_i \in R, l_i \in A, i = 1, \ldots, N_e\} \tag{13.6}$$

where e_i shows the value of the ith event and l_i denotes its respective label (or appliance). N_e stands for the total number of events in the aggregated signal.

Learning-based algorithms extract a model based on the events and their assigned labels. Then, based on the extracted model (learned model), a proper label (or appliance) is assigned to each event of the aggregated test signal.

13.4 Review on NILM Algorithms

The existing optimization-based and learning-based approaches to NILM will be reviewed in this section.

13.4.1 Optimization-Based NILM Algorithms Review

The NILM problem is formulated as a convex quadratic programming in Lin et al. (2016). In this paper, the whole training dataset is used to train the model which increases the computation complexity and time. Furthermore, the proposed method is highly dependent to the training dataset. Therefore, it is not scalable and has low accuracy in disaggregating the consumption of other consumers. Authors in Singh and Majumdar (2017) proposed sparse optimization NILM problem which is solved by dictionary learning-based algorithm. Six different metaheuristic optimization techniques were evaluated in Egarter and Elmenreich (2015) by considering different datasets. NILM is formulated as a stochastic-based optimization problem in Shahroz et al. (2020), and the probability of usage of appliances is defined as the regularization term.

Authors in Suzuki et al. (2008) defined NILM problem as an integer programming (IP) to detect the simultaneously ON appliances and increase the disaggregation accuracy in disaggregating the high number of appliances with multi-operation modes. NILM problem is formulated as a mixed-integer programming (MIP) in Piga et al. (2016), and the results are compared with a learning-based one. Authors in Bhotto et al. (2017) modified the formulation of IP by defining novel boundaries for the appliances. To reduce the computation burden in MIP-based NILM, a window-based algorithm is utilized in Wittmann et al. (2018). To increase the disaggregation accuracy, ILP and clustering algorithm are merged in Ayub et al. (2018). Authors in Azizi et al. (2020) defined constraints based on the transitions between different modes multi-mode appliances to decrease the computation time of the MIP-based NILM method.

All the aforementioned optimization-based NILM algorithms require pre-information about appliances, such as the total number of appliances, the consumption of each operation mode of appliances, their consumption pattern, etc. This information is obtained via involving and asking consumers, the manual datasheet of appliances, or collecting a training dataset which causes new challenges in this field.

13.4.2 Learning-Based NILM Review

Based on the presence of the training dataset, learning-based NILM can be divided into two main groups, supervised NILM algorithms and unsupervised NILM algorithms (Zhao et al., 2020). A brief review of these methods is discussed below.

(1) Supervised NILM Methods
Supervised NILM algorithms aim to extract a proper model from the training dataset which maps each sample or event of the aggregated signal to a specific appliance. Then, based on the extracted model, the test aggregated signal is disaggregated. One of the well-known supervised NILM algorithms is the classification technique which attracts a lot of attention in recent years. NILM is formulated as KNN problem in Schirmer and Mporas (2019) which reduces the burden of computation. Authors in Tabatabaei et al. (2017) and Massidda et al. (2020) proposed multi-label classification-based NILM which results in high accuracy in disaggregating the multi-mode appliances. Deep-learning-based approach is utilized in Singhal et al. (2019) which requires a high volume of a training dataset. To disaggregate the consumption of multi-mode appliances, authors in Shahroz et al. (2020) utilized a novel deep convolutional neural networks-based algorithm.

All supervised NILM methods proved to be accurate in disaggregating high number of appliances and appliances with overlapping power values (Singhal et al., 2019). However, the main challenge in this field is the requirement for a high volume of training dataset that is not in general feasible to collect (Yang et al., 2019). In the last few years, researchers focused on extracting more features and signatures of appliances from small training datasets to address this challenge (Dash et al., 2020).

(2) Unsupervised NILM Methods
In contrast with supervised methods, unsupervised NILM algorithms do not require a training dataset and prior information about appliances. One of the common methods of unsupervised methods is clustering which is used in this field frequently. Authors in Henao et al. (2015) utilized the subtractive clustering method to the events of the aggregated signal to extract a proper number of appliances as shown in Fig. 13.6. In Kong et al. (2016), *k*-means method is applied on the events of the aggregated signal. The principal component analysis method as an unsupervised method is utilized in Moradzadeh et al. (2020) to reduce the dimension of data and extract the profile of each appliance.

Despite some success in the case where all appliances are type I, these algorithms have been ineffective in dealing with multi-mode appliances (Kong et al., 2016; Dinesh et al., 2019). Moreover, these methods, extract different groups of appliances without assigning any appliances to them.

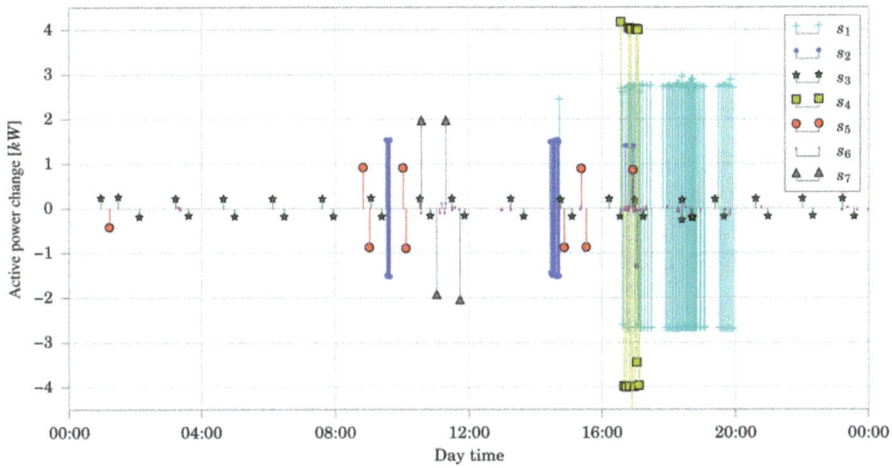

Fig. 13.6 Grouping events based on substractive clustering (Henao et al., 2015)

13.5 Types of Appliances

The accuracy of disaggregation of the consumption of an appliance is highly dependent on its category. Generally speaking, appliances are divided into four main groups based on their number of operation modes and their consumption pattern as discussed below (Zoha et al., 2012):

1. Type I: This type of appliances has two operation modes (ON/OFF modes). Examples of these appliances are table lamp, toaster, etc.
2. Type II: Appliances such as washing machine, dishwasher, stove burner, etc., which have multi-operation modes are included in the type II category. This type of appliances is entitled as a finite state machines (FSM). The switching pattern of these appliances makes them more distinguishable and increases the disaggregation accuracy.
3. Type III: Appliances such as the power drill and dimmer lights are included in type III category that have not fixed number of operation modes which are entitled as continuously variable appliances. Disaggregating the consumption of these appliances is one of the main challenges in this field.
4. Type IV: This type of appliances consumes power 24 h of a day. Examples of type IV appliances are hard-wired smoke alarm and external power supplies.

Figure 13.7 illustrates the power pattern of the first three types of appliances. Among different types of appliances, Type III ones are rarely used by consumers and type IV appliances consume low power. Therefore, disaggregating the consumption of type I and II appliances gains a great deal of attention among researchers.

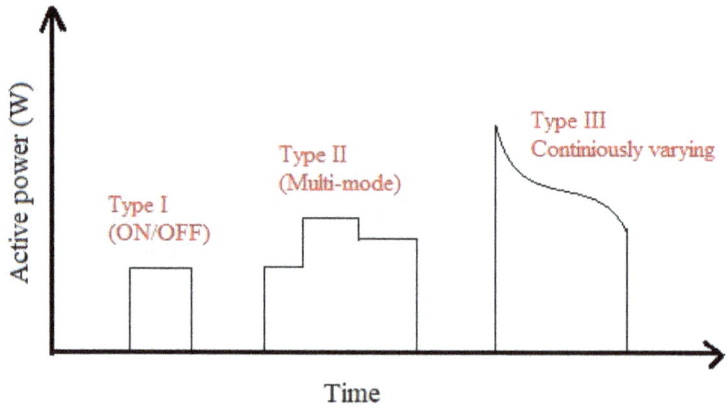

Fig. 13.7 Different types of appliances

13.6 Features/Signatures of Appliances

To disaggregate the aggregated signal, various specific features or signatures of appliances are used. These features or signatures are unique to the appliances and make them more distinguishable from each other. The different types of these features are described below.

13.6.1 Steady-State Features

Steady-state features of appliances refer to longer-lasting changes in power signal when at least one appliance switches its operating mode. The most well-known steady-state features include active and reactive power, current, and current and voltage waveforms (Azizi et al., 2021).

13.6.2 Transient Features

Transient features are short-term fluctuations in power signal before settling to a steady-state condition which include shape, size, duration, and harmonics of the transient. The transient signature of the majority of appliances is pronounced, which makes it suitable for load identification and increases the disaggregation accuracy. However, these features require a high-frequency sampling dataset and specific hardware to collect these data.

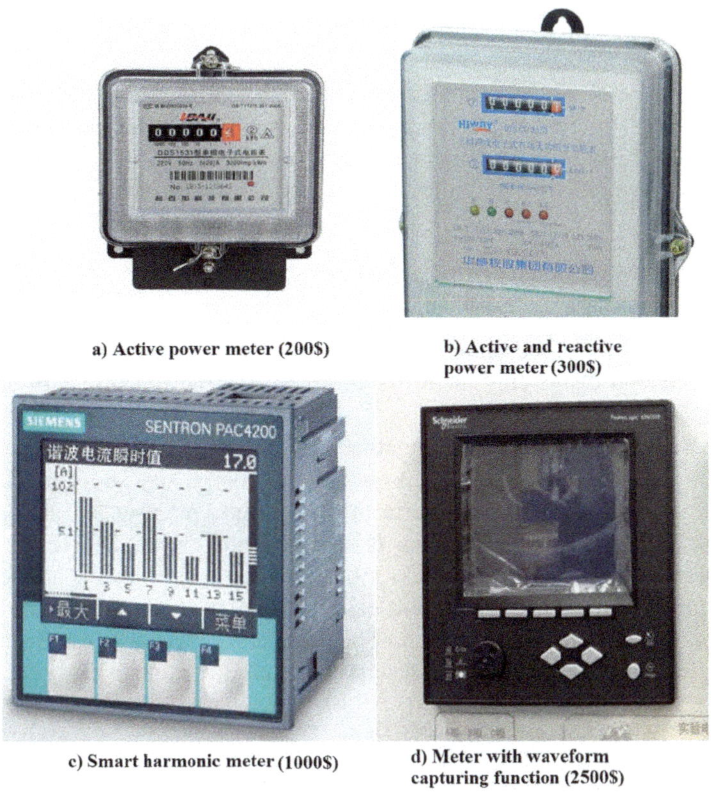

Fig. 13.8 Different types of smart meters vs their cost (Xu et al., 2018)

13.6.3 *Non-traditional Features*

In addition to the aforementioned signatures of appliances, in recent years non-traditional signatures such as temperature, light, consumers' special usage behaviour, time of day, etc. are increasingly used in disaggregation methods to improve the disaggregation accuracy.

Figure 13.8 shows different kinds of smart meters. Since smart meters of active power have the lowest price, the majority of studies focused on steady-state active power-based NILM algorithms.

13.7 Classification of NILM Methods

Active power-based NILM studies can be broadly divided into two main classes, sample-based algorithms and event-based algorithms which are described below.

13.7.1 Sample-Based Approaches

In sample-based NILM algorithms, by assuming a finite-state machine for each appliance, the total consumption signal is disaggregated based on the pre-learned model of state transitions of appliances (Zhao et al., 2018).

13.7.2 Event-Based Approaches

Event-based methods focused on detecting significant variations in the power signal called events and classifying them based on the specific transitions of appliances (Lu & Li, 2019).

Compared to the sample-based methods, event-based methods are more direct and comprehensive and have low computational complexity and time. Therefore, they have gained a lot of attention in the last years (Azizi et al., 2020). One of the main challenges in these methods is detecting real events. Due to the presence of various noises as shown in Fig. 13.9, event detection plays a crucial role in the disaggregation accuracy of event-based NILM methods (Zhao et al., 2018). Therefore, not detecting any real event or mistakenly considering a fluctuation as an event decreases the accuracy of disaggregation in these methods.

13.8 Evaluation Metrics

Two main groups of evaluation metrics exist in the NILM problem: (1) event-detection evaluation metric and (2) load disaggregation accuracy.

13.8.1 Event-Detection Accuracy Metric

Three famous metrics for measuring the accuracy of event detection can be listed as:

1. False-positive rate (FPR): Ratio of FP to the actual negatives

$$FPR = \frac{FP}{FP + TN}, \tag{13.7}$$

   where FP and TN show the false-positive events (detected non-event as event) and true-negative events (detected non-event as non-event), respectively.
2. F_β: Trade-off between precision and recall

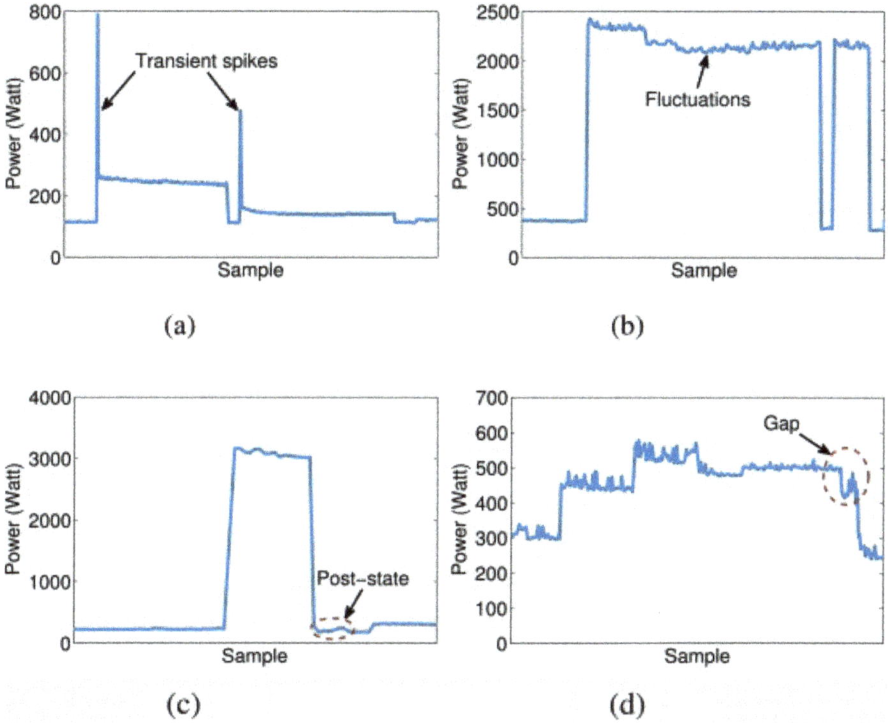

Fig. 13.9 Different types of noises in the measured power signal (Zhao et al., 2018)

$$F_{\beta} = \frac{(1 + \beta^2) \times TRP \times RC}{(\beta^2 \times TRP) + RC}, \tag{13.8}$$

$$TPR = \frac{TP}{TP + FN}, \ RC = \frac{FP}{FP + TN}, \tag{13.9}$$

where TRP and RC show the precision and recall; TP is true positive, FP is false positive, TN is true negative, and FN is false negative. The weighting factors is shown with β. F_1 score is specific type of F_{β} in which $\beta = 1$.

3. Overall accuracy:

$$OA = \frac{N_{dis}}{N_{true}} \tag{13.10}$$

where N_{dis} shows number of events and N_{true} stands for the non-events that are detected correctly.

13.8.2 Energy Disaggregation Accuracy

The common energy disaggregation accuracy metrics based on the difference between the estimated consumption of each appliance $\hat{E}_i$ and its actual consumption E_i, for N appliances can be listed as:

1. Relative error (RE): The ratio of the error of disaggregation to the actual power consumption which is formulated as

$$RE = \frac{\sum_{i=1}^{N} E_i - \sum_{i=1}^{N} \hat{E}_i}{\sum_{i=1}^{N} E_i} \tag{13.11}$$

2. Root mean square error (RMSE): This parameter measures the standard deviation of the disaggregation error based on

$$RMSE = 1 - \sqrt{\frac{1}{N}\sum_{i=1}^{N}(E_i - \hat{E}_i)} \tag{13.12}$$

3. Mean absolute error (MAE): MAE measures the average estimated error based on

$$MAE = \frac{1}{N}\sum_{i=1}^{N}|E_i - \hat{E}_i| \tag{13.13}$$

4. Mean absolute percentage error (MAPE): MAPE defines the accuracy as the following ratio

$$MAPE = \frac{1}{N}\sum_{i=1}^{N}\left|\frac{E_i - \hat{E}_i}{E_i}\right| \tag{13.14}$$

5. R^2: This accuracy metric is defined based on the variance of the data and the residual sum of squares as follows

$$R^2 = 1 - \frac{\sum_i^N (E_i - \tilde{E}_i)^2}{\sum_i^N (E_i - \hat{E}_i)^2} \tag{13.15}$$

where $\tilde{E}_i$ is the average of E_i.

13.9 Sampling Frequency

Depending on the sampling frequency of the power signals, NILM approaches are divided into two main groups:

1. High-frequency sampling data: To obtain higher-resolution data and monitor small changes ($<50\,\text{W}$), the consumption is sampled with high frequency ($>1\,\text{Hz}$). This sampling rate requires specific hardware and contains a high volume of dataset. However, different signatures (steady and transient signatures) of appliances can be easily extracted from this dataset.
2. Low-frequency sampling data: In this case, the data is sampled with low frequency ($<1\,\text{Hz}$) with existing meters. Therefore, this type of sampling omits the cost of additional hardware. Steady signatures must be extracted from these datasets.

Due to the low cost and volume of low-frequency sampling data, in recent years researchers have been focused on these dataset in NILM problem.

13.10 NILM Datasets

With the importance of NILM came the need for standardized datasets to benchmark the wide variety of published algorithms. The most popular datasets in this field and their characteristics are listed in Table 13.1.

13.11 Companies of NILM Products

Over the last decade, different companies have developed different NILM products that analyse the total building electricity consumption and extract the consumption of individual appliances based on complex learning or optimization approaches. Based on these results, some of them suggest different energy management and cost-saving strategies to consumers. Table 13.2 shows the list of known companies in this field.

13.12 Application of NILM in Energy Flexibility Potential Extraction

Shiftable appliances, electrical vehicles, and energy storage devices in the residential sector provide flexibility in the power grid. Generally speaking, the energy flexibility potential of each consumer is computed based on the earliest

Table 13.1 Different benchmarking datasets for the NILM problem

Dataset	Sampling rate	Duration	Houses	Ground truth	Country
REDD (Kolter & Johnson, 2011)	1 Hz	Several months	6	Submeter channels	US
UK-DALE (Kelly & Knottenbelt, 2015)	16,000 Hz/1 Hz	2 years	6	Submeter channels	UK
AMPDs (Makonin et al., 2013)	1 min	2 years	1	Submeter channels	Canada
ECO (Kleiminger et al., 2015)	1 s	8 months	6	Submeter channels	Switzerland

Table 13.2 Different NILM companies

Company	Country	Founded year	Description
Bidgely	USA	2010	Load disaggregation and recommendations to save energy
Powerly	USA	2015	Load disaggregation, appliance health monitoring and smart device management
You know what	Belgium	2013	Load disaggregation and anomaly detection
Chai energy	USA	2012	Load disaggregation and identify saving energy oppurtiunities
Watty	Sweden	2013	Load disaggregation

start time and latest finished time for shiftable/flexible appliance (D'hulst et al., 2015). Mathematically, assuming N_{app} shiftable/flexible appliances, the total energy consumption is obtained based on:

$$E_{tip}(t) = \sum_{i=1}^{N_{app}} \int_{t_0}^{t} P^i(n)dn \tag{13.16}$$

where,

$$P^i \triangleq [P^i(1), \ldots, P^i(T)], \;\; i \in [1, \ldots, N_{app}] \tag{13.17}$$

in which P^i shows the power of each flexible appliance i.

To extract the energy flexibility potential two parameters are defined as

$$\begin{aligned} P^i_{max} &\triangleq [P^i(1), \ldots, P^i(T)], P^i(t_\alpha) \geq 0; \\ & P^i(t_\beta) = 0; \;\; \forall t_\alpha < t_\beta \end{aligned} \tag{13.18}$$

$$\begin{aligned} P^i_{min} &\triangleq [P^i(1), \ldots, P^i(T)], P^i(t_\alpha) = 0; \\ & P^i(t_\beta) \geq 0; \;\; \forall t_\alpha < t_\beta \end{aligned} \tag{13.19}$$

where P_{max} and P_{min} show the power signal when appliance i starts at earliest time and finished at the latest time, respectively.

The energy patterns considering earliest start and latest finished time for appliances is computed based on

$$E_{max}(t) = \sum_{i=1}^{N} \int_{t_0}^{t} P^i_{max}(n)dn \tag{13.20}$$

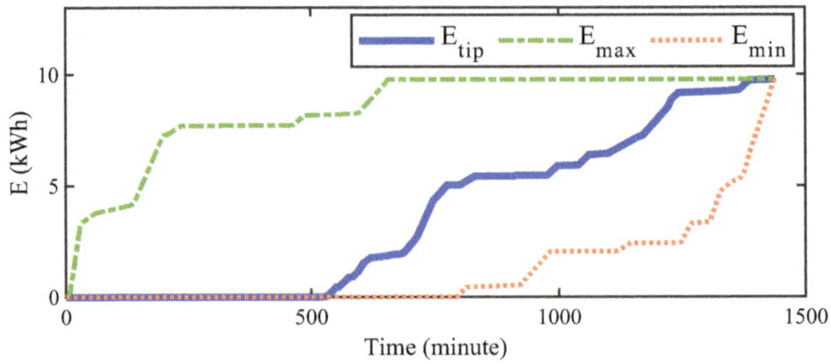

Fig. 13.10 E_{max} and E_{min} vs typical energy pattern

$$E_{min}(t) = \sum_{i=1}^{N} \int_{t_0}^{t} P_{min}^{i}(n)dn \tag{13.21}$$

It should be noted that E_{max} and E_{min} show the boundaries of the energy flexibility of the considered appliances' set as shown in Fig. 13.10.

To compute the energy flexibility of consumers the power signal of shiftable/flexible appliances are required. To obtain them, installing a meter on each appliance (ILM technique) is time-consuming and expensive. Therefore, NILM method must be applied to the aggregated signal of each consumer.

13.12.1 Numerical Results on AMPds

To evaluate the efficiency of the NILM method in the energy flexibility potential extraction the AMPds which consists of minutely measured data is considered (Makonin et al., 2013). Four appliances were considered in this case which include a heat pump, a clothes dryer, a dishwasher, as flexible appliances, and a refrigerator as the most frequent ON/OFF appliance. Figure 13.11 shows the power signal of these appliances and their aggregated one in a typical day.

The event-based optimization NILM method of Azizi et al. (2020) is applied to this dataset which consists of three main steps:

1. **Pre-processing:** In this step, three-point method is applied to the consumption signal of each appliance and the aggregated signal to omit the outliers such as spikes and overshoots, filter the signal and extract events. Then, k-means clustering is applied to events of each appliance to extract the number of their operation modes and their consumption in each mode.
2. **Optimization-based NILM:** Based on the results of the previous step, MINLP is utilized to solve NILM problem in this stage.

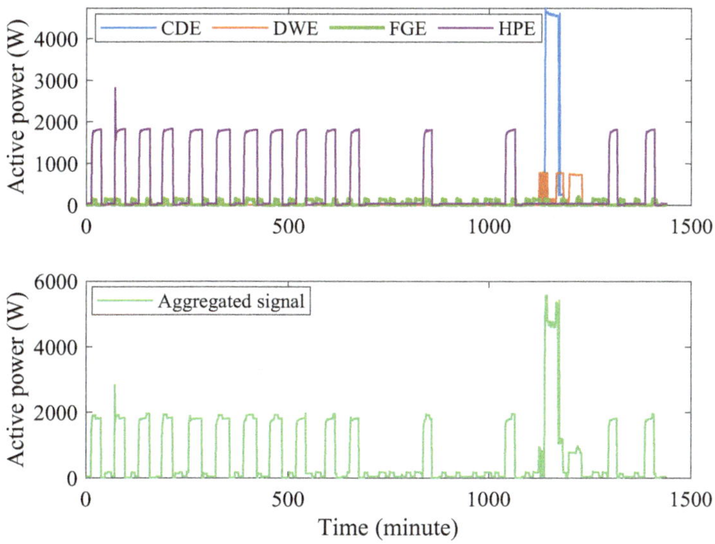

Fig. 13.11 Total power consumption of appliances

Table 13.3 Evaluation metrics of the NILM method

	Apps			
	HP	DW	CW	FRG
RE	0.02	0.12	0.03	0.05
R^2	0.96	0.88	0.97	0.9

3. **Post-processing:** Finally, considering non-fixed consumption values for each operation mode of appliances, the consumption signal of appliances are reconstructed.

Table 13.3 shows the accuracy of the estimated power signals based on (13.11). Based on the extracted power signals of shiftable/flexible appliances (HP, DW, and CW), the boundaries of the energy flexibility are computed which is illustrated in Fig. 13.12. The error between the estimated boundaries and the real ones is less than 5% which shows the effectiveness of NILM methods in energy flexibility extraction of consumers.

13.13 Status Quo, Challenges, and Outlook

In recent years, large amounts of datasets on power profiles have become available, leading to various methods being proposed for the NILM problem. However, there are still some challenges (such as (1) the need for a training dataset for supervised methods, (2) achieving high accuracy in disaggregating the multi-mode appliances, (3) proposing scalable approaches, (4) limiting human contribution

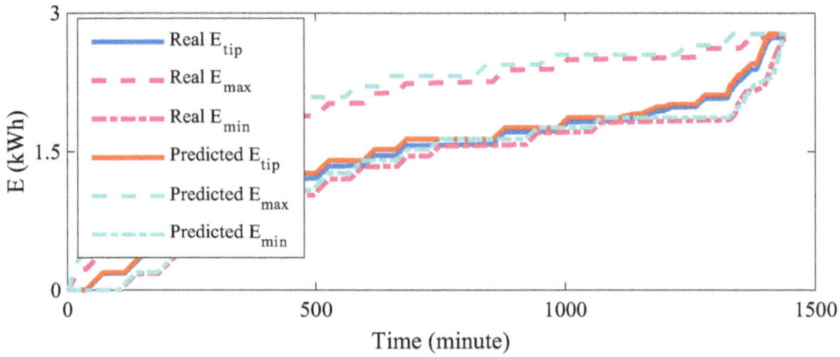

Fig. 13.12 Boundaries of energy flexibility based on (13.20) and (13.21)

during configuration, (5) minimizing hardware complexity and cost, etc.) that have not been addressed yet. Therefore, a practical NILM algorithm should have the following guidelines:

1. The proposed methods should be applicable on low-frequency sampled data (1 Hz or less) which are reported with existing smart meters. The proposed NILM approaches on the high-frequency sample data require specific meters which increases the cost.
2. The accuracy of disaggregation should be more than 80% for each appliance. However, the presence of multi-mode appliances increases the complexity of algorithms and decreases accuracy. Developing a NILM algorithm that could work well in disaggregating all types of appliances regardless of their number of modes is still an unsolved challenge in this field.
3. The majority of proposed NILM algorithms require pre-information about appliances or training dataset. However, to address the privacy concerns of consumers, no training dataset should be required which is still one of the main challenges in this field.
4. The proposed NILM techniques may have high accuracy for specific consumers, while they may have low accuracy in disaggregating the aggregated signal of other consumers. However, the algorithm must be scalable to apply to different residential buildings and consumers.

13.14 Conclusion

Using energy resources responsibly, protecting the environment, and reducing CO2 emissions are key objectives in the development of smart sustainable cities. In this regard, energy management strategies are integral parts of smart sustainable cities. The major goals of energy management programs are extracting information about the consumption behaviour of consumers, proposing proper strategies based on consumer's behaviour in order to peak shaving and valley filling, and finally, increasing the balance between demand and generation. Since more than 30% of total energy consumption belongs to the residential sector, energy management in this sector plays a crucial role to attain the aforementioned goals.

One of the effective tools in residential energy management is non-intrusive load monitoring (NILM). NILM algorithms extract the consumption pattern of each appliance and provide feedback to the consumer about appliances such as their operation condition (normal/faulty), the cost of consumption of each appliance, etc. Furthermore, based on this feedback the energy flexibility potential of each consumer can also be characterized.

In this chapter, we discussed the basic concepts of NILM. To illustrate its application in energy flexibility characterization, four appliances of AMPds are considered, and an optimization-based NILM algorithm is utilized to disaggregate their consumption signal. Then, considering the earliest start time and latest finished time, the energy flexibility of this dataset is extracted.

In this chapter, the energy flexibility of shiftable appliances is computed. Extracting the electrical vehicle consumption pattern using NILM methods and characterizing its energy flexibility potential are one of the main future directions for the researchers who consider reading this chapter. Furthermore, proposing a NILM algorithm that does not require any training dataset and results in high accuracy in disaggregation high number of appliances with multi-mode operation modes can be considered as another future direction for researchers.

References

Abed-Meraim, K., Qiu, W., & Hua, Y. (1997). Blind system identification. *Proceedings of the IEEE, 85*(8), 1310–1322.

Aydin, E., Brounen, D., & Kok, N. (2018). Information provision and energy consumption: Evidence from a field experiment. *Energy Economics, 71*, 403–410.

Ayub, M. A., Hassan, N. U., & Yuen, C. (2018). Hybrid iterative algorithm for non-intrusive load disaggregation. In *2018 IEEE International Conference on Communications (ICC)* (pp. 1–6).

Azizi, E., Shotorbani, A. M., Hamidi-Beheshti, M.-T., Mohammadi-Ivatloo, B., & Bolouki, S. (2020). Residential household non-intrusive load monitoring via smart event-based optimization. *IEEE Transactions on Consumer Electronics, 66*(3), 233–241.

Azizi, E., Beheshti, M. T., & Bolouki, S. (2021). Event matching classification method for non-intrusive load monitoring. *Sustainability, 13*(2), 693.

Bhotto, M. Z. A., Makonin, S., & Bajić, I. V. (2017). Load disaggregation based on aided linear integer programming. *IEEE Transactions on Circuits and Systems II: Express Briefs, 64*(7), 792–796.

Bulut, M. B., Odlare, M., Stigson, P., Wallin, F., & Vassileva, I. (2016). Active buildings in smart grids—exploring the views of the Swedish energy and buildings sectors. *Energy and Buildings, 117*, 185–198.

Clarke, J. (2021). Designing active buildings. In *Emerging Research in Sustainable Energy and Buildings for a Low-Carbon Future* (pp. 11–24). Springer.

D'hulst, R., Labeeuw, W., Beusen, B., Claessens, S., Deconinck, G., & Vanthournout, K. (2015). Demand response flexibility and flexibility potential of residential smart appliances: Experiences from large pilot test in Belgium. *Applied Energy, 155*, 79–90.

Darby, S., et al. (2006). The effectiveness of feedback on energy consumption. In *A Review for DEFRA of the Literature on Metering, Billing and direct Displays* (Vol. 486, p. 26).

Dash, S., Sodhi, R., & Sodhi, B. (2019). A semi-intrusive load monitoring approach for demand response applications. In *2019 8th International Conference on Power Systems (ICPS)* (pp. 1–6). IEEE.

Dash, S., Sodhi, R., & Sodhi, B. (2020). An appliance load disaggregation scheme using automatic state detection enabled enhanced integer-programming. *IEEE Transactions on Industrial Informatics, 17*(2), 1176–1185.

Devlin, M. A., & Hayes, B. P. (2019). Non-intrusive load monitoring and classification of activities of daily living using residential smart meter data. *IEEE Transactions on Consumer Electronics, 65*(3), 339–348.

Dinesh, C., Makonin, S., & Bajić, I. V. (2019). Residential power forecasting using load identification and graph spectral clustering. *IEEE Transactions on Circuits and Systems II: Express Briefs, 66*(11), 1900–1904.

Egarter, D., & Elmenreich, W. (2015). Load disaggregation with metaheuristic optimization. In *Energieinformatik* (pp. 1–12).

Erkin, Z., Troncoso-Pastoriza, J. R., Lagendijk, R. L., & Pérez-González, F. (2013). Privacy-preserving data aggregation in smart metering systems: An overview. *IEEE Signal Processing Magazine, 30*(2), 75–86.

Esa, N. F., Abdullah, M. P., & Hassan, M. Y. (2016). A review disaggregation method in non-intrusive appliance load monitoring. *Renewable and Sustainable Energy Reviews, 66*, 163–173.

Faustine, A., Pereira, L., & Klemenjak, C. (2020). Adaptive weighted recurrence graphs for appliance recognition in non-intrusive load monitoring. *IEEE Transactions on Smart Grid, 12*(1), 398–406.

Hafemeister, D. W. (2009). Review of the 2008 APS energy study, energy future: Think efficiency. *Physics and Society, 38*(2), 16.

Hart, G. W. (1992). Nonintrusive appliance load monitoring. *Proceedings of the IEEE, 80*(12), 1870–1891.

Henao, N., Agbossou, K., Kelouwani, S., Dubé, Y., & Fournier, M. (2015). Approach in nonintrusive type I load monitoring using subtractive clustering. *IEEE Transactions on Smart Grid, 8*(2), 812–821.

Kelly, J., & Knottenbelt, W. (2015). The UK-DALE dataset, domestic appliance-level electricity demand and whole-house demand from five UK homes. *Scientific Data, 2*(1), 1–14.

Kleiminger, W., Beckel, C., & Santini, S. (2015). Household occupancy monitoring using electricity meters. In *Proceedings of the 2015 ACM International Joint Conference on Pervasive and Ubiquitous Computing (UbiComp 2015), Osaka, Japan.*

Kolter, J. Z., & Johnson, M. J. (2011). REDD: A public data set for energy disaggregation research. In *Workshop on Data Mining Applications in Sustainability (SIGKDD), San Diego, CA* (Vol. 25, pp. 59–62).

Kong, W., Dong, Z. Y., Ma, J., Hill, D. J., Zhao, J., & Luo, F. (2016). An extensible approach for non-intrusive load disaggregation with smart meter data. *IEEE Transactions on Smart Grid, 9*(4), 3362–3372.

Li, D., & Dick, S. (2018). Residential household non-intrusive load monitoring via graph-based multi-label semi-supervised learning. *IEEE Transactions on Smart Grid, 10*(4), 4615–4627.

Lin, S., Zhao, L., Li, F., Liu, Q., Li, D., & Fu, Y. (2016). A nonintrusive load identification method for residential applications based on quadratic programming. *Electric Power Systems Research, 133*, 241–248.

Liu, Q., Kamoto, K. M., Liu, X., Sun, M., & Linge, N. (2019). Low-complexity non-intrusive load monitoring using unsupervised learning and generalized appliance models. *IEEE Transactions on Consumer Electronics, 65*(1), 28–37.

Lu, M., & Li, Z. (2019). A hybrid event detection approach for non-intrusive load monitoring. *IEEE Transactions on Smart Grid, 11*(1), 528–540.

Makonin, S., Popowich, F., Bartram, L., Gill, B., & Bajić, I. V. (2013). AMPds: A public dataset for load disaggregation and eco-feedback research. In *2013 IEEE Electrical Power & Energy Conference* (pp. 1–6). IEEE.

Massidda, L., Marrocu, M., & Manca, S. (2020). Non-intrusive load disaggregation by convolutional neural network and multilabel classification. *Applied Sciences, 10*(4), 1454.

Moradzadeh, A., Sadeghian, O., Pourhossein, K., Mohammadi-Ivatloo, B., & Anvari-Moghaddam, A. (2020). Improving residential load disaggregation for sustainable development of energy via principal component analysis. *Sustainability, 12*(8), 3158.

Piga, D., Cominola, A., Giuliani, M., Castelletti, A., & Rizzoli, A. E. (2016). Sparse optimization for automated energy end use disaggregation. *IEEE Transactions on Control Systems Technology, 24*(3), 1044–1051.

Schirmer, P. A., & Mporas, I. (2019). Statistical and electrical features evaluation for electrical appliances energy disaggregation. *Sustainability, 11*(11), 3222.

Shahroz, M., Younis, M. S., & Nasir, H. A. (2020). A scenario-based stochastic optimization approach for non-intrusive appliance load monitoring. *IEEE Access, 8*, 142205–142217.

Singh, S., & Majumdar, A. (2017). Analysis co-sparse coding for energy disaggregation. *IEEE Transactions on Smart Grid, 10*(1), 462–470.

Singhal, V., Maggu, J., & Majumdar, A. (2019). Simultaneous detection of multiple appliances from smart-meter measurements via multi-label consistent deep dictionary learning and deep transform learning. *IEEE Transactions on Smart Grid, 10*(3), 2969–2978.

Suzuki, K., Inagaki, S., Suzuki, T., Nakamura, H., & Ito, K. (2008). Nonintrusive appliance load monitoring based on integer programming. In *IEEE, SICE Annual Conference, Tokyo, Japan* (pp. 20–22)

Tabatabaei, S. M., Dick, S., & Xu, W. (2017). Toward non-intrusive load monitoring via multi-label classification. *IEEE Transactions on Smart Grid, 8*(1), 26–40.

Wang, Y., Shukla, A., & Liu, S. (2017). A state of art review on methodologies for heat transfer and energy flow characteristics of the active building envelopes. *Renewable and Sustainable Energy Reviews, 78*, 1102–1116.

Wittmann, F. M., López, J. C., & Rider, M. J. (2018). Nonintrusive load monitoring algorithm using mixed-integer linear programming. *IEEE Transactions on Consumer Electronics, 64*(2), 180–187.

Xu, F., Huang, B., Cun, X., Wang, F., Yuan, H., Lai, L. L., & Vaccaro, A. (2018). Classifier economics of semi-intrusive load monitoring. *International Journal of Electrical Power & Energy Systems, 103*, 224–232.

Yang, C. C., Soh, C. S., & Yap, V. V. (2019). A systematic approach in load disaggregation utilizing a multi-stage classification algorithm for consumer electrical appliances classification. *Frontiers in Energy, 13*(2), 386–398.

Zhang, J., Chen, X., Ng, W. W., Lai, C. S., & Lai, L. L. (2019). New appliance detection for nonintrusive load monitoring. *IEEE Transactions on Industrial Informatics, 15*(8), 4819–4829.

Zhao, B., He, K., Stankovic, L., & Stankovic, V. (2018). Improving event-based non-intrusive load monitoring using graph signal processing. *IEEE Access, 6*, 53944–53959.

Zhao, B., Ye, M., Stankovic, L., & Stankovic, V. (2020). Non-intrusive load disaggregation solutions for very low-rate smart meter data. *Applied Energy, 268*, 114949.

Zhou, Z., Xiang, Y., Xu, H., Yi, Z., Shi, D., & Wang, Z. (2020). A novel transfer learning based intelligent non-intrusive load monitoring with limited measurements. *IEEE Transactions on Instrumentation and Measurement, 70*, 1–8.

Zoha, A., Gluhak, A., Imran, M. A., & Rajasegarar, S. (2012). Non-intrusive load monitoring approaches for disaggregated energy sensing: A survey. *Sensors, 12*(12), 16838–16866.

Chapter 14
Active Buildings Demand Response: Provision and Aggregation

Omid Sadeghian, Arash Moradzadeh, Behnam Mohammadi-Ivatloo, and Vahid Vahidinasab

Status Quo, Challenges, and Outlook
The increasing energy demand and environmental issues should be mitigated in future in different energy sectors, especially the building sector. For this aim, the conventional passive buildings are moving toward active buildings. The major feature of active buildings is their demand flexibility, which is feasible by demand response programs. Such programs are designed by aggregators, which have an important role in future power systems. Therefore, the first step is to answer the question that what is the role of aggregators in implementing such programs? What are the characteristics of demand response programs? What kind of demand response programs could be implemented in active buildings? and what are the enabling technologies for implementing such programs?

O. Sadeghian · A. Moradzadeh (✉)
Faculty of Electrical and Computer Engineering, University of Tabriz, Tabriz, Iran
e-mail: arash.moradzadeh@ieee.org

B. Mohammadi-Ivatloo
Department of Electrical and Electronics Engineering, Muğla Sıtkı Koçman University, Muğla, Turkey
e-mail: mohammadi@ieee.org

V. Vahidinasab
Department of Engineering, School of Science and Technology, Nottingham Trent University, Nottingham, UK
e-mail: vahid.vahidinasab@ntu.ac.uk

V. Vahidinasab, B. Mohammadi-Ivatloo (eds.), *Active Building Energy Systems*, Green Energy and Technology, https://doi.org/10.1007/978-3-030-79742-3_14

14.1 Introduction

Active buildings attracted considerable attention in recent years due to the energy challenges and climate change issues (Vahidinasab et al., 2021; Mehrtash et al., 2020). The most important feature of the active building is demand flexibility (Bulut et al., 2016). There are several solutions for increasing demand flexibility of buildings such as demand response (DR) programs. Aggregators are critical for the development of future active buildings by contributing to the implementation of such programs (Golmohamadi et al., 2019). This contribution might include the activation of enabling technologies along with the management of DR programs.

DR has been widely investigated in recent years due to the increasing trend of energy consumption and undesirable peak loads caused by the building sector, which is responsible for a large amount of energy consumption (Rahman et al., 2018). Environmental issues and depletion of fossil fuel reservoirs have accelerated the interests in such researches. A wide range of researches has been performed to contribute to the mentioned challenges using DR programs. DR contains price-based and incentive-based DR programs. In the price-based programs, the time-varying electricity prices are considered to motivate the consumers to change their load habits, whereas, in incentive-based programs, an incentive is paid to the consumers in exchange for participation in DR based on predetermined programs.

Implementation of DR programs in buildings can contribute to electricity system economic, reliability, and security aspects. These programs have been researched in different aspects such as modeling, enabling technologies, and drivers. Some other researchers have discussed their benefits, challenges, and barriers. The aggregators have an important role in providing infrastructures and effective implementation of DR programs to benefit the end-users and electricity systems. DR can be implemented for every load in a building, which includes appliances, lighting loads, heating loads, cooling loads, ventilation loads, and also heating, ventilation, and air conditioning (HVAC) systems.

An appropriate assessment and implementation of DR in buildings can benefit the consumers by reducing the electricity bill (Yoon et al., 2020) and improving the reliability of the power supply (Kim et al., 2020). Load peak reduction by load curtailment of buildings' electricity consumers and, accordingly, cost reduction by turning off costly generators in such hours are the benefit of DR for the power system. The mentioned advantages are achieved through load management by consumers through curtailment, reduction, or shifting their demand. When the load curtailment or reduction is targeted in DR (and not load displacement), CO_2 emissions are reduced which is environmentally beneficial (Leerbeck et al., 2020).

This chapter deals with the role of aggregators in DR programs in the active buildings as a large consumer in the electricity grid. In the first stage, aggregators and the retail electricity market are described in this chapter. This chapter also involves the benefits, barriers, and drivers of implementing DR programs. Implementing DR needs designing efficient mechanisms to motivate the consumers to participate in intended programs, which is discussed in this chapter. Furthermore, both the incentive-based and price-based DR programs are described where one of them or both can be effective according to the existing conditions. Enabling technologies for implementing DR is also addressed which can facilitate DR participation by providing real-time information and monitoring for the data centers and is a necessity for implementing DR.

The structure of this chapter is organized into five sections. Aggregators and the retail electricity markets are discussed in Sect. 14.2. Section 14.3 deals with benefits, barriers, and drivers for implementing DR. DR mechanisms and programs are discussed in Sect. 14.4. Enabling technologies is described in Sect. 14.5. Section 14.6 as the last section makes the main conclusions.

14.2 Aggregators and the Retail Electricity Market

Aggregators are important entities in future power systems such as demand flexibility of buildings, which retails the electricity and implements DR programs. For implementing efficient DR programs, aggregators should accomplish the needed actions like providing the enabling technologies for the consumers in buildings and managing the DR programs. For this aim, knowing about the aggregators and retail electricity market is critical. Figure 14.1 illustrates the conceptual diagram of the role of aggregator in DR of active buildings. In the following, these two concepts are discussed.

14.2.1 Aggregators

Building as a part of the electricity system is currently subjecting to considerable alterations due to advances in data and communication technologies, power electronics devices, and dispersed generations. Samples for dispersed energy resources

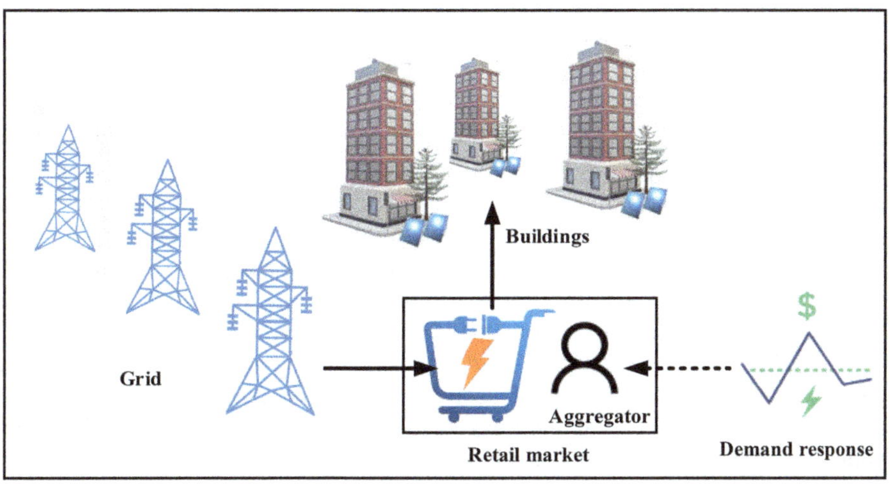

Fig. 14.1 Conceptual diagram of the role of aggregator in demand response of active buildings

are diesel generators, solar photovoltaics, small wind farms, electric vehicles, fuel cells, energy storage systems, and DR programs (Burger & Chaves-ávila, 2016). An active building needs some of these elements to provide demand flexibility. Unlike conventional centralized generating units, distributed energy resources commonly have small capacities and are connected to low- and medium-voltage electricity grids (Ackermann et al., 2001; El-Khattam & Salama, 2004; Pepermans et al., 2005). An example of low-voltage electricity grids is active buildings.

Aggregators enable distributed generations to provide these valuable electricity services appropriately to improve the energy flexibility of buildings (Burger et al., 2017). Aggregators play other roles in power markets. For instance, aggregators possess the technology to perform DR and are accountable for the installation of the communication, manage, and control devices such as smart meters at end-user premises to achieve direct control of their appliances like air conditioning in critical conditions such as peak-demand emergency condition (Gkatzikis et al., 2013). Although providing the needed technologies by aggregators is dependent on the energy policy of governments, aggregators are an intermediate entity between the power system and the end-users situated in buildings.

There are various definitions provided for the aggregator concept in literature. In Ikäheimo et al. (2010), three aggregator types including retailer aggregator, demand aggregator, and generation aggregator are introduced. Therefore, the definition of an aggregator is proposed as an interface company between electricity end-users, which provides distributed energy resources and participants in the power system. Gkatzikis et al. (2013) has suggested that aggregators are novel entities in the power markets that act as mediators/brokers between users and the utility operator. Another definition has been given in (Burger et al., 2017), where the aggregation is defined as the act of grouping distinct agents in a power system to act as a single entity when participating in electricity markets or selling services to the system operator.

Furthermore, (Burger & Chaves-ávila, 2016) has defied the aggregator as third-party intermediates between power market participants and residential, industrial, and commercial end-consumers. In Momber et al. (2016), the researchers have suggested that the aggregator acts between wholesale and retail markets. The aggregator is defined in Kim and Thottan (2011) as a company that collects power generated from microgrids and sells it to the utility. Some references have claimed that the direct participation of individual microgrids in the retail market may not be scalable since a utility faces the problem of dealing with a large number of small power sources, and hence, the concept of aggregator has been introduced (Kim & Thottan, 2011; Chowdhury et al., 2009) according to this issue.

Two types of economic value are described for aggregation in Burger and Chaves-ávila (2016) containing system and private values. System value is achieved if aggregation increases the economic efficiency of the electricity system as a whole. On the other hand, private value is attained for the economic improvement of a single agent or subset of agents. Three broad categories of aggregation are addressed in Burger et al. (2017) containing aggregations with "fundamental" or "intrinsic" value, aggregations with "transitory" value, and aggregations with "opportunistic" value. They are defined as follows:

1. Aggregations with "fundamental" or "intrinsic" value: which is independent of the market context as well as the regulatory context
2. Aggregations with "transitory" value: contribute to the better operation of the electricity system under the present and near-term conditions. However, it may be decreased by regulatory or technological developments
3. Aggregations with "opportunistic" value: emerge as a result of regulatory or market design flaws and may endanger the economic efficiency of the power system

Another aggregator has been defined in literature, namely, the electric vehicles aggregator. In Momber et al. (2016), an electric vehicle aggregator has two different aims in the wholesale-side market and client-side retail market. On the wholesale side, the competition is between electric vehicle aggregators against other demand-side market participants for optimal electricity purchases as well as the optimal profit of ancillary services. On the client-side retail market, the aggregators contribute end-users against other agents, which provide alike products and services. In Guille and Gross (2009), the aggregator is defined as an interface company that aggregates the electric vehicles to act as the dispersed energy resource to make the vehicle-to-grid concept implementable. The aggregator also seeks to provide an interface with the independent system operator, regional transmission organization, and the energy service providers who provide the electricity supply to customers through the distribution grid.

In the conducted researches, the aggregator plays different roles in the system. In Gkatzikis et al. (2013), the goal of every aggregator is to maximize the net profit, i.e., the revenue received from the operator minus its compensation to home users. In this research, a set of competing aggregators act as intermediaries between the utility operator, and the home users is considered in which the operator seeks to minimize the smart grid operational cost and offers rewards to aggregators toward this goal. The aggregator implements the indirect load control program by determining the energy retail prices in Momber et al. (2016).

Among the price signals, which final customers are exposed, the aggregator is only responsible for optimizing the wholesale energy component of the electricity bill. In Kim and Thottan (2011), aggregators are an entity, which enables microgrids to sell their surplus power to a utility in which aggregators collect distributed generations of microgrids and sell it to the utility. In this research, a two-stage market is considered where the first stage market is between the utility and the aggregators, and the second stage market is between the aggregator and the microgrids, and the profit of the aggregator is obtained based on the first and second stage prices. The obtained results have revealed that the participation of aggregators considerably impacts the market according to the supply elasticity and, hence, the cost structure of microgrids. In some other research (Papavasiliou & Oren, 2014; Ponds et al., 2018), large-scale integration of DR and renewable energy sources for participating in the market is accomplished. In another attempt (Golmohamadi et al., 2019), an agent-based structure to integrate the flexibility potential of industrial and residential demands has been introduced.

14.2.2 *Retail Electricity Market*

Deregulation and restructuring of electricity systems have attracted much attention for the past decades, due to promote competition among retail electricity providers, cause the electricity price to be more inexpensive, marginalize high-cost providers, and result in lower electricity bills for customers in buildings (Yu et al., 2004). In order to limit the aggregator's monopolistic profitability, an adequate competition in retail markets is necessary (Momber et al., 2016). Additionally, traditional problems like congestion, voltage and frequency derivations, and line losses of a power sector including buildings can be improved using distributed generation sources and inter-ties, which add complexity from both technical and economic perspectives, and hence, the emergence of new entities within the distribution network domain, namely, retailers, is sensed (Algarni & Bhattacharya, 2009; Teimourzadeh et al., 2021).

A simplified categorization of the participants in the retail electricity market is as (Yu et al., 2004) (a) independent power plants, (b) power exchange companies, (c) retail companies, and (d) power users. Therefore, the end-users in buildings can be the participants in a retail market. With the establishment of a competitive electricity market, demand-side participation by consumers in the buildings plays an important role in improving power market efficiency (Caves et al., 2000; Boogert & Dupont, 2008). Customers, however, have no considerable influence on the design of such markets. This is because they do not have adequate financial incentive or level of knowledge to actively participate in such a complex and time-consuming task (Nguyen et al., 2009). Designing efficient incentive-based mechanisms based on the local condition can be effective on the active participation of buildings in demand flexibility programs.

Three alternative market settings are considered: islanding mode, interaction with the power market, and interactions of aggregators/retailers with the power market (Ilieva et al., 2016). By deregulation, in addition, to increase competition among retailers/suppliers, the consumers in a building will enable them to have more choices among providers (Yang, 2014).

A wide number of researches have considered the retail markets. In Liang et al. (2019), an optimal DR scheme enables demand-side appliances that are basically too small to participate in a retail electricity market by using local controllers. In this reference, the appliances include HVAC systems, plug-in electric vehicles, and electric water heaters. A centralized decision-making architecture is introduced in Javadi et al. (2018) for multiple home microgrids in the retail electricity market. A retail electricity market is proposed in Marzband et al. (2018) for active distribution systems and home microgrids. In Marzband et al. (2016), a strategy for implementing a retail market structure is introduced by considering a high penetration of distributed energy resources and demand-side management. Evaluating switching behavior in the household sector in the Danish retail electricity market is studied in Yang (2014). In Lehto (2011), the effective factors on electricity pricing in the Finnish retail market and the impact of ownership structure on retail prices are investigated for a real case. In this research, the influence of low-cost power sources on retail prices is considered. In Babar et al. (2015), a new framework is proposed for bidding in a retail market using demand elasticity. Benefits of customers from electricity retail competition of US states are discussed in Su (2015). In another attempt (Kuleshov et al., 2012), the advancement of reforms in the Russian retail electricity market is studied in which the social, political, and technological challenges are discussed as well as market liberalization. The results indicate that the retail market is currently divided into inactive and sub-active markets.

Researchers in Yu et al. (2004) have focused on developing an agent-based retail electricity market model to simulate the trading mechanism and behavior of the participants. A number of business and price models are offered in Bae et al. (2014) which reflect participants' viewpoints when a retail power market is established. An electricity retail market model is presented in Sekizaki et al. (2016) in which elastic demands of consumers in a distribution network are sold through a retailer by offering such elastic demands in that market. The investment for a retail company in the electricity market is optimized in He et al. (2015). Liberalization of the Japanese household retail market and its positive impact on consumer satisfaction are studied in Shin and Managi (2017) where consumers are enabled to choose an electricity provider. Investigation on the fall of residential electricity rates after the retail competition is investigated in Swadley and Yücel (2011). This research has revealed that an increase in participation rates, more price controls, larger markets, and higher shares of hydro in power generation decrease the retail prices, whereas the rates are increased by increases in coal and natural gas prices. A survey on the next generation of US retail electricity market and recent technology developments is presented in Chen et al. (2018a) considering customers and prosumers.

A techno-economic strategy is presented in Chen et al. (2017) to encourage typical load aggregators, like parking lots with high penetration of vehicles, to directly participate in the retail power market. In Flores and Waddams Price (2013), the behavior of customers in the British retail electricity market is studied. Moving toward retail competition in the Portuguese electricity market and success of the market liberalization has been studied in Ghazvini et al. (2016) in which the consumers' behavior, changes in the retail electricity rates, and, additionally, the market concentration are taken into account. In another research (Naing & Miranda, 2005), multi-energy retail markets have been investigated.

Some research studies have investigated the risk management for participation in retail markets. Striking the right balance between risk and profit of retailing in the power market is investigated in Golmohamadi et al. (2015). Risk-constrained electricity retail contracting has been studied in Downward et al. (2016). In Kharrati et al. (2016), a risk-aversion equilibrium model is proposed for the mid-term decision-making of retailers considering the strategy of other retailers. An effective risk-constrained scheme for retail electricity providers is introduced in Fotouhi Ghazvini et al. (2015) to evaluate the uncertainties in the day-ahead market to hedge against financial losses in the electricity market.

14.3 Benefits, Barriers, and Drivers of Implementing Demand Response

Demand response programs are critical in future power systems, especially the building sector, which is a large energy consumer (Chen et al., 2018b). According to the US Department of Energy report, electricity consumption in the USA has increased by 2.5% annually for the past two decades (Ding et al., 2014), and the electricity grid may expose its limits in the upcoming years (Palensky & Dietrich, 2011). Additionally, by increasing the penetration of renewable energy sources, the need for demand flexibility has been increased. Furthermore, by assuming the demand to be inelastic, the hourly and daily load variations should be supported by the generation side which is undesirable due to limitation in generation facilities (Paterakis et al., 2017). Also, energy crisis and environmental issues are existence challenges of today's world (Sadeghian et al., 2019a). The activation mechanism of the demand side for energy management is called demand-side management (Warren, 2014). Demand-side management solutions include energy efficiency and load management techniques. Load management and especially DR strategies have attracted much attention for improving the power system operations (Paterakis et al., 2017).

DR is an important tool for electric utilities to overcome the growing demand for electricity (Eissa, 2011) and improve the flexibility of the power system (Bayer, 2015). Moreover, according to the benefits of DR to attain reliable electricity markets, it is also a key element on the reliable development path (Su & Kirschen, 2009). DR may satisfy the reliability needs of active and dynamic power markets by overcoming the uncertainties (Martinez & Rudnick, 2012). Identifying the load pattern of a typical consumer is effective to implement the DR in a more appropriate and beneficial manner. There are various methods for load forecasting such as principal component analysis (Moradzadeh et al., 2020a; Moradzadeh & Pourhossein, 2020), artificial neural networks (Moradzadeh et al., 2020b; Early detection of turn-to-turn faults in power transformer winding: An experimental study, 2019; Moradzaeh & Khaffafi, 2018), deep learning (Moradzadeh et al., 2020c; Moradzadeh & Pourhossein, 2019a; Moradzadeh et al., 2021a, 2021b), and support vector machines (Moradzadeh & Pourhossein, 2019b; Moradzadeh et al., 2020d; Mansour-Saatloo et al., 2020). DR can improve different aspects of the electricity grid, and hence, some see DR as a strategy to improve the economic operation of markets, while others see it as a new control variable that can improve power system reliability and security (Mathieu et al., 2013). In the viewpoint of the loads, DR is performed by load reduction, load curtailment, and load shifting (Mirzaei et al., 2020). DR is an essential characteristic of the smart grid to diminish the generation of expensive generators and further to defer the capacity addition in long-term periods (Deng et al., 2015).

The US Department of Energy (DoE) has defined DR as "changes in electric usage by end-use customers from their normal consumption patterns in response to changes in the price of electricity over time, or to incentive payments designed to induce lower electricity use at times of high wholesale market prices or when system reliability is jeopardized." It includes price-based and incentive-based programs, which are discussed in U. department of Energy (2005). DR has been defined also in Albadi and El-Saadany (2008) as all intentional electricity consumption pattern alterations by end-users that are intended to alter the timing, level of instantaneous, or total electricity consumption. In another reference (Razmara et al., 2017), DR is defined as ancillary services in which the consumers adjust their electricity consumption in response to a signal from the grid called DR signal. This signal may contain information about the power price, system reliability emergencies, or incentives. The needed infrastructures for implementing DR programs have been described in Paterakis et al. (2017). In the following, benefits, barriers, and drivers of implementing DR are discussed.

14.3.1 Benefits

Both the buildings and power supply can benefit from implementing DR programs (Gao & Sun, 2016). In a report presented by the International Energy Agency, $10–$15 billion annual saving can be achieved for US commercial buildings by peak demand management (Sadineni & Boehm, 2012). Stavrakas and Flamos (2020) has focused on self-consumption of electricity as a benefit of DR. The advantage of DR on the electricity bill of an educational office building with a responsive district heating system is investigated in Vand et al. (2020).

Various positive aspects have been mentioned for DR in the literature. In Strbac (2008), several benefits have been described for DR programs. U. department of Energy (2005) has studied the potential benefits of DR and gives some recommendations for achieving them. As mentioned in Zhang and Li (2012), DR can be an alternative for investment in constructing new power plants that may be underutilized aim to provide capacity reserves. The economic benefits of DR have been discussed in Conchado and Linares (2012). In some other works (Oconnell et al., 2014; Mirzaei et al., 2018), researchers have focused on the contribution of DR for balancing the fluctuations caused by renewable energy source generations. Researchers in Gyamfi and Krumdieck (2011) investigate voluntary load shedding by costumers in a real case in New Zealand in response to real-time data about the critical peak demand costs, power supply security, and emission profile. 10% reduction in critical peak load has been achieved in which results have shown that the customers are more sensitive to security information compared to that of emission. In another research (Bradley et al., 2013), economic welfare effects of DR in the UK are considered, while in previous researches, the only size of investments and certain form returns of DR in isolation had been investigated.

The advantages of residential DR in a real-time distribution energy market are investigated in Siano and Sarno (2016). In Conchado et al. (2016), both the supply and demand considerations are taken into account in an integrated evaluation of a DR program benefits in Spanish households. In Shafie-Khah et al. (2016), it has been revealed that electric vehicle parking lots can benefit from selective participation in DR programs. Potential electricity market efficiencies from implementing DR have been discussed in Spees and Lave (2007). A review of the residential price-based DR studies is performed in Yan et al. (2018), in which the advantages and disadvantages of different price-based DR programs are discussed. Concerns and future research challenges related to price-based DR programs are also addressed in this reference. This reference has mentioned various advantages of DR by using the price signal such as carbon emissions and energy costs reduction, risk and stability management, and also shaving of peak demand.

The influence of implementing DR on CO_2 emission in peak electricity demand days in the residential sector of New York City is studied in Gilbraith and Powers (2013). Yao et al. (2015) also has concentrated on CO_2 emission improvement as one of the benefits of DR. Integration of DR with renewable energy sources has been reviewed in Aghaei and Alizadeh (2013), and a comprehensive benefit and cost evaluation of DR is presented. In Parvania et al. (2012), different advantages of DR participation in wholesale markets are addressed. In Joung and Kim (2013), the long-term benefits of DR in viewpoints of electricity price and system reliability are evaluated. DR can contribute to the system flexibility in renewable-based systems (Sadeghian et al., 2020a) and contribute to increasing the renewable energy share in power supply (Zhang et al., 2015) and, hence, be considered as a renewable energy enabling technology (Nikolic et al., 2016) since the renewable energy sources such as wind turbines and solar photovoltaics have intermittent and uncertain nature (Sadeghian et al., 2020b).

In Sadeghian et al. (2020a) and Sadeghian et al. (2019b), curtailable loads are considered aims to cost minimization of grid-connected virtual power plants. A comprehensive study on the potential of DR on the main characteristics of residential distribution network operation is performed in Safdarian et al. (2016). It has been revealed in this research that the realization of 10% of DR potentials can improve the peak load, losses, and reliability by 5.6%, 1.3%, and 1.7%, respectively. In another research (Safdarian et al., 2014a), DR has been adopted to flatten the overall load profile. In Safdarian et al. (2014b), reliability improvement of the distribution network has been achieved using DR, while such programs have been used in Goel et al. (2006) for reliability improvement of a deregulated power system. An event-driven emergency DR scheme is introduced in Wang et al. (2011) for improving the power system security. Rapid frequency response using DR is investigated in Chakravorty et al. (2017) for Great Britain Power System. Frequency control of the power system is also performed by considering the DR in heat pumps and electric vehicles in Khezri et al. (2018) and electric vehicles in Oshnoei (2017). In some other researches, DR has been used for voltage control of power system (Rabiee et al., 2014), voltage and frequency support of MV microgrids (Bayat et al., 2016), under-frequency load shedding (Rafinia et al., 2020), and real-time voltage control in a distribution system (Zakariazadeh et al., 2014), and hence, the need for reactive compensation by conventional shunt capacitors (Sadeghian et al., 2020c) is decreased.

The benefits of DR are summarized as follows:

14.3.1.1 End-users

Participation of customers in DR leads to reliability improvement of power supply due to load management in peak hours and also electricity bill reduction due to energy consumption reduction resulting from reduction, curtailment, or displacement of demand based on the DR program that a typical consumer has participated in it.

14.3.1.2 System

In short-term horizons, reduction of unexpected failures and, hence, lower need for mandatory maintenance (Sadeghian et al., 2019c, 2019d) especially in peak hours as well as cost reduction due to turning off the costly generators in peak hours are achieved by DR. Energy losses can also be improved by implementing DR. On the other hand, in the long-term horizons, decreasing the need for constructing new power plants is the advantage of DR for the power system.

14.3.1.3 Environment

Mitigate the reliance on fossil fuels, and hence, reduction of CO_2 emissions and lower air pollution are the short-term benefit of DR from the environmental aspect. Solving the challenge of global warming and climate changes resulting from CO_2 emission in long-term periods is another potential benefit for the environment.

14.3.2 Barriers

Despite the mentioned benefits of DR implementation in the building sector, some barriers and issues exist in the DR implementation. The main reasons for negative attitudes toward DR are discussed in Yamaguchi et al. (2020), in which "scheduling" has been mentioned as a major reason for consumers in residential buildings. Understanding the potential DR value is further highlighted as a barrier to DR in the previous researches (Nolan & O'Malley, 2015). Another possible barrier to DR could be consumers' deficit of expertise in market functions and lack of interest (Kim & Shcherbakova, 2011). Low consumers' confidence in the electricity market functions is another barrier (Council of European Energy Regulators, 2011). If customers do not have confidence in the electricity market, this could be a potential barrier and diminish the willingness for participation. The long-term negative effect of the lack of customers' confidence in the electricity market for participation in DR programs is studied in Torstensson and Wallin (2015) in which it has suggested that without confidence that the market acts impartially, it might reduce the households' interest to contribute with more flexible electricity loads. The lack of suitable mechanism in the structure of current market is one of the greatest barriers to DR (Eric et al., 2012; Cappers et al., 2013). A further barrier for DR is associated with current tariff and regulatory mechanisms, especially for residential consumers (Oconnell et al., 2014). The installation of enabling devices such as smart meters is costly and is a challenge for implementing DR (Eid et al., 2016). The lack of information and communication technology infrastructures is another barrier (Strbac, 2008). Low electricity prices resulting from high reserve margin and few reliability concerns are barriers to the customer's participation in DR programs in typical regions that have those features (Smith & Hledik, 2011). The biggest barriers to establishing a policy to support the DR development are uncertainty of costs and true benefits related to DR in competitive electricity markets compared to other schemes of meteorological change mitigation and energy efficiency (Oconnell et al., 2014). From the physical aspects' point of view, another barrier is corresponding to limitations related to network capacity (Haider et al., 2016). The DR barriers are categorized as fundamental and secondary aspects (Good et al., 2017):

- Fundamental barriers: related to the social, technological, and economic aspects
- Secondary barriers: associated with regulatory aspects, markets design, physical problems (electrical grid)

Regulatory barriers are discussed in detail in Van Dievel et al. (2014). Researchers in Vallés et al. (2016) revise the principle regulatory barriers that prevent the prosperous development related to DR mechanisms for minor consumers and management of active distribution network across Europe. Rahimi and Ipakchi (2010) reviews the barriers and potential schemes for the implementation of DR under the smart grid and market paradigms.

14.3.3 Drivers/Motivators

The introduction of smart metering and the availability of bidirectional communications are two main technical drivers for incorporating DR into smart grids (Deng et al., 2015). Economic savings are widely considered as a major driver for consumers to change their energy use habits and are the most common driver in research projects (Verbong et al., 2013).

Electricity price, especially during peak hours, is considered another driver for DR. Regions that incur high electricity costs based on high power consumption can benefit from such programs and pursue more programs and resources accordingly. Similarly, customers in high-priced systems should have a stronger incentive for participation in getting higher rewards through reduction/shifting their loads (Smith & Hledik, 2011). However, some exceptions may be in which they may have high electricity prices but no significant DR resources arising from customer attributes or the policy supporting evolution in these states (Smith & Hledik, 2011), for instance, the Hawaii region.

In a conducted research in Santinelli et al. (2016), 9% of customers are interested to participate in the DR programs for economic benefits. 85% are interested to understand more about their domestic energy consumption, to reduce their electricity bills. Interest in environmental issues also seemed to be a strong driver for 90% of consumers.

Drivers of implementing smart DR in China have been discussed in Guo et al. (2017). Sharma and Sharma (2019) has investigated also some drivers for DR implementation.

14.4 Demand Response Mechanisms and Programs

DR mechanisms have received considerable attention from researchers since the efficient implementation of DR programs in buildings needs appropriate mechanisms. Additionally, DR programs including price-based and incentive-based programs should be assessed in advance to identify their implications and quantitative benefits for the building sector. The mechanisms and programs are discussed further.

14.4.1 Designing Efficient Demand Response Mechanisms

An efficient and appropriate DR mechanism is essential to achieve the benefits of DR by consumers in a building. An ideal mechanism for achieving the electricity consumption efficiency through economic incentives has been considered in Barreto et al. (2013). The suggested incentives might be seen as an indirect mechanism, in which users cannot see private information about their preferences. This research has concluded that the success of the suggested mechanism requires subsidies from external institutions, at least during the transition between an inefficient outcome and the efficient equilibrium. Researchers in Martinez and Rudnick (2012) have formulated a novel perspective of DR in emerging countries, based on the US and European experience, with important programs, instruments, and mechanisms of demand management and integration of such programs in smart grids. It reviews strategies, opportunities, and challenges for incorporating DR in emerging countries, by concentrating on designs for Latin America. In Mathieu et al. (2013), researchers have reviewed many options for the exploitation of residential loads for DR in which engineering and economic consequences for three specific cases including real-time pricing, dispatch-based control via an aggregator that participates in markets, and direct participation in markets are investigated. It has been founded that the suitable selection of DR program design depends on the goal. For instance, economic goals may be achieved via bidding mechanisms and well-designed pricing. Sustainability is best achieved through programs of dispatch. In the quest for a realistic implementation of mechanism design in DR, a two-stage strategy is suggested in Mhanna et al. (2014) for sharing the cost of electricity among strategic, rational, and selfish household retailers that have private information about their preferences. A residential power network is considered in Cao et al. (2012), where consumers are asked to send their information about electricity usage to the service provider and then the provider determines the optimal power allocations based on social welfare maximization. The proposed scheme may be beneficial for both the service provider and consumers.

14.4.2 *Incentive-Based vs. Price-Based Programs*

DR programs for implementation in buildings are divided into two main programs including price-based and incentive-based programs (Albadi & El-Saadany, 2007). These programs are mainly implemented aim of peak shaving. In the time-based programs, the electricity price is changed to impact the electricity consumption pattern. In such programs, electricity price is increased in high-price hours and decreased in low-price hours. On the other hand, in the incentive-based programs, incentives are paid to participants in terms of change in their consumption. In Imani et al. (2018), the results of incentive-based and price-based programs have been compared, while in Mirzaei et al. (2019) and Sadeghian et al. (2020d), only the time-based program is employed. In Albadi and El-Saadany (2008) and Yan et al. (2018), DR programs are categorized. The programs are classified as Fig. 14.2.

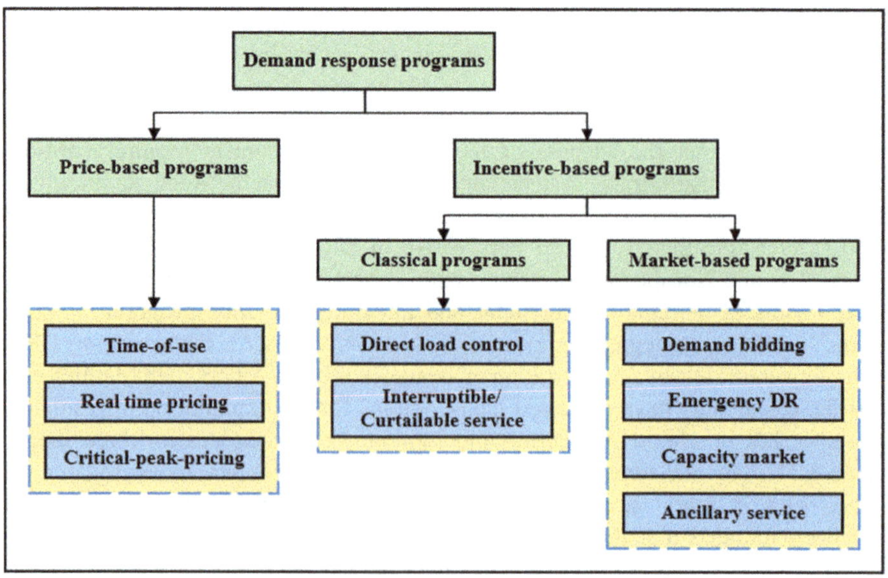

Fig. 14.2 Demand response categories

In the following, the description of each one is presented as follows:

(a) Price-based programs:

- Time of use: In this program, the 24-h horizon is divided into three different categories of time including night hours, off-peak hours, and on-peak hours. In on-peak hours, the electricity price is increased, while in night hours, the electricity price is decreased. This work is an accomplished aim to flatten the load profile.
- Real-time pricing: In this program, each hour has its related price. Peak hours have the most electricity price, while the night hours have lower electricity price, compared to the other hours of the day. Also, days with more load have more electricity price. Therefore, the electricity price varies by day and hour.
- Critical peak pricing: This program is implemented only in peak hours of the year by increasing the price to a very high value to peak shaving due to a limit in generation capacity or electricity grid. The security and reliability of the power system are attained by satisfying the mentioned physical limits.

(b) Incentive-based programs:

- Direct load control: In this program, the related utility has the ability to shut down the predetermined appropriate appliances, which have, in turn, lower impact on customer comforts such as air conditioner, refrigerator, or water heater or a short time period.
- Interruptible/curtailable service: This program is similar to direct load control with the difference that, in the interruptible load program, participants are asked to reduce their programs for predetermined values.
- Demand bidding: In this program, participants bid a portion of their load in the market which the bid should be lower than the market price. When a bid is accepted, the participants should curtail the contracted load; otherwise, they face a penalty.
- Emergency DR: In emergency DR, utility pay incentives are paid to participants that curtail their loads during determined peak hours.
- Capacity market: This program is for contingency conditions. In such situations, participants should curtail their predetermined loads; otherwise, they face a penalty.

Ancillary service: In this program, the spot market price is paid to participants that have participated in this service with their determined bid.

14.5 Enabling Technologies for Demand Response

For implementing efficient DR programs in buildings, enabling technologies should be provided for consumers by aggregators. The various US DR pilots have indicated that by peak power reduction without enabling technologies, 16% improvement is achieved, whereas with enabling technologies, 28% improvement is achieved, especially if thermostats are employed (Ivanov et al., 2013). The availability and development of enabling technologies give many systems and societal potential benefits of DR (Balijepalli et al., 2011). Advances in integrated electronic circuits, information and communications technologies, and control systems have appropriately improved the operation of advanced metering and technologies of DR (Siano, 2014). Enabling technologies are effective factor to determine the capability and potential of energy efficiency and DR at a typical facility used by a customer (Siano, 2014). The most known enabling technology is smart meters, which provide real-time bidirectional information and communication infrastructure between the appliance and the service provider. Smart meters record the energy consumption data at intervals (every minute) and send them to the utility. Enabling technologies consist of, but are not limited, the following (Siano, 2014):

- Optimized demand reduction strategies to meet the objectives
- Two-way communications of interval meters
- Communication technologies to inform customers to act based on the DR schedule
- Energy-information tools to reflect the load data in near real time
- Control systems related to building energy management and load controllers
- On-site generation appliance used either for back-up in emergency conditions

Some smart technologies, such as smart meters, employed to monitor energy usage and automatically regulate the load based on price changes or remote signals from system operators and represent a key requirement for most DR programs (Assessment of Demand Response and Advanced Metering, 2006). The innovative enabling technologies permit both better participation of the customers and greater confidence of the utility (Siano, 2014).

In Ghatikar (2014), enabling technologies from viewpoint of data centers are discussed. The enabling technologies need to link data center operational needs with the supply-side systems and provide integration for visualization of recorded DR information (Ghatikar, 2014). In this research, enabling technologies for information technology equipment and control systems and the need to enable DR automation are described.

Enabling technologies generally include smart thermostats, energy management systems, peak load control, and on-site generators. A response mechanism needs to be implementable in case of an event (Albadi & El-Saadany, 2007). Utility-side and customer-side enabling technologies for energy management in smart homes have been discussed in Zipperer et al. (2013). In Arens et al. (2006), developments of enabling technologies for DR have been discussed. In Albadi and El-Saadany (2007), different DR costs are depicted in Fig. 14.3. The location of enabling technology cost is seen in this figure.

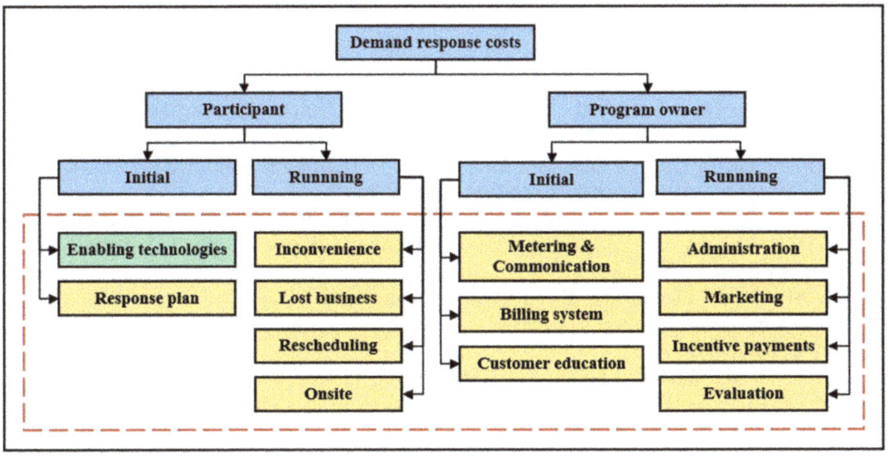

Fig. 14.3 The status of enabling technology cost among demand response costs

14.6 Conclusions

Active buildings have received significant attraction in recent years due to the pollution issues and energy crisis. The major feature of the active building is demand flexibility rather than self-generation of electricity. One of the solutions for demand flexibility of buildings is implementing demand response programs. Aggregators as an intermediate entity between the power system and end-users play the role of implementing demand response issues and also proving to enable technologies for consumers situated in the buildings. This chapter discussed the role of aggregators in implementing demand response programs in the buildings as a large consumer in the electricity grid. Aggregators and the retail electricity market were described in this chapter. This chapter involved the benefits, challenges, and drivers of implementing demand response programs. Furthermore, both the incentive-based and price-based demand response programs were described. Enabling technologies for implementing demand response were also addressed which can facilitate demand response participants by providing real-time information and monitoring for the data centers and are necessary for implementing demand response. This chapter shows the different roles of aggregators in local electricity grids such as implementing demand response programs. Different enabling technologies were discussed, which should be provided by aggregators. However, policymakers play a decisive role in the development of demand response programs by the removal of barriers and especially proving enabling technologies. The selection of demand response programs for implementation depends on the interests of the consumers. For instance, economic saving may be the aim of consumers for participation in such programs.

References

Ackermann, T., Andersson, G., & Söder, L. (2001). Distributed generation: A definition. *Electric Power Systems Research, 57*(3), 195–204. https://doi.org/10.1016/S0378-7796(01)00101-8

Aghaei, J., & Alizadeh, M. I. (2013). Demand response in smart electricity grids equipped with renewable energy sources: A review. *Renewable and Sustainable Energy Reviews, 18*, 64–72. https://doi.org/10.1016/j.rser.2012.09.019

Albadi, M. H., & El-Saadany, E. F. (2007). Demand response in electricity markets: An overview. In *2007 IEEE power engineering society general meeting, PES*, pp. 1–5. https://doi.org/10.1109/PES.2007.385728

Albadi, M. H., & El-Saadany, E. F. (2008). A summary of demand response in electricity markets. *Electric Power Systems Research, 78*(11), 1989–1996. https://doi.org/10.1016/j.epsr.2008.04.002

Algarni, A. A. S., & Bhattacharya, K. (2009). A generic operations framework for discos in retail electricity markets. *IEEE Transactions on Power Systems, 24*(1), 356–367. https://doi.org/10.1109/TPWRS.2008.2007001

Arens, E. A., Auslander, D., Culler, D., & Al, E. (2006). Demand response enabling technology development. *UC Berkeley Controls and Information Technology*. https://doi.org/10.11436/mssj.15.250

"Assessment of Demand Response and Advanced Metering". (2006). *Staff report Federal Energy Regulatory Commission (FERC).*

Babar, M., Nguyen, P. H., Cuk, V., & Kamphuis, I. G. (2015). The development of demand elasticity model for demand response in the retail market environment. In *2015 IEEE Eindhoven PowerTech, PowerTech 2015*. https://doi.org/10.1109/PTC.2015.7232789

Bae, M., Kim, H., Kim, E., Chung, A. Y., Kim, H., & Roh, J. H. (2014). Toward electricity retail competition: Survey and case study on technical infrastructure for advanced electricity market system. *Applied Energy, 133*, 252–273. https://doi.org/10.1016/j.apenergy.2014.07.044

Balijepalli, V. S. K. M., Pradhan, V., Khaparde, S. A., & Shereef, R. M. (2011). Review of demand response under smart grid paradigm. In *2011 IEEE PES International conference on innovative smart grid technologies-India, ISGT India 2011*, pp. 236–243. https://doi.org/10.1109/ISET-India.2011.6145388

Barreto, C., Mojica-Nava, E., & Quijano, N. (2013). Design of mechanisms for demand response programs. In *Proceedings of the IEEE conference on decision and control*, pp. 1828–1833. https://doi.org/10.1109/CDC.2013.6760148

Bayat, M., Sheshyekani, K., Hamzeh, M., & Rezazadeh, A. (2016). Coordination of distributed energy resources and demand response for voltage and frequency support of MV microgrids. *IEEE Transactions on Power Systems, 31*(2), 1506–1516. https://doi.org/10.1109/TPWRS.2015.2434938

Bayer, B. (2015). Current practice and thinking with integrating demand response for power system flexibility in the electricity markets in the USA and Germany. *Current Sustainable/Renewable Energy Reports, 2*(2), 55–62. https://doi.org/10.1007/s40518-015-0028-7

Boogert, A., & Dupont, D. (2008). When supply meets demand: The case of hourly spot electricity prices. *IEEE Transactions on Power Systems, 23*(2), 389–398. https://doi.org/10.1109/TPWRS.2008.920731

Bradley, P., Leach, M., & Torriti, J. (2013). A review of the costs and benefits of demand response for electricity in the UK. *Energy Policy, 52*, 312–327. https://doi.org/10.1016/j.enpol.2012.09.039

Bulut, M. B., Odlare, M., Stigson, P., Wallin, F., & Vassileva, I. (2016). Active buildings in smart grids – Exploring the views of the Swedish energy and buildings sectors. *Energy and Buildings, 117*, 185–198. https://doi.org/10.1016/j.enbuild.2016.02.017

Burger, S., & Chaves-ávila, J. P. (2016). The value of aggregators in electricity systems. In C. Batlle, & I. J. Pérez-Arriaga (Eds,), *MIT Center for Energy and Environmental Policy Research*, 2016, vol. CEEPR WP 2, no. January.

Burger, S., Chaves-Ávila, J. P., Batlle, C., & Pérez-Arriaga, I. J. (2017). A review of the value of aggregators in electricity systems. *Renewable and Sustainable Energy Reviews, 77*(February), 395–405. https://doi.org/10.1016/j.rser.2017.04.014

Cao, J., Yang, B., Chen, C., & Guan, X. (2012). Optimal demand response using mechanism design in the smart grid. In *Chinese Control Conference, CCC*, pp. 2520–2525.

Cappers, P., MacDonald, J., Goldman, C., & Ma, O. (2013). An assessment of market and policy barriers for demand response providing ancillary services in U.S. electricity markets. *Energy Policy, 62*, 1031–1039. https://doi.org/10.1016/j.enpol.2013.08.003

Caves, D., Eakin, K., & Faruqui, A. (2000). Mitigating price spikes in wholesale markets through market-based pricing in retail markets. *Electricity Journal, 13*(3), 13–23. https://doi.org/10.1016/S1040-6190(00)00092-0

Chakravorty, D., Chaudhuri, B., & Hui, S. Y. R. (2017). Rapid frequency response from smart loads in Great Britain power system. *IEEE Transactions on Smart Grid, 8*(5), 2160–2169. https://doi.org/10.1109/TSG.2016.2517409

Chen, T., Pourbabak, H., Liang, Z., & Su, W. (2017). An integrated eVoucher mechanism for flexible loads in real-time retail electricity market. *IEEE Access, 5*(c), 2101–2110. https://doi.org/10.1109/ACCESS.2017.2659704

Chen, T., Alsafasfeh, Q., Pourbabak, H., & Su, W. (2018a). The next-generation U.S. retail electricity market with customers and prosumers-A bibliographical survey. *Energies, 11*(1). https://doi.org/10.3390/en11010008

Chen, Y., Xu, P., Gu, J., Schmidt, F., & Li, W. (2018b). Measures to improve energy demand flexibility in buildings for demand response (DR): A review. *Energy and Buildings, 177*, 125–139. https://doi.org/10.1016/j.enbuild.2018.08.003

Chowdhury, S., Chowdhury, S. P., & Crossley, P. (2009). *Microgrids and active distribution networks*.

Conchado, A., & Linares, P. (2012). The economic impact of demand-response programs on power systems. A survey of the state of the art. *Handbook of Networks in Power Systems, 1*, 281–301. https://doi.org/10.1007/978-3-642-23193-3_11

Conchado, A., Linares, P., Lago, O., & Santamaría, A. (2016). An estimation of the economic and environmental benefits of a demand-response electricity program for Spain. *Sustainable Production and Consumption, 8*(March), 108–119. https://doi.org/10.1016/j.spc.2016.09.004

Council of European Energy Regulators. (2011). *Advice on the take-off of a demand response electricity market with smart meters*.

Deng, R., Yang, Z., Chow, M. Y., & Chen, J. (2015). A survey on demand response in smart grids: Mathematical models and approaches. *IEEE Transactions on Industrial Informatics, 11*(3), 570–582. https://doi.org/10.1109/TII.2015.2414719

Ding, Y. M., Hong, S. H., & Li, X. H. (2014). A demand response energy management scheme for industrial facilities in smart grid. *IEEE Transactions on Industrial Informatics, 10*(4), 2257–2269. https://doi.org/10.1109/TII.2014.2330995

Downward, A., Young, D., & Zakeri, G. (2016). Electricity retail contracting under risk-aversion. *European Journal of Operational Research, 251*(3), 846–859. https://doi.org/10.1016/j.ejor.2015.11.040

Early detection of turn-to-turn faults in power transformer winding: An experimental study (2019). In *2019 international aegean conference on electrical machines and power electronics (ACEMP) & 2019 international conference on optimization of electrical and electronic equipment (OPTIM)*. https://doi.org/10.1109/acemp-optim44294.2019.9007169

Eid, C., Koliou, E., Valles, M., Reneses, J., & Hakvoort, R. (2016). Time-based pricing and electricity demand response: Existing barriers and next steps. *Utilities Policy, 40*, 15–25. https://doi.org/10.1016/j.jup.2016.04.001

Eissa, M. M. (2011). Demand side management program evaluation based on industrial and commercial field data. *Energy Policy, 39*(10), 5961–5969. https://doi.org/10.1016/j.enpol.2011.06.057

El-Khattam, W., & Salama, M. M. A. (2004). Distributed generation technologies, definitions and benefits. *Electric Power Systems Research, 71*(2), 119–128. https://doi.org/10.1016/j.epsr.2004.01.006

Eric, C., Fredrich, K., & Andy, T. (2012). Maximizing the value of demand response. *The Electricity Journal, 25*(7), 6–16. https://doi.org/10.1016/j.tej.2012.08.004

Flores, M., & Waddams Price, C. (2013). Consumer behaviour in the British retail electricity market. *Centre for Competition Policy*. no. 10.

Fotouhi Ghazvini, M. A., Faria, P., Ramos, S., Morais, H., & Vale, Z. (2015). Incentive-based demand response programs designed by asset-light retail electricity providers for the day-ahead market. *Energy, 82*, 786–799. https://doi.org/10.1016/j.energy.2015.01.090

Gao, D. C., & Sun, Y. (2016). A GA-based coordinated demand response control for building group level peak demand limiting with benefits to grid power balance. *Energy and Buildings, 110*, 31–40. https://doi.org/10.1016/j.enbuild.2015.10.039

Ghatikar, G. (2014). Demand response opportunities and enabling technologies for data centers: Findings from field studies. *Lawrence Berkeley National Laboratory*.

Ghazvini, M. A. F., Ramos, S., Soares, J., Vale, Z., & Castro, R. (2016). Toward retail competition in the Portuguese electricity market. In *International conference on the European energy market, EEM*, vol. 2016-July. https://doi.org/10.1109/EEM.2016.7521209

Gilbraith, N., & Powers, S. E. (2013). Residential demand response reduces air pollutant emissions on peak electricity demand days in New York City. *Energy Policy, 59*, 459–469. https://doi.org/10.1016/j.enpol.2013.03.056

Gkatzikis, L., Koutsopoulos, I., & Salonidis, T. (2013). The role of aggregators in smart grid demand response markets. *IEEE Journal on Selected Areas in Communications, 31*(7), 1247–1257. https://doi.org/10.1109/JSAC.2013.130708

Goel, L., Qiuwei, W., & Peng, W. (2006). Reliability enhancement of a deregulated power system considering demand response. In *2006 IEEE power engineering society general meeting, PES*, pp. 1–6. https://doi.org/10.1109/pes.2006.1708965

Golmohamadi, H., Keypour, R., Hassanpour, A., & Davoudi, M. (2015). Optimization of green energy portfolio in retail market using stochastic programming. In *2015 North American power symposium, NAPS 2015*. https://doi.org/10.1109/NAPS.2015.7335233

Golmohamadi, H., Keypour, R., Bak-Jensen, B., & Pillai, J. R. (2019). A multi-agent based optimization of residential and industrial demand response aggregators. *International Journal of Electrical Power and Energy Systems, 107*(December 2018), 472–485. https://doi.org/10.1016/j.ijepes.2018.12.020

Good, N., Ellis, K. A., & Mancarella, P. (2017). Review and classification of barriers and enablers of demand response in the smart grid. *Renewable and Sustainable Energy Reviews, 72*(January 2016), 57–72. https://doi.org/10.1016/j.rser.2017.01.043

Guille, C., & Gross, G. (2009). A conceptual framework for the vehicle-to-grid (V2G) implementation. *Energy Policy, 37*(11), 4379–4390. https://doi.org/10.1016/j.enpol.2009.05.053

Guo, P., Li, V. O. K., & Lam, J. C. K. (2017). Smart demand response in China: Challenges and drivers. *Energy Policy, 107*(December 2015), 1–10. https://doi.org/10.1016/j.enpol.2017.04.019

Gyamfi, S., & Krumdieck, S. (2011). Price, environment and security: Exploring multi-modal motivation in voluntary residential peak demand response. *Energy Policy, 39*(5), 2993–3004. https://doi.org/10.1016/j.enpol.2011.03.012

Haider, H. T., See, O. H., & Elmenreich, W. (2016). A review of residential demand response of smart grid. *Renewable and Sustainable Energy Reviews, 59*, 166–178. https://doi.org/10.1016/j.rser.2016.01.016

He, J., Cai, L., Cheng, P., & Fan, J. (2015). Optimal investment for retail company in electricity market. *IEEE Transactions on Industrial Informatics, 11*(5), 1210–1219. https://doi.org/10.1109/TII.2015.2475215

Ikäheimo, J., Evens, C., & Kärkkäinen, S. (2010). DER Aggregator business: The Finnish case. *Technical Research Centre of Finland (VTT)*.

Ilieva, I., Bremdal, B., Ottesen, S. Ø., Rajasekharan, J., & Olivella-Rosell, P. (*2016*). Design characteristics of a smart grid dominated local market. *CIRED Workshop – Helsinki, 2016.*

Imani, M. H., Ghadi, M. J., Ghavidel, S., & Li, L. (2018). Demand response modeling in microgrid operation: A review and application for incentive-based and time-based programs. *Renewable and Sustainable Energy Reviews, 94*(June), 486–499. https://doi.org/10.1016/j.rser.2018.06.017

Ivanov, C., Getachew, L., Fenrick, S. A., & Vittetoe, B. (2013). Enabling technologies and energy savings: The case of EnergyWise smart meter pilot of Connexus energy. *Utilities Policy, 26*, 76–84. https://doi.org/10.1016/j.jup.2012.10.001

Javadi, M., Marzband, M., Akorede, M. F., Godina, R., Al-Sumaiti, A. S., & Pouresmaeil, E. (2018). A centralized smart decision-making hierarchical interactive architecture for multiple home microgrids in retail electricity market. *Energies, 11*(11), 1–22. https://doi.org/10.3390/en11113144

Joung, M., & Kim, J. (2013). Assessing demand response and smart metering impacts on long-term electricity market prices and system reliability. *Applied Energy, 101*, 441–448. https://doi.org/10.1016/j.apenergy.2012.05.009

Kharrati, S., Kazemi, M., & Ehsan, M. (2016). Equilibria in the competitive retail electricity market considering uncertainty and risk management. *Energy, 106*, 315–328. https://doi.org/10.1016/j.energy.2016.03.069

Khezri, R., Oshnoei, A., Hagh, M. T., & Muyeen, S. M. (2018). Coordination of heat pumps, electric vehicles and AGC for efficient LFC in a smart hybrid power system via SCA-based optimized FOPID controllers. *Energies, 11*(2). https://doi.org/10.3390/en11020420

Kim, J. H., & Shcherbakova, A. (2011). Common failures of demand response. *Energy, 36*(2), 873–880. https://doi.org/10.1016/j.energy.2010.12.027

Kim, H., & Thottan, M. (2011). A two-stage market model for microgrid power transactions via aggregators. *Bell Labs Technical Journal, 16*(3), 101–107. https://doi.org/10.1002/bltj.20524

Kim, S., Mowakeaa, R., Kim, S. J., & Lim, H. (2020). Building energy management for demand response using kernel lifelong learning. *IEEE Access, 8*, 82131–82141. https://doi.org/10.1109/ACCESS.2020.2991110

Kuleshov, D., Viljainen, S., Annala, S., & Gore, O. (2012). Russian electricity sector reform: Challenges to retail competition. *Utilities Policy, 23*, 40–49. https://doi.org/10.1016/j.jup.2012.05.001

Leerbeck, K., Bacher, P., Junker, R. G., Tveit, A., Corradi, O., & Madsen, H. (2020). Control of heat pumps with CO_2 emission intensity forecasts. *Energies, 13*(11), 1–19. https://doi.org/10.3390/en13112851

Lehto, E. (2011). Electricity prices in the Finnish retail market. *Energy Policy, 39*(4), 2179–2192. https://doi.org/10.1016/j.enpol.2011.02.007

Liang, Z., Bian, D., Zhang, X., Shi, D., Diao, R., & Wang, Z. (2019). Optimal energy management for commercial buildings considering comprehensive comfort levels in a retail electricity market. *Applied Energy, 236*, 916–926. https://doi.org/10.1016/j.apenergy.2018.12.048

Mansour-Saatloo, A., Moradzadeh, A., Mohammadi-Ivatloo, B., Ahmadian, A., & Elkamel, A. (2020). Machine learning based PEVs load extraction and analysis. *Electronics (Switzerland), 9*(7), 1–15. https://doi.org/10.3390/electronics9071150

Martinez, V. J., & Rudnick, H. (2012). Design of demand response programs in emerging countries. In *2012 IEEE international conference on power system technology, POWERCON 2012*, pp. 1–6. https://doi.org/10.1109/PowerCon.2012.6401387

Marzband, M., Javadi, M., Domínguez-García, J. L., & Moghaddam, M. M. (2016). Non-cooperative game theory based energy management systems for energy district in the retail market considering DER uncertainties. *IET Generation, Transmission and Distribution, 10*(12), 2999–3009. https://doi.org/10.1049/iet-gtd.2016.0024

Marzband, M., Javadi, M., Pourmousavi, S. A., & Lightbody, G. (2018). An advanced retail electricity market for active distribution systems and home microgrid interoperability based on game theory. *Electric Power Systems Research, 157*, 187–199. https://doi.org/10.1016/j.epsr.2017.12.024

Mathieu, J. L., Haring, T., Ledyard, J. O., & Andersson, G. (2013). Residential Demand Response program design: Engineering and economic perspectives. In *International conference on the European energy market, EEM*. https://doi.org/10.1109/EEM.2013.6607296

Mehrtash, M., Capitanescu, F., Heiselberg, P. K., Gibon, T., & Bertrand, A. (2020). An enhanced optimal PV and battery sizing model for zero energy buildings considering environmental impacts. *IEEE Transactions on Industry Applications, 56*(6), 6846–6856. https://doi.org/10.1109/TIA.2020.3022742

Mhanna, S., Verbič, G., & Chapman, A. C. (2014). Towards a realistic implementation of mechanism design in demand response aggregation. In *Proceedings – 2014 power systems computation conference, PSCC 2014*.https://doi.org/10.1109/PSCC.2014.7038379

Mirzaei, M. A., Yazdankhah, A. S., & Mohammadi-Ivatloo, B. (2018). Integration of demand response and hydrogen storage system in security constrained unit commitment with high penetration of wind energy. In *26th Iranian conference on electrical engineering, ICEE 2018*, pp. 1203–1208. https://doi.org/10.1109/ICEE.2018.8472631

Mirzaei, M. A., Sadeghi Yazdankhah, A., & Mohammadi-Ivatloo, B. (2019). Stochastic security-constrained operation of wind and hydrogen energy storage systems integrated with price-based demand response. *International Journal of Hydrogen Energy, 44*(27), 14217–14227. https://doi.org/10.1016/j.ijhydene.2018.12.054

Mirzaei, M. A., et al. (2020). An integrated energy hub system based on power-to-gas and compressed air energy storage technologies in presence of multiple shiftable loads. *IET Generation, Transmission & Distribution*. https://doi.org/10.1049/iet-gtd.2019.1163

Momber, I., Wogrin, S., & Gomez San Roman, T. (2016). Retail pricing: A bilevel program for PEV aggregator decisions using indirect load control. *IEEE Transactions on Power Systems, 31*(1), 464–473. https://doi.org/10.1109/TPWRS.2014.2379637

Moradzadeh, A., & Pourhossein, K. (2019a). Short circuit location in transformer winding using deep learning of its frequency responses. In *Proceedings 2019 international aegean conference on electrical machines and power electronics, ACEMP 2019 and 2019 international conference on optimization of electrical and electronic equipment, OPTIM 2019*, pp. 268–273. https://doi.org/10.1109/ACEMP-OPTIM44294.2019.9007176

Moradzadeh, A., & Pourhossein, K. (2019b). Application of support vector machines to locate minor short circuits in transformer windings. In *2019 54th international universities power engineering conference, UPEC 2019 – Proceedings*, pp. 1–6. https://doi.org/10.1109/UPEC.2019.8893542

Moradzadeh, A., & Pourhossein, K. (2020). PCA-assisted location of small short circuit in transformer winding. In *2020 28th Iranian conference on electrical engineering, ICEE 2020*, pp. 1–6. https://doi.org/10.1109/ICEE50131.2020.9260815

Moradzadeh, A., Sadeghian, O., Pourhossein, K., Mohammadi-Ivatloo, B., & Anvari-Moghaddam, A. (2020a). Improving residential load disaggregation for sustainable development of energy via principal component analysis. *Sustainability (Switzerland), 12*(8), 3158. https://doi.org/10.3390/SU12083158

Moradzadeh, A., Mansour-Saatloo, A., Mohammadi-Ivatloo, B., & Anvari-Moghaddam, A. (2020b). Performance evaluation of two machine learning techniques in heating and cooling loads forecasting of residential buildings. *Applied Sciences (Switzerland), 10*(11), 3829. https://doi.org/10.3390/app10113829

Moradzadeh, A., Zakeri, S., Shoaran, M., Mohammadi-Ivatloo, B., & Mohammadi, F. (2020c). Short-term load forecasting of microgrid via hybrid support vector regression and long short-term memory algorithms. *Sustainability (Switzerland), 12*(17), 7076. https://doi.org/10.3390/su12177076

Moradzadeh, A., Zeinal-Kheiri, S., Mohammadi-Ivatloo, B., Abapour, M., & Anvari-Moghaddam, A. (2020d). Support vector machine-assisted improvement residential load disaggregation. In *2020 28th Iranian conference on electrical engineering (ICEE)*, pp. 1–6. https://doi.org/10.1109/ICEE50131.2020.9260869

Moradzadeh, A., Mohammadi-Ivatloo, B., Abapour, M., Anvari-Moghaddam, A., Gholami Farkoush, S., & Rhee, S.-B. (2021a). A practical solution based on convolutional neural network for non-intrusive load monitoring. *Journal of Ambient Intelligence and Humanized Computing*. https://doi.org/10.1007/s12652-020-02720-6

Moradzadeh, A., Moayyed, H., Zakeri, S., Mohammadi-Ivatloo, B., & Aguiar, A. P. (2021b). Deep learning-assisted short-term load forecasting for sustainable management of energy in microgrid. *Inventions, 6*(1), 15. https://doi.org/10.3390/inventions6010015

Moradzaeh, A., & Khaffafi, K. (2018). Comparison and evaluation of the performance of various types of neural networks for planning issues related to optimal management of charging and discharging electric cars in intelligent power grids. *Emerging Science Journal, 1*(4). https://doi.org/10.28991/ijse-01123

Naing, W. O., & Miranda, V. (2005). Multi-energy retail market simulation with intelligent agents. In *2005 IEEE Russia Power Tech, PowerTech*. https://doi.org/10.1109/PTC.2005.4524755

Nguyen, D. T., Negnevitsky, M., De Groot, M., & Wang, C. (2009). Demand response in the retail market: Benefits and challenges. In *AUPEC'09 – 19th Australasian universities power engineering conference: Sustainable energy technologies and systems*, pp. 1–6.

Nikolic, D., Negnevitsky, M., De Groot, M., Gamble, S., Forbes, J., & Ross, M. (2016). Fast demand response as an enabling technology for high renewable energy penetration in isolated power systems. *CIGRE Session, 46*.

Nolan, S., & O'Malley, M. (2015). Challenges and barriers to demand response deployment and evaluation. *Applied Energy, 152*, 1–10. https://doi.org/10.1016/j.apenergy.2015.04.083

Oconnell, N., Pinson, P., Madsen, H., & Omalley, M. (2014). Benefits and challenges of electrical demand response: A critical review. *Renewable and Sustainable Energy Reviews, 39*, 686–699. https://doi.org/10.1016/j.rser.2014.07.098

Oshnoei, A. (2017). Application of IPSO and fuzzy logic methods in electrical vehicles for efficient frequency control of multi-area power systems. In *2017 Iranian conference on electrical engineering (ICEE)*, pp. 1349–1354.

Palensky, P., & Dietrich, D. (2011). Demand side management: Demand response, intelligent energy systems, and smart loads. *IEEE Transactions on Industrial Informatics, 7*(3), 381–388. https://doi.org/10.1109/TII.2011.2158841

Papavasiliou, A., & Oren, S. S. (2014). Large-scale integration of deferrable demand and renewable energy sources. *IEEE Transactions on Power Systems, 29*(1), 489–499. https://doi.org/10.1109/TPWRS.2013.2238644

Parvania, M., Fotuhi-Firuzabad, M., & Shahidehpour, M. (2012). Demand response participation in wholesale energy markets. In *IEEE power and energy society general meeting*, pp. 1–4. https://doi.org/10.1109/PESGM.2012.6344591

Paterakis, N. G., Erdinç, O., & Catalão, J. P. S. (2017). An overview of demand response: Key-elements and international experience. *Renewable and Sustainable Energy Reviews, 69*(July 2016), 871–891. https://doi.org/10.1016/j.rser.2016.11.167

Pepermans, G., Driesen, J., Haeseldonckx, D., Belmans, R., & D'haeseleer, W. (2005). Distributed generation: Definition, benefits and issues. *Energy Policy, 33*(6), 787–798. https://doi.org/10.1016/j.enpol.2003.10.004

Ponds, K. T., Arefi, A., Sayigh, A., & Ledwich, G. (2018). Aggregator of demand response for renewable integration and customer engagement: Strengths, weaknesses, opportunities, and threats, *11*(*Energies*, 9). https://doi.org/10.3390/en11092391

Rabiee, A., Soroudi, A., Mohammadi-Ivatloo, B., & Parniani, M. (2014). Corrective voltage control scheme considering demand response and stochastic wind power. *IEEE Transactions on Power Systems, 29*(6), 2965–2973. https://doi.org/10.1109/TPWRS.2014.2316018

Rafinia, A., Moshtagh, J., & Rezaei, N. (2020). Towards an enhanced power system sustainability: An MILP under-frequency load shedding scheme considering demand response resources. *Sustainable Cities and Society*, 102168. https://doi.org/10.1016/j.scs.2020.102168

Rahimi, F., & Ipakchi, A. (2010). Demand response as a market resource under the smart grid paradigm. *IEEE Transactions on Smart Grid, 1*(1), 82–88. https://doi.org/10.1109/TSG.2010.2045906

Rahman, A., Srikumar, V., & Smith, A. D. (2018). Predicting electricity consumption for commercial and residential buildings using deep recurrent neural networks. *Applied Energy, 212*, 372–385. https://doi.org/10.1016/j.apenergy.2017.12.051

Razmara, M., Bharati, G. R., Hanover, D., Shahbakhti, M., Paudyal, S., & Robinett, R. D. (2017). Building-to-grid predictive power flow control for demand response and demand flexibility programs. *Applied Energy, 203*, 128–141. https://doi.org/10.1016/j.apenergy.2017.06.040

Sadeghian, O., Nazari-Heris, M., Abapour, M., Taheri, S. S., & Zare, K. (2019a). Improving reliability of distribution networks using plug-in electric vehicles and demand response. *Journal of Modern Power Systems and Clean Energy, 7*(5), 1189–1199. https://doi.org/10.1007/s40565-019-0523-8

Sadeghian, O., Shotorbani, A. M., & Mohammadi-Ivatloo, B. (2019b). Generation maintenance scheduling in virtual power plants. *IET Generation, Transmission and Distribution, 13*(12), 2584–2596. https://doi.org/10.1049/iet-gtd.2018.6751

Sadeghian, O., Oshnoei, A., Nikkhah, S., & Mohammadi-Ivatloo, B. (2019c). Multi-objective optimisation of generation maintenance scheduling in restructured power systems based on global criterion method. *IET Smart Grid, 2*(2), 203–213. https://doi.org/10.1049/iet-stg.2018.0140

Sadeghian, O., Mohammadpour Shotorbani, A., & Mohammadi-Ivatloo, B. (2019d). Risk-based stochastic short-term maintenance scheduling of GenCos in an oligopolistic electricity market considering the long-term plan. *Electric Power Systems Research, 175*, 105908. https://doi.org/10.1016/j.epsr.2019.105908

Sadeghian, O., Oshnoei, A., Khezri, R., & Muyeen, S. M. (2020a). Risk-constrained stochastic optimal allocation of energy storage system in virtual power plants. *Journal of Energy Storage, 31*(January), 101732. https://doi.org/10.1016/j.est.2020.101732

Sadeghian, O., Oshnoei, A., Kheradmandi, M., Khezri, R., & Mohammadi-Ivatloo, B. (2020b). A robust data clustering method for probabilistic load flow in wind integrated radial distribution networks. *International Journal of Electrical Power and Energy Systems, 115*, 105392. https://doi.org/10.1016/j.ijepes.2019.105392

Sadeghian, O., Oshnoei, A., Kheradmandi, M., & Mohammadi-ivatloo, B. (2020c). Optimal placement of multi-period-based switched capacitor in radial distribution systems. *Computers and Electrical Engineering*, 1–25.

Sadeghian, O., Moradzadeh, A., Mohammadi-Ivatloo, B., Mohammadi-Ivatloo, B., Abapour, M., & Marquez, F. P. G. (2020d). Generation units maintenance in combined heat and power integrated systems using the mixed integer quadratic programming approach. *Energies, 13*(11), 2840. https://doi.org/10.3390/en13112840

Sadineni, S. B., & Boehm, R. F. (2012). Measurements and simulations for peak electrical load reduction in cooling dominated climate. *Energy, 37*(1), 689–697. https://doi.org/10.1016/j.energy.2011.10.026

Safdarian, A., Fotuhi-Firuzabad, M., & Lehtonen, M. (2014a). A distributed algorithm for managing residential demand response in smart grids. *IEEE Transactions on Industrial Informatics, 10*(4), 2385–2393. https://doi.org/10.1109/TII.2014.2316639

Safdarian, A., Degefa, M. Z., Lehtonen, M., & Fotuhi-Firuzabad, M. (2014b). Distribution network reliability improvements in presence of demand response. *IET Generation, Transmission and Distribution, 8*(12), 2027–2035. https://doi.org/10.1049/iet-gtd.2013.0815

Safdarian, A., Fotuhi-Firuzabad, M., & Lehtonen, M. (2016). Benefits of demand response on operation of distribution networks: A case study. *IEEE Systems Journal, 10*(1), 189–197. https://doi.org/10.1109/JSYST.2013.2297792

Santinelli, G., Beillan, V., Monteverdi, I., Jalmain, I., Decorme, R., & Tatibouet, M. (2016). 'Good causes' crowdfunding as a driver for behavioural change in demand response scenarios. In *4th European conference on behaviour and energy efficiency (Behave 2016)*.

Sekizaki, S., Nishizaki, I., & Hayashida, T. (2016). Electricity retail market model with flexible price settings and elastic price-based demand responses by consumers in distribution network. *International Journal of Electrical Power and Energy Systems, 81*, 371–386. https://doi.org/10.1016/j.ijepes.2016.02.029

Shafie-Khah, M., et al. (2016). Optimal behavior of electric vehicle parking lots as demand response aggregation agents. *IEEE Transactions on Smart Grid, 7*(6), 2654–2665. https://doi.org/10.1109/TSG.2015.2496796

Sharma, A., & Sharma, H. (2019). Demand side response: Drivers, challenges, and opportunities. In *International conference on advancements in computing & management (ICACM-2019)*, pp. 754–759. https://doi.org/10.2139/ssrn.3462969

Shin, K. J., & Managi, S. (2017). Liberalization of a retail electricity market: Consumer satisfaction and household switching behavior in Japan. *Energy Policy, 110*(January), 675–685. https://doi.org/10.1016/j.enpol.2017.07.048

Siano, P. (2014). Demand response and smart grids – A survey. *Renewable and Sustainable Energy Reviews, 30*, 461–478. https://doi.org/10.1016/j.rser.2013.10.022

Siano, P., & Sarno, D. (2016). Assessing the benefits of residential demand response in a real time distribution energy market. *Applied Energy, 161*, 533–551. https://doi.org/10.1016/j.apenergy.2015.10.017

Smith, K., & Hledik, R. (2011). Drivers of Demand response adoption: Past, present, and future. *Institute for Building Efficiency*.

Spees, K., & Lave, L. B. (2007). Demand response and electricity market efficiency. *The Electricity Journal, 20*(3), 69–85. https://doi.org/10.1016/j.tej.2007.01.006

Stavrakas, V., & Flamos, A. (2020). A modular high-resolution demand-side management model to quantify benefits of demand-flexibility in the residential sector. *Energy Conversion and Management, 205*(August 2019), 112339. https://doi.org/10.1016/j.enconman.2019.112339

Strbac, G. (2008). Demand side management: Benefits and challenges. *Energy Policy, 36*(12), 4419–4426. https://doi.org/10.1016/j.enpol.2008.09.030

Su, X. (2015). Have customers benefited from electricity retail competition? *Journal of Regulatory Economics, 47*(2), 146–182. https://doi.org/10.1007/s11149-014-9263-x

Su, C. L., & Kirschen, D. (2009). Quantifying the effect of demand response on electricity markets. *IEEE Transactions on Power Systems, 24*(3), 1199–1207. https://doi.org/10.1109/TPWRS.2009.2023259

Swadley, A., & Yücel, M. (2011). Did residential electricity rates fall after retail competition? A dynamic panel analysis. *Energy Policy, 39*(12), 7702–7711. https://doi.org/10.1016/j.enpol.2011.09.014

Teimourzadeh, H., Moradzadeh, A., Shoaran, M., Mohammadi-Ivatloo, B., & Razzaghi, R. (2021). High impedance single-phase faults diagnosis in transmission lines via deep reinforcement learning of transfer functions. *IEEE Access, 9*, 15796–15809. https://doi.org/10.1109/ACCESS.2021.3051411

Torstensson, D., & Wallin, F. (2015). Potential and barriers for demand response at household customers. *Energy Procedia, 75*, 1189–1196. https://doi.org/10.1016/j.egypro.2015.07.570

U. department of Energy. (2005). Benefits of demand response in electricity markets and recommendations for achieving them. In *A report to The United States Congress*. https://doi.org/10.1503/cmaj.1070122

Vahidinasab, V., Ardalan, C., Mohammadi-Ivatloo, B., Giaouris, D., & Walker, S. L. (2021). Active building as an energy system: Concept, challenges, and outlook. *IEEE Access*, 1–1. https://doi.org/10.1109/ACCESS.2021.3073087

Vallés, M., Reneses, J., Cossent, R., & Frías, P. (2016). Regulatory and market barriers to the realization of demand response in electricity distribution networks: A European perspective. *Electric Power Systems Research, 140*, 689–698. https://doi.org/10.1016/j.epsr.2016.04.026

Van Dievel, P., De Vos, K., & Belmans, R. (2014). Demand response in electricity distribution grids: Regulatory framework and barriers. In *International conference on the European energy market, EEM*. https://doi.org/10.1109/EEM.2014.6861286

Vand, B., Martin, K., Jokisalo, J., Kosonen, R., & Hast, A. (2020). Demand response potential of district heating and ventilation in an educational office building. *Science and Technology for the Built Environment, 26*(3), 304–319. https://doi.org/10.1080/23744731.2019.1693207

Verbong, G. P. J., Beemsterboer, S., & Sengers, F. (2013). Smart grids or smart users? Involving users in developing a low carbon electricity economy. *Energy Policy, 52*, 117–125. https://doi.org/10.1016/j.enpol.2012.05.003

Wang, Y., Pordanjani, I. R., & Xu, W. (2011). An event-driven demand response scheme for power system security enhancement. *IEEE Transactions on Smart Grid, 2*(1), 23–29. https://doi.org/10.1109/TSG.2011.2105287

Warren, P. (2014). A review of demand-side management policy in the UK. *Renewable and Sustainable Energy Reviews, 29*, 941–951. https://doi.org/10.1016/j.rser.2013.09.009

Yamaguchi, Y., et al. (2020). An integrated approach of estimating demand response flexibility of domestic laundry appliances based on household heterogeneity and activities. *Energy Policy, 142*(March), 111467. https://doi.org/10.1016/j.enpol.2020.111467

Yan, X., Ozturk, Y., Hu, Z., & Song, Y. (2018). A review on price-driven residential demand response. *Renewable and Sustainable Energy Reviews, 96*(August), 411–419. https://doi.org/10.1016/j.rser.2018.08.003

Yang, Y. (2014). Understanding household switching behavior in the retail electricity market. *Energy Policy, 69*, 406–414. https://doi.org/10.1016/j.enpol.2014.03.009

Yao, M., Hu, Z., Zhang, N., Duan, W., & Zhang, J. (2015). Low-carbon benefits analysis of energy-intensive industrial demand response resources for ancillary services. *Journal of Modern Power Systems and Clean Energy, 3*(1), 131–138. https://doi.org/10.1007/s40565-015-0102-6

Yoon, A. Y., Kang, H. K., & Moon, S. (2020). Optimal price based demand response of HVAC systems in commercial buildings considering peak load reduction. *Energies, 13*(4). https://doi.org/10.3390/en13040862

Yu, J., Zhou, J. Z., Yang, J. J., Wu, W., Fu, B., & Liao, R. T. (2004). Agent-based retail electricity market: Modeling and analysis. In *Proceedings of 2004 international conference on machine learning and cybernetics*, Vol. 1, no. August, pp. 95–100. https://doi.org/10.1109/icmlc.2004.1380618

Zakariazadeh, A., Homaee, O., Jadid, S., & Siano, P. (2014). A new approach for real time voltage control using demand response in an automated distribution system. *Applied Energy, 117*, 157–166. https://doi.org/10.1016/j.apenergy.2013.12.004

Zhang, Q., & Li, J. (2012). Demand response in electricity markets: A review. In *9th international conference on the european energy market, EEM 12*. https://doi.org/10.1109/EEM.2012.6254817

Zhang, X., Hug, G., Kolter, Z., & Harjunkoski, I. (2015). Industrial demand response by steel plants with spinning reserve provision. In *2015 North American power symposium, NAPS 2015*. https://doi.org/10.1109/NAPS.2015.7335115

Zipperer, A., et al. (2013). Electric energy management in the smart home: Perspectives on enabling technologies and consumer behavior. *Proceedings of the IEEE, 101*(11), 2397–2408. https://doi.org/10.1109/JPROC.2013.2270172

Index

A

Active buildings (AB), 52, 220, 222–225, 273–275
- ancillary services, 13–14
- and active homes, 28
- challenges, 10–11
- comparison of energy storages in, 64–66
- cybersecurity research roadmap for, 238
  - efficiency codes, 238–239
  - emergent techniques, 241
  - end-user related, 241
  - incident response, 241
  - legal aspects, 241
  - market features, 239
  - mentioning, 242
  - modelling and simulation, 240–241
  - power-based attacks, 240
  - risk assessments, 238
  - smart grid components, 239
- data security and privacy, 29
- decarbonization, 8
- definition, 2, 26
- demand-side response programs, 5–6
  - aggregate loads, 16
  - autonomous demand-side response programs, 17–18
  - energy storage-based demand-side response programs, 17
- digitalisation of buildings and energy services, 28
- distributed generation in, 8–10
- enabling technologies, 13
- energy management system, role of, 288–290
- energy sharing, 7–8, 16
- energy storage systems, 9–10
  - electric vehicles, 14
  - stationary energy storage systems, 14
  - thermal energy storage systems, 15
- flexibility services, 278
  - DSO-level flexibility services, 284–287
  - TSO-level flexibility services, 278–284
- flexible appliances and devices, 275
  - high-flexible appliances and devices, 276–277
  - low-flexible appliances and devices, 278
  - medium-flexible appliances and devices, 277
- hybrid AC/DC electrical power grids in, 71–72, 74–75
  - advantages of, 75–76
  - contribution and organization, 73–74
  - DC-DC, 82–84, 88–89
  - load-shift systems for, 79
  - single-phase AC-DC converter, 80–81, 86–88
  - smart home with, 89–92
  - status quo, challenges, 72–73
  - structures of, 78–79
  - three-phase AC-DC converter, 81–82, 84–86
  - unipolar and bipolar, 76–77
- load forecasting
  - load disaggregation techniques, 12–13
  - model predictive control approaches, 12
- main keywords of previous works, 4
- major features, 5
- passive and active energy management systems, 15

V. Vahidinasab, B. Mohammadi-Ivatloo (eds.), *Active Building Energy Systems*, Green Energy and Technology, https://doi.org/10.1007/978-3-030-79742-3

Active buildings (AB) (*cont.*)
  and predecessor conceptualisations, standards and certificates, 27
  providing ancillary services by, 6–7
  renewable energy sources, 18
  schematic diagram, 3
  self-generation of electricity, 2
  vs smart building, 3
Active energy management system (AEMS), 328
Active homes
  active buildings as, 26–29
  active housing developments, 36–40
  challenges, 43–44
  contextual landscape, 30
  dynamic socio-technical configurations, 26
  energy and homes in transition, 29
  lived experience, 30
  location, multi-site
    energy sufficient, 33–34
    environmental issues, 34
    health and wellbeing, 34–35
    renewable energy production, 32
    system-change, 33
  multimethod research, 42–43
  narratives and subjectivity, 30
  outlook, 43–44
  people, 30
  qualitative longitudinal research, multistage, 35, 41, 42
  research areas, 31
  research project, 32
Advanced metering infrastructure (AMI), 223
Advanced persistent threats (APT), 224, 240
Aggregators
  conceptual diagram of role of, 357
  definition, 358
  electric vehicle aggregator, 359
  "fundamental" or "intrinsic" value, 359
  indirect load control program, 359
  microgrids, 359
  "opportunistic" value, 359
  in power markets, 358
  private value, 358
  system value, 358
  "transitory" value, 359
  types, 358
Air-source heat pumps, 106
Airbnb, 295
AMPds dataset, 348–349
Ancillary services, 13–14, 296
Artificial intelligence, 264, 267
Attack surface, 225–226
Attack trees, 235
Automatic frequency restoration reserve (aFRR), 283–284
Autonomous demand-side response programs, 17–18

B
Back-end DC-DC converter, 83, 84, 88, 91
Battery energy storage systems (BESS), 57
Battery storage systems (BSS), 302–303
Bidding strategy, 290
Biophilia, 34
Building Controls Virtual Test Bed (BCVTB), 102
Building energy management system (BEMS), 252
Building management systems (BMS), 223
Building-to-grid strategy, 17

C
Clustering algorithm, 337–339, 348
Co-design
  closed-loop control, 132
  optimisation, 131–132
  predictive control
    model predictive control, 133
    residential buildings, 135–138
    stochastic model predictive control, 134
  residential buildings, 130
  simultaneous optimisation, 131
  software tools, 134–135
    continuous-time optimisation problem, 135
    finite-dimensional optimisation problem, 135
    ICLOCS2, 134, 135
Coefficient of performance (COP), 105
Collaborative economy, 295, 296
Combined heat and power (CHP) systems, 304–305
Common Vulnerability and Exposures (CVE) database, 232
Common Vulnerability Scoring System (CVSS), 232
Congestion management, 285–286
Control approaches
  energy services provision, 161–162
  poor quality control, 163
  smart grid control, 162
  system non-linearity, 162

Control methods and techniques
data driven control
artificial neural networks, 173–175
backpropogation, 174
deep neural networks, 173
genetic programming, 175–176
machine learning (ML), 17
overfitting, 174
parameter synthesis, 175
recurrent neural networks, 173
reinforcement learning, 173
game theoretic approaches
best response dynamics, 177
cooperative game theory, 176
fictitious play, 177
Nash bargaining strategy, 176
noncooperative game theory, 176
regret matching, 177
model predictive control, 169–170
multi-agent system approaches, 170–171
intelligent agent, 171
learning agent, 171
maintenance cost, 172
proactivity of system, 172
reactive agent, 171
PID control, 168
receding horizon control (RHC), 169–170
supervisory control and data acquisition (SCADA), 166–168
control system architecture, 167
human-machine interface, 167
Internet of Things (IoT), 167
Control of population of buildings
aggregate models, 194–195
coordinated control (*See* Coordinated control of active buildings)
coordinated volt/var control, 197–200
coordinated control, 197
corrective control, 197
distribution network, 198–199
preventative control, 197
transmission network, 197–198
transmission/distributed system operator coordination, 199–200
demand response, 194
energy storage systems (ESS), 197
frequency instability, 195
inertia emulation, 196
maximum power point (MPP), 197
peak load shifting, 196
rate of change of frequency, 196
virtual synchronous generators, 196
Controlled stress testing, 240
Coordinated control of active buildings
centralised, 178–180
frequency response, 180
load-frequency control (LFC) mode, 178
low frequency demand disconnection, 179
model for participation, 178
simple LFC model, 179
step load disturbance, 178
system values for simulation, 178
cyber-physical system, 201–203
access control, 202
authentication, 202
availability, 202
confidentiality, 202
integrity, 202
non-repudiation, 202
SMA Solar Technology, 202
decentralised, 180–185
base controller, 183
frequency response, 181
frequency response improvements, 184
global frequency measurements, 185
simple LFC model, 182
system values for simulation, 182
distributed, 185–188
frequency measurements, 186, 188
frequency response, 189, 190
global frequency measurements, 187
frequency response services, 213–215
symbolic control, 215
TCL reference tracking, 214
TCL temperature setpoint, 214
multi-vector nano energy hub, 200–201
black box model, 201
grey box model, 201
white box model, 201
Coordinated structures, energy systems
energy perspective, 165–166
lens of security, 166
system perspective, 163–165
centralised scheme, 164
decentralised scheme, 165
distributed scheme, 165
Cost minimization, 254–255
Current source converter (CSC), 54
Cyber-attacks, 221, 229
Cyber-physical counterparts, 240
Cyber-physical system (CPS), 219
Cybersecurity practitioners, 232
Cybersecurity roadmap
active buildings, 222–225
attack surface, 225–226

Cybersecurity roadmap (*cont.*)
interfacing with transactive energy market, 226–227
cyber-physical security, 227
business stakeholders and cybersecurity, 238
cybersecurity perspective, 237
intrusion detection, 233–235
modelling and simulation, 236–237
risk assessments, 230–231
security metrics, 232–233
smart grid cybersecurity, 228–230
threat modelling, 235–236
roadmap for
efficiency codes, 238–239
emergent techniques, 241
end-user related, 241
incident response, 241
legal aspects, 241
market features, 239
mentioning, 242
modelling and simulation, 240–241
power-based attacks, 240
risk assessments, 238
smart grid components, 239
Cybersecurity-oriented simulation, 236

D
Data-driven control schemes, 190
DC-DC converter, 82–84, 88–89
Decarbonisation, 8, 29, 30, 44, 328
Demand response (DR) programs
aggregators
conceptual diagram of role of, 357
definition, 358
electric vehicle aggregator, 359
"fundamental" or "intrinsic" value, 359
indirect load control program, 359
microgrids, 359
"opportunistic" value, 359
in power markets, 358
private value, 358
system value, 358
"transitory" value, 359
types, 358
barriers, 365–366
benefits
CO_2 emission improvement, 364
economic benefits, 363
end-users, 365
environment, 365
event-driven emergency DR scheme, 364
power system, 365
renewable energy, 364
residential DR, 363–364
designing efficient DR mechanism, 367–368
drivers/motivators, 366–367
enabling technologies, 370–371
incentive-based programs, 369
price-based programs, 368, 369
retail electricity market, 360–361
Demand side response (DSR), 5–6, 16–18, 193, 297, 298, 300, 301, 304, 307, 312
Demand-side management, 79, 360, 362
Demand-side storage systems, 79
Department for Business, Energy and Industrial Strategy (BEIS), 43
Direct current (DC) storage, 53
Distributed energy resources (DER), 193, 258, 273, 295, 297–302, 304, 305, 307, 309–311, 313
Distributed generation (DG), 285
Distribution system operator (DSO), 273, 298
Domestic hot water (DHW), 108
DSO-level flexibility services, 284–287
congestion management, services related to, 285–286
voltage control, services related to, 286–287
Dynamic voltage restorer (DVR), 56

E
Electric double-layer capacitors (ELDC), 55
Electric vehicles, 14, 62–64
Electrical grids, 71, 72
Electrical power grids (EPG), 222
Electricity network service provider, active building, 273–275
energy management system, role of, 288–290
flexibility services, 278
DSO-level flexibility services, 284–287
TSO-level flexibility services, 278–284
flexible appliances and devices, 275
high-flexible appliances and devices, 276–277
low-flexible appliances and devices, 278
medium-flexible appliances and devices,–277
Electrification, 99, 105

Electrochemical storage systems, 57
advanced batteries
lithium-ion (Li-ion), 60–61
sodium sulphur, 61
conventional batteries
lead-acid battery (Pb-acid), 57–58
nickel cadmium, 59
nickel metal hydride, 59–60
flow batteries, 61–62
Enabling technologies, 13
End Users, 30
Energy disaggregation accuracy metrics, 344
Energy flexibility, 345–348
Energy management system (EMS), 252–253, 274, 288–290
constraints of
controllable load appliances' constraints, 260–261
grid exchange limitation, 263
thermal comfortness, 262
cost minimization, 254–255
flexibility services, 257–259
modeling energy management of, 263–264
artificial intelligence and machine learning approaches, 267
optimization-based approaches, 264–266
rule-based approaches, 266
multi-objective function, 259–260
peak shaving, 255–257
Energy sharing, among buildings, 16
Energy storage systems (ESS), 51–52, 193, 220
ABs, comparison of energy storages in, 64–66
electric vehicles, 14, 62–64
static storages
electrical storage, 52–56
electrochemical storage systems, 57–63
stationary energy storage systems, 14
thermal energy storage systems, 15
Energy storage-based demand-side response programs, 17
Energy sustainability, 18
Energy-sufficient developments, 33–34
EnergyPlus, 236
European Network and Information Security Agency (ENISA), 231
European Programme for Critical Infrastructure Protection (EPCIP), 225
European Telecommunications Standards Institute's (ETSI's), 231
Event-based NILM algorithms, 342
Event-detection accuracy metric, 342–343

F
False-positive rate (FPR), 342
Fast frequency reserve (FFR), 282, 284
Fault-Induced Delayed Voltage Recovery (FIDVR), 198
Federated power plants, 307
Finnish distribution network, 17
Flexibility service, 257–259, 278, 289–290
DSO-level flexibility services, 284–287
TSO-level flexibility services, 278–284
Flexible appliances, 275
high-flexible appliances and devices, 276–277
low-flexible appliances and devices, 278
medium-flexible appliances and devices,–277
Flexible power system, 66
Flow batteries, 61–62
Formal control synthesis, 204–206
behaviour, 206
determinism, 204
equivalence property, 204–205
of fast-response elements, 207–210
baseline controller, 208
control strategy, 209
formal specification, 207
SCOTS software tool, 208
symbolic controller, 207, 209, 210
symbolic model, 207
infinite-state systems, 204
non-determinism, 205
symbolic model, 206
system specifications, 205–206
Frequency containment reserve in disturbance (FCR-D), 283
Frequency containment reserve in normal condition (FCR-N), 279–281
Front-end AC-DC converter, 84–86
Fuel cell-combined heat and power (FC-CHP), 305
Full P2P market, 313–314

G
Gas storage systems (GSS), 305–306
Grid flexibility, 26, 28, 36, 63, 274
Grid-connected active buildings, energy management systems, 252–253
constraints of
controllable load appliances' constraints, 260–261
grid exchange limitation, 263
thermal comfortness, 262
cost minimization, 254–255

Grid-connected active buildings, energy management systems (*cont.*)
flexibility services, 257–259
modeling energy management of, 263–264
artificial intelligence and machine learning approaches, 267
optimization-based approaches, 264–266
rule-based approaches, 266
multi-objective function, 259–260
peak shaving, 255–257
Grid-connected virtual power plants, 364
Grid-responsive or grid-aware buildings, 2
Grid-to-vehicle mode (G2V), 90, 92, 256
GridWise Architecture Council (GWAC), 296, 298, 299
Ground-source (GSHP) heat pumps, 106

H
Hankel singular values (HSVs), 113, 114
Hardware-in-the-Loop (HIL) simulation, 236, 240
Health and wellbeing, 34–35
Heating, ventilation, and air-conditioning system (HVAC) systems, 304
Heuristic optimisation algorithm, 198
High-frequency sampling data, 345
Home appliances, 257
Home energy management (HEM), 331
Hottel–Whillier–Bliss model, 106, 107
Hybrid AC/DC electrical power grids, 71–72, 74–75
advantages of, 75–76
contribution and organization, 73–74
load-shift systems for, 79
power electronics and operation modes, experimental validation of, 84
DC-DC converter, 88–89
single-phase AC-DC converter, 86–88
three-phase AC-DC converter, 84–86
power electronics for, 80
DC-DC, 82–84
single-phase AC-DC, 80–81
three-phase AC-DC, 81–82
smart home with, 89–92
status quo, challenges, 72–73
structures of, 78–79
unipolar and bipolar, 76–77
Hybrid P2P market, 316
Hybrid Threat Modelling Method (hTMM), 235

I
Incentive-based programs, 369
Information and Communication Technologies (ICT), 220, 288, 291, 296, 300, 310, 314
Information Security Risk Assessment (ISRA), 231
Information systems (IS), 220
Interpretive phenomenology, 31
Intrusive load monitoring, 333–334
IoT security, 228–229

L
Laboratory prototypes, 84, 91, 93
Lead-acid battery (Pb-acid), 57–58
Learning-based NILM problem statement, 336–337
Line voltage regulators (LVRs), 199
Linear programming (LP) problem, 265
Lithium-ion (Li-ion), 60–61
Lived experience, 30, 32, 35
Load disaggregation techniques, 12–13
Load forecasting
load disaggregation techniques, 12–13
model predictive control approaches, 12
Load frequency control (LFC), 227
Load monitoring (LM)
benefits to consumers and grid, 332, 333
framework, 332, 333
intrusive load monitoring, 333–334
non-intrusive load monitoring, 334
semi-intrusive load monitoring, 334–335
Load shedding mechanisms, 223
Load-shift systems, 79
Local P2P market, 314–315
Low-frequency sampling data, 345
Low-order modelling, MPC
data-driven pathway
air-handling unit (AHU), 115
Matlab's system identification toolbox, 116
outputs for different model orders, 117
parameter identification, 116
simulation environment, 116–117
model-based pathway
balanced truncation, 113–114
Hankel singular values (HSVs), 113, 114
outputs for different model orders, 115
simulation environment, 114–115

M

Machine learning, 267
Manual frequency restoration reserve (mFRR), 281–282
Markov decision processes (MDPs), 173
Mean absolute error (MAE), 344
Mean absolute percentage error (MAPE), 344
Mixed-integer linear programming problem (MILP), 265
Mixed-integer nonlinear programming (MINLP) problem, 266
Model Predictive Control (MPC), 100, 130
  average zone temperatures, 122, 123
  building models within, 117–125
  Building Resistance-Capacitance Modelling (BRCM), 102
  centralised MPC
    building control strategy, 118
    low-complexity model matrices, 118
    optimisation times, 125
    zone-level control, 118–119
  data-driven techniques, 102
  decentralised MPC
    building control strategy, 120
    multi-input-multi-output (MIMO), 120
    optimisation times, 125
    single-input-single-output (SISO), 120
    system-level control, 119–120
  demand-side flexibility, 100
  low-order modelling, 112–117
  modelling workflow, 100, 101
  parameter selection considerations, 124–125
  price-awareness, 122–123
  simulated zone temperatures, 121
  simulation model implementation, 121–125
  space heating performance, 122, 123
  time-step for different model orders, 124
Modelling scale, 240
MPC-enabled heating/cooling systems, 210–213
  continuous space model, 210
  Markov chain, 211
  Markov space model, 210
  modified Kalman filter, 210, 212
  stochastic hybrid system (SHS), 211
  TCL tracking outcome, 213
  total power consumption, 212
Multi-Input Multi-Output (MIMO), 168
Multi-Level Perspective (MLP), 33
Multi-objective function, 259–260
Multiagent systems (MAS), 287

N

National energy grids, 30
Net-zero energy buildings (nZEB), 220
Niche-innovation, 33
Nickel cadmium (NiCd), 59
Nickel metal hydride (NiMH), 59–60
Non-intrusive load monitoring (NILM)
  appliances' consumption extraction, 335
  challenges, 349–350
  companies of NILM products, 345, 347
  datasets, 345, 346
  energy flexibility potential extraction
    E_{max} and E_{min} vs typical energy pattern, 348
    energy patterns, 347
    numerical results on AMPds, 348–349
    total energy consumption, 347
  evaluation metrics
    energy disaggregation accuracy metrics, 344
    event-detection accuracy metric, 342–343
  event-based approaches, 342
  features/signatures of appliances
    non-traditional features, 341
    steady-state features, 340
    transient features, 340
  guidelines, 350
  learning-based NILM review
    supervised NILM algorithms, 338
    unsupervised NILM algorithms, 338
  learning-based problem statement, 336–337
  optimization-based formulation, 335–336
  optimization-based NILM algorithms review, 337
  sample-based approaches, 342
  sampling frequency
    high-frequency sampling data, 345
    low-frequency sampling data, 345
  schematic representation, 334
  status quo, 349–350
  types of appliances, 339, 340
Non-traditional features of appliances, 341
North American Electric Reliability Corporation (NERC), 225

O

Occupants, 27–42
On-load tap changers (OLTCs), 274
Open Platform for Energy Networks (OPEN), 237
Operation control centre (OCC), 220
Optimal technology sizing, 129, 143–149

Optimization-based approaches, 264–266
Optimization-based formulation, 335–336

P
Passive and active energy management systems, 15
Peak shaving, 255–257
Peer-to-Peer (P2P) energy market
for active buildings
full P2P market, 313–314
hybrid P2P market, 316
local P2P market, 314–315
ancillary services, 306–308
basic P2P electricity market, 317–318
challenges, 328
congestion management, 320–321
electricity
battery storage systems, 302–303
electric vehicles, 303–304
photovoltaic storage systems, 302–303
frequency response, 312
future directions, 328
gas and hydrogen
FC-CHP systems, 305
GSS, 305–306
heat and electricity trading, 326–328
heating and cooling
CHP systems, 304–305
HVAC systems, 304
peak load shaving, 309–310
power and heat trading, 321–323
power losses management, 318–320
resilience response, 310–311
simulation results
congestion management, 325–326
11-bus radial distribution network test case, 323–324
power losses management, 324–325
transactions in the basic model, 324
voltage and reactive power control, 310
Peer-to-peer energy trading, 8, 16
Photovoltaic (PV) panels, 301
Photovoltaic (PV) technologies, 220
Photovoltaic storage systems, 302–303
Physics-based modelling
auxiliary electrical loads, 109
domestic hot water modelling, 108–109
heat pump modelling, 105–106
occupancy modelling, 107–108
photovoltaic generation, 106–107
ambient temperature, 107
heat removal factor, 107
Hottel–Whillier–Bliss model, 107
inlet flow temperature, 107
overall loss coefficient, 107
thermal efficiency calculation, 107
resistance-capacitance model, 103–105
software tools, 110
three-story dwelling
building structure, 111
construction details, 111
Energy Performance Certificate (EPC), 110
linear state-space model, 110
simulated heat demand, 112
simulated zone temperatures, 112
Point of common coupling (PCC), 263
Power electronics, 72, 80
DC-DC, 82–84
and operation modes, experimental validation of, 84
DC-DC converter, 88–89
single-phase AC-DC converter, 86–88
three-phase AC-DC converter, 84–86
single-phase AC-DC, 80–81
three-phase AC-DC, 81–82
Power-based attacks, 240
Power-to-gas (P2G), 200, 305
Prediction under uncertainty, 240
Price-based programs, 368, 369
Primary frequency response, 207
Process for Attack Simulation and Threat Analysis (PASTA), 235
Prosumers, 220, 296, 298, 300, 306, 307, 309, 311, 361

Q
Qualitative longitudinal (QLL) research, 35, 41, 42

R
Redox-flow battery (RFB), 61–62
Relative error (RE), 344
Renewable energy sources (RES), 2–4, 7–9, 11, 14, 16, 18, 52, 71, 220, 295, 299, 300
Reserve markets, 14
Reserve products, 282
Residential buildings, 135–138
co-design
annual electricity bill, 137
on-site PV generation, 136
optimization framework, 136

modelling, 139–143
asset parameters, 140
dwelling parameters, 139
energy generation, 142–143
energy storage, 141–142
energy supply, 140
temperature modelling, 139
MPC
battery storage, 137–138
seasonal storage, 138
optimal technology sizing with seasonal storage, 143–149
annual state of charge of seasonal storage, 151
annual trajectories, 150
annual trajectories for $CAPEX_1$, 147
asset investment parameters, 145
$CAPEX_1$, 146
$CAPEX_2$, 146
co-design problem with capital cost, 145
Nordpool data, 144
physical limitation, 143
time-varying electricity prices, 144
total system cost, 145
trajectories for $CAPEX_2$, 149
weekly electricity and carbon prices, 154
weekly electricity consumption for $CAPEX_1$, 155
weekly trajectories for $CAPEX_1$, 148, 152, 156
Residential demand-side management (RDSM), 331
Residential energy management (REM), 331
Residential sector, 331, 351
Retail electricity market, 360–361
Risk assessment (RA) methodology to, 221
Root mean square error (RMSE), 344
Rule-based approaches, 266

S
Sample-based NILM algorithms, 342
Sampling frequency, 345
Security Card methodology approach, 236
Security Equipment Inc. (SEI), 235
Security for Smart Electricity Grids (SEGRID), 231
Security Quality Requirements Engineering (SQUARE), 235
SEGRID Risk Management Methodology (SRMM), 231
Semi-intrusive load monitoring (SILM) technique, 334–335
Sensitivity analysis, 241
Sharing economy, *see* Collaborative economy
Sick Building Syndrome, 34
Single-Input Single-Output (SISO), 168
Single-phase AC-DC, 80–81
Single-phase AC-DC converter, 86–88
Smart buildings (SB), 2, 3, 12, 13, 16, 17, 225
Smart city, 161
Smart grids, 12, 71, 161, 189, 215
Smart home, 89–92, 229
Socio-ecological developments, 34
Socio-technical landscape, 33
Socio-technical regime, 33
Sodium sulphur (NaS), 61
Solid-state transformers (SSTs), 78–79
Static storages
electrical storage, 52–56
electrochemical storage systems, 57–63
Stationary energy storage systems, 14
Steady-state features of appliances, 340
Storage devices, 99
Supercapacitors (SCs), 52, 55–56
Superconducting magnetic energy storage (SMES), 53–55
Supervised NILM algorithms, 338
Supervisory control and data acquisition (SCADA) system, 166–168, 228, 287
Sustainable buildings, 27, 32
Sustainable energy, 71, 73
Synchronised attacks, 229–230
Synchronised cyber incidents, 240

T
Telecommunication, 225
Testbeds, 240
Thermal energy storage systems (TES), 15, 220
Threat modelling (TM), 235–236
Threat Vulnerability and Risk Analysis (TVRA) method, 231
Three-phase AC-DC, 81–82, 84–86
Time Use Survey (TUS), 108
Tradable energy packet (ePKT), 328
Transactive energy (TE)
challenges, 300
definition, 296
DERs, 297–300
DSR, 297, 298
local layer, 299

Transactive energy (TE) (*cont.*)
microgrid layer, 299
P2P market, 300–301
participants of, 298
power system, 298, 299
regional layer, 299
residential layer, 298
Transactive energy market, 226–227
Transient features of appliances, 340
Transmission system operator (TSO), 199, 273, 274
TSO-level flexibility services, 278–284
disturbances and low-inertia situations, services for, 282
aFRR, 283–284
FCR-D, 283
FFR, 284
services for normal operation, 279
FCR-N, 279–281
mFRR, 281–282

U
Uber, 295
UK, 26, 29, 43
Ultra-capacitors (UC), 55
Unsupervised NILM algorithms, 338

V
Vehicle-to-Grid (V2G) concept, 303
Virtual power plant (VPP), 226
Voltage control, 274
Voltage source converter (VSC), 54

W
Wind turbine power converter systems (WTPCS), 220

Z
Zero-day attacks, 228

Lightning Source UK Ltd.
Milton Keynes UK
UKHW051231160223
416809UK00004B/15

9 783030 797416